ALS

Advances in Life Sciences

Marked Individuals in the Study of Bird Population

Edited by
J.-D. Lebreton
Ph. M. North

Birkhäuser Verlag
Basel · Boston · Berlin

Editors

Dr. Jean-Dominique Lebreton
Centre d'Ecologie Fonctionnelle
et Evolutive
C.N.R.S., BP 5051
F-34033 Montpellier Cedex 1
France

Dr. Philipp M. North
Applied Statistics Research Unit
Institute of Mathematics and Statistics
University of Kent
Canterbury, CT2 7NF
U.K.

A CIP catalogue record for this book is available from the Library of Congress, Washington D.C., USA

Deutsche Bibliothek Cataloging-in-Publication Data
Marked individuals in the study of bird population / ed. by J.-D. Lebreton ; Ph. M. North.
– Basel ; Boston ; Berlin : Birkhäuser, 1993
(Advances in life sciences)
ISBN 3-7643-2780-4 (Basel...)
ISBN 0-8176-2780-4 (Boston)
NE: Lebreton, Jean-Dominique [Hrsg.]

Camera-ready copy prepared by the authors
Printed on acid-free paper produced from chlorine-free pulp
Printed in Germany
ISBN 3-7643-2780-4
ISBN 0-8176-2780-4

9 8 7 6 5 4 3 2 1

CONTENTS

Merging different sources of information

Recruitment and dispersal

Survival in relation with habitat

Reviews

Foreword

The present threats to biodiversity are both a direct and an indirect result of man's activity (Erhlich, 1988). These threats are exemplified in a striking way by the status of bird species: out of about 9000 bird species in the world, 1029 are considered as threatened (Collar and Andrew, 1988) and there is evidence of continuing declines in bird populations, especially in insular habitats (Noon and Young, 1991). In such a context, the study of bird populations is of particular interest. Moreover birds are fairly easy to see and to count (Ralph and Scott, 1981). It is thus not surprising that bird populations have been traditionally the subject of many studies.

One of the earliest techniques, individual marking, became a standard tool during the last forty years (see e.g. Dorst, 1956 p.30, and Tautin, this volume). Presently, each year, more than 4 Million birds are ringed in Europe (Wassenaar, Euring data bank, pers.comm.) and more than 1 Million birds in North America (Tautin, U.S.Fish and Wildlife Service, pers. comm.). In North America, one third of the birds which are ringed each year belong to game species (Tautin, pers.comm.).

While information on ringed birds has been long used for understanding migration patterns (see e.g. Dorst, 1956), it has been progressively more and more widely used for studying demographic processes. The roots of this line of research can be traced back to seminal papers by Lack (1946), and Haldane (1955), devoted to the quantitative analysis of recoveries of dead ringed birds. The latter paper was the first example of iterative estimation of a demographic parameter. In a parallel way impressive progress took place in the analysis of live recapture data, culminating in the Cormack-Jolly-Seber model for the estimation of survival probabilities, of capture probabilities, and, possibly, of population size (Cormack, 1964; Jolly, 1965; Seber 1965). The main development of survival models based on the analysis of recovery data started in the 1970s (Cormack, 1970; Seber, 1971, 1972; Brownie et al., 1976, 1985; see reviews by North, 1987; Seber, 1992). Then an explosive development of live recapture analysis took place again (see reviews by Pollock, 1991; Nichols, 1992; North, 1990; Lebreton et al., 1992; Seber, 1992). This development was accompanied by a progressive change in the main emphasis, which moved from the estimation of population size to that of survival parameters (Brownie, 1987; Clobert and Lebreton, 1987; Burnham et al., 1987). The characteristic common to all live recapture models is that the intermittence of capture of individually marked animals should be accounted for. While many developments took place on purely statistical grounds, fish population studies (see e.g. Burnham et al., 1987) and bird population studies remained the two main areas waiting for applications but also raising methodological developments. In parallel, rapid developments took place from 1970 onwards in survival models designed for the study of human health problems, in which, in principle at least, individuals are followed continously (see e.g. Cox and Oakes, 1984). Such models are increasingly relevant for the study of bird populations with the spread of radio-tracking techniques.

The role of bird population studies in the progress of statistical methodology for animal populations has been reflected by several international meetings. The first one took place in London on May 4, 1982, at the initiative of the Mathematical Ecology Group (MEG) of the British Ecological Society. The papers presented, together with a few other ones, appeared as a collection in book form (Morgan and North, 1985). The second meeting, based in part on the same group of experts, was organised at the initiative of the MEG and of EURING, the organisation of the bird ringing centres of European countries, in Wageningen, The Netherlands, in February 1986. This meeting was largely devoted to advances in the analysis of recovery data. The proceedings appeared as a special issue of the journal "Acta Ornithologica" (Vol.23, N°1, 1987). The next conference, still under the leadership of the MEG and EURING, took place in Sempach, Switzerland, in April 1989. The topics treated were broadened, with more emphasis on the rapid progress being made in the analysis of live recapture data. The proceedings appeared as an issue of the journal "The Ring" (Vol.13, N° 1 and 2, 1990).

The rapid advances in the analysis of variation in demographic parameters of animal populations, and the leading role of bird population studies in the development of new methodology, had thus made this series of conferences necessary. They have been the only interdisciplinary forum on the subject.

The present book is the result of a further conference in this series. It took place in Montpellier (France) from April 7 to April 11, 1992, at the Centre d'Ecologie Fonctionnelle et Evolutive (Centre National de la Recherche Scientifique) and was named EURING 92 for short. Its full title "The use of marked individuals in the study of bird populations" emphasised that the main goal of the conference was to establish the state-of-the-art of the methods for analysing demographic parameters in bird populations. Another objective was to gather together representatives of ringing centres, applied statisticians, and bird population biologists to discuss in a multidisciplinary way the quality and the relevance of demographic methods. Thirty-one speakers were invited by five session organisers. Sixty participants (including the speakers) attended the conference and enjoyed its social programme in a relaxed and friendly atmosphere.

After peer review, twenty-eight papers were retained for publication and form the present book. A list of the referees who participated in this review process is given below.

George M. Jolly kindly agreed to give an introductory address in which he traced back to Quetelet the use of "instinctive statistics" that played a major role in the development of the present methodology.

Several papers provided an overview of recent refinements in the analysis of recoveries, live recaptures, and radio-tracking data. Most of these developments concern the ability to take into account complex effects (Skalski et al.; Pradel;

Cormack; Bunck and Pollock; Rinne et al.; Oatley and Underhill; White). Several other papers test these sophisticated approaches in practice (Peach; Baillie and McCulloch; Bell et al.; Kenward; Francis and Cooke), or discuss their role for ringing centres (Tautin). Among these papers, several compared the quality and amount of information brought on demographic parameters and processes by various types of data. This was in particular the case for Kenward, and Francis and Cooke.

Another group of papers attempted to merge different sources of information and to encourage more refined designs: recaptures and recoveries (Burnham; Catchpole et al.), recaptures and resightings (Nichols and Hines), short term and long term live recaptures (Pollock et al.).

Even more complicated sources of variation were considered by several other authors, working on models of recruitment or of dispersal (Schwarz;, Nichols et al.; Clobert et al.; Cooke and Francis; Manly and Chatterjee; Van Noordwijk). The move from pattern to processes and the potential relevance to evolutionary biology are particularly striking there. One may link to this line papers studying relationships between survival and habitat use (Conroy; Aebischer et al.).

Finally, papers provided reviews of software (Lebreton et al.) and of foreseeable developments (Anderson et al.).

From the overall scene developed by this set of texts, two main points seem to arise:

1) a complementarity and a balance between various sources of information (recaptures, recoveries, radio-tracking) can be achieved in the design of bird population studies;

2) one may model individual fates, in terms of demographic parameters subject to various sources of variation. This now seems feasible with great statistical soundness, the intermittence of capture and the way various events compete being more and more adequately represented. More relevant procedures now seem to be within the reach of population biologists and evolutionists. We expect such progress in future will be linked in turn with the ad hoc approaches inherent in Lifetime Reproductive Success calculations (Clutton-Brock, 1988; Newton, 1989) and attract more and more interest from evolutionary biologists.

The EURING 92 conference received the support of many organisations. The Centre National de la Recherche Scientifique (C.N.R.S.) provided the largest contribution to funding, via four different sources: The Interdisciplinary Research Program on Environment (P.I.R.E.), the Department of Life Sciences, the Centre d'Ecologie Fonctionnelle et Evolutive, and the Délégation Régionale Languedoc-Roussillon. The Société Francaise de Biométrie (French region of the Biometric Society), the Office National de la Chasse, the Biometric Unit (run jointly by University Montpellier II, E.N.S.A.M., and I.N.R.A.), and the computer service company AVENIX provided further funding for travel of speakers and participants.

This book and the EURING 92 conference would not have been possible without the participation of many people. Maryse Gautier worked efficiently on the local organisation, together with Josy Trauchessec, Rosa Soto and Claudette

Trinquier. The organisation of sessions was dealt with by Philippe Perret, Paula Dias, Roger Pradel, German Gonzalès-Dàvila and Erica Taube, that of field trips by Jacques Blondel and Max Debussche. Software demonstrations were organised by Anne-Marie Reboulet and Gérard Banco. We wish to thank also the members of the Centre d'Ecologie Fonctionnelle et Evolutive, in particular its director Frank Warembourg, its drawing team, Alain Dervieux and Jean-Louis Martin. For their help in various ways, we thank also M. Lafon, A. Gibert and Sylvie Bénéfice-Malouet, Délégation Régionale Languedoc-Roussillon du C.N.R.S., and Gill Warr and her secretarial team, Applied Statistics Research Unit, University of Kent at Canterbury.

Our warmest thanks go also to the scientific committee of the Conference (The editors, K.P. Burnham, J. Clobert, W. Peach, F. Spina, L. Underhill), and to the session organisers (K.P. Burnham, J. Clobert, R. Kenward, B.J.T. Morgan, W. Peach) to whom we are strongly indebted for the scientific quality of this book.

J.-D.Lebreton, P.M.North
Montpellier, Canterbury
August 1992

References

Brownie, C. (1987).- Recent models for mark-recapture and mark-resighting data. Biometrics, 43: 1017-1019.

Brownie, C., Anderson, D.R., Burnham, K.P. and Robson, D.S. (1976).- Statistical inference from band recovery data - a handbook. U.S. Fish and Wildlife Service, Resource publication n°156.

Brownie, C., Anderson, D.R., Burnham, K.P. and Robson, D.S. (1985).- Statistical inference from band recovery data - a handbook. U.S. Fish and Wildlife Service, Resource publication n°156, 2nd edition.

Burnham, K.P., Anderson, D.R., White, G.C., Brownie, C. and Pollock, K.H.(1987).- Design and analysis methods for fish survival experiments based on release-recapture. American Fisheries Society Monograph n° 5, 437 p.

Clobert, J. and Lebreton, J.D. (1987).- Recent models for mark-recapture and mark resighting data. A response to C.Brownie. Biometrics, 43: 1019-1022.

Clutton-Brock, T.H. (Ed.)(1988).- Reproductive success; studies of individual variation in contrasting breeding systems. University of Chicago Press, Chicago and London.

Collar, N. and Andrew, P.(1988).- Birds to watch; the ICBP world checklist of threatened birds. ICBP technical publication n°8, Smithsonian Institution Press, Washington, D.C..

Cormack, R.M.(1964).- Estimates of survival from the sightings of marked animals. Biometrika, 51 : 429-438.

Cormack, R.M.(1970).- Statistical appendix to R.Fordham's paper. Journal of Animal Ecology, 39: 24-27.

Cox, D.R. and Oakes, D.(1984).- Analysis of survival data. Chapman and Hall, London., 201 p.

Dorst, J.(1956).- Les migrations des oiseaux. Payot, Paris.

Erhlich, P. (1988).- The loss of diversity: causes and consequences. pp. 21-27 in Wilson, E.O. (Ed.).- Biodiversity. Nat.Acad. Press, Washington, D.C..

Haldane, J.B.S.(1955).- The calculation of mortality rate from ringing data. Proc.11th Int.Orn.Congress, Basel, 11: 454-458.

Jolly, G.M.(1965).- Explicit estimates for capture-recapture rates with both death and immigration - stochastic model. Biometrika, 52: 225-247.

Lack, D.(1946).- Do juvenile birds survive less well than adults ? British Birds, 39: 258-264.

Lebreton, J.D., Burnham, K.P., Clobert, J. and Anderson, D.R.(1992).- Modelling survival and testing biological hypotheses using marked animals : a unified approach with case studies. Ecological Monographs, 62: 67-118.

Morgan, B.J.T. and North, P.M.(Eds.)(1985).- Statistics in Ornithology. Springer Verlag, New-York.

Newton, I. (Ed.)(1989).- Lifetime reproduction in Birds. Academic Press, London.

Nichols, J.D. (1992).- Capture-Recapture Models; using marked animals to study population dynamics. BioScience, 42: 94-102.

Noon, B.R. and Young, K. (1991).- Evidence of continuing worldwide declines in bird populations: insights from an international conference in New-Zealand. Conservation Biology: 5, 141-143.

North, P.M.(1987).- Ring recovery analytical methods: a brief review. Acta Ornithologica, 23: 7-12.

North, P.M.(1990).- Analysis of avian ring recovery and live recapture data: where have we come from Wageningen to Sempach? The Ring, 13: 11-22.

Pollock, K.H.(1991).- Modeling capture, recapture, and removal statistics for estimation of demographic parameters for fish and wildlife populations: past, present, and future. Journal of the American Statistical Association, 86 : 225-238.

Ralph, C.J. and Scott, J.M. (Eds.)(1981).- Estimating numbers of terrestrial birds. Studies in avian biology, vol.6 Allen Press, Laurence, Kansas, 630 p.

Seber, G.A.F.(1965).- A note on the multiple-recapture census. Biometrika, 52 : 249-259.

Seber, G.A.F.(1971).- Estimating age-specific survival rates from bird-band returns when the reporting rate is constant. Biometrika, 58: 491-497.

Seber, G.A.F.(1972).- Estimating survival rates from bird-band returns. Journal of Wildlife Management, 36: 405-413.

Seber, G.A.F.(1992).- A review of estimating animal abundance II. International Statistical Review, 60:129-166.

List of Contributors

Aebischer N.J.
The Game Conservancy, Fordingbridge, Hants SP6 1EF, United Kingdom

Anderson, D.R.
Colorado Cooperative Fish & Wildlife Research Unit, U.S. Fish & Wildlife Service, Fort Collins, CO 80523, U.S.A.

Baillie S.R.
British Trust for Ornithology, The Nunnery, Thetford, Norfolk IP24 2PU, United Kingdom

Banco G.
Centre d'Ecologie Fonctionnelle et Evolutive, C.N.R.S., BP 5051, 34033 Montpellier cedex 1, France

Bell M.C.
The Wildfowl & Wetlands Trust, Slimbridge, Gloucester, GL2 7BT, United Kingdom

Black J.M.
The Wildfowl & Wetlands Trust, Slimbridge, Gloucester, GL2 7BT, United Kingdom

Brownie C.
Department of Statistics, North Carolina State University, Raleigh, NC 27695-8203, U.S.A.

Bunck C.M.
U.S. Fish & Wildlife Service, Patuxent Wildlife Research Center, Laurel, MD 20708, U.S.A.

Burnham K.P.
Colorado Cooperative Fish and Wildlife Research Unit, U.S. Fish & Wildlife Service, Fort Collins, Colorado 80523, U.S.A.

Catchpole E.A.
University College, University of New South Wales, Canberra, ACT 2600, Australia

Chatterjee C.
Department of Mathematics and Statistics, University of Otago, P.O. Box 56, Dunedin, New Zealand

Clobert J.
Laboratoire d'Ecologie, Ecole Normale Supérieure 46 rue d'Ulm 75230 Paris cedex 05, France

Conroy M.J.
U.S. Fish and Wildlife Service, Georgia Cooperative Fish and Wildlife Research Unit, School of Forest Resources, University of Georgia, Athens, GA 30602, U.S.A.

Cooke F.
Biology Department, Queen's University, Kingston, Ontario, Canada K7L 3N6

Cormack R.M.
Department of Mathematical and Computational Sciences, University of St Andrews, St Andrews, Fife KY16 9SS, United Kingdom

Fox A.D.
The Wildfowl & Wetlands Trust, Slimbridge, Gloucester, GL2 7BT, United Kingdom

Francis C.M.
Center for Tropical Conservation, , Duke University, Durham, NC 27705, U.S.A.

Freeman S.N.
Renewable Resources Assessment Group, Centre for Environmental Technology, Imperial College of Science, Technology and Medicine, 8 Princes Gardens, London, SW7 INA, United Kingdom

Hestbeck J.B.
U.S. Fish & Wildlife Service, Massachusetts Cooperative Fish & Wildlife Research Unit, University of Massachusetts, Amherst, MA 010031, U.S.A.

Hines J.E.
U.S. Fish and Wildlife Service, Patuxent Wildlife Research Center, Laurel MD 20708, U.S.A.

Hoffmann A.
Center for Quantitative Science, HR-20, University of Washington, Seattle, WA 98195, U.S.A.

Jolly G. M.
38, Silverknowes Terrace, Edinburgh, EH4 5HJ, United Kingdom

Julliard R.
Laboratoire d'Ecologie, Ecole Normale Supérieure 46 rue d'Ulm 75230 Paris cedex 05, France

Karlbom M.
Institute of Zoophysiology, Box 560, 751 22 Uppsala 1, Sweden

Kendall W.L.
Office of Migratory Bird Management, U.S. Fish and Wildlife Service, Laurel, MD 20708, U.S.A.

Kenward R.E.
Institute of Terrestrial Ecology, Furzebrook road, Wareham, Dorset BH20 5AS, United Kingdom

Lebreton J.D.
Centre d'Ecologie Fonctionnelle et Evolutive, C.N.R.S., BP 5051, 34033 Montpellier cedex 1, France

Lokki H.
Department of Computer Science, University of Helsinki, Teollisuuskatu 23, SF-00510 Helsinki, Finland

Manly B.F.J.
Department of Mathematics and Statistics, University of Otago, P.O. Box 56, Dunedin, New Zealand

Marcström V.
Institute of Zoophysiology, Box 560, 751 22 Uppsala 1, Sweden

McCleery R.H.
Edward Grey Institute, Department of Zoology, Oxford University, South Parks Road, OX1 3PS Oxford, United Kingdom

McCulloch N.
British Trust for Ornithology, The Nunnery, Thetford, Norfolk IP24 2PU, United Kingdom

Morgan B.J.T.
Institute of Mathematics and Statistics, University of Kent, Canterbury, Kent, CT2 7NF, United Kingdom

Nichols J.D.
U.S. Fish and Wildlife Service, Patuxent Wildlife Research Center, Laurel MD 20708, U.S.A.

Oatley T.B.
Avian Demography Unit, Department of Statistical Sciences, University of Cape Town, Rondebosch, 7700 South Africa.

Owen M.
The Wildfowl & Wetlands Trust, Slimbridge, Gloucester, GL2 7BT, United Kingdom

Peach W.J.
British Trust for Ornithology, The Nunnery, Thetford, Norfolk IP24 2PU, United Kingdom

Pollock K.H.
Department of Statistics, North Carolina State University, Raleigh, NC 27695-8203, U.S.A.

Pradel R.
Centre d'Ecologie Fonctionnelle et Evolutive, C.N.R.S., B. P. 5051, 34033 Montpellier cedex 1, France

Reboulet A.M.
Centre d'Ecologie Fonctionnelle et Evolutive, C.N.R.S., BP 5051, 34033 Montpellier cedex 1, France

Rexstad E.A.
Institute of Artic Biology and Department of Biology & Wildlife, University of Alaska, Fairbanks, AK 99775 U.S.A.

Rinne J.
Department of Meteorology, University of Helsinki, Hallituskatu 11-13, SF-00100 Helsinki, Finland

Saurola P.
Ringing Centre, Zoological Museum, University of Helsinki, Pohjoinen Rautatiekatu 13, SF-00100 Helsinki, Finland

Schwarz C.J.
Department os Statistics, University of Manitoba, Winnipeg, Manitoba, Canada R3T 2N2

Skalski J. R.
Center for Quantitative Science, HR-20, University of Washington, Seattle, WA 98195, U.S.A.

Smith S.G.
Center for Quantitative Science, HR-20, University of Washington, Seattle, WA 98195, U.S.A.

Tautin J.
Bird Banding Laboratory, U.S. Fish and Wildlife Service, Laurel, Maryland 20708 U.S.A.

Underhill L.G.
Avian Demography Unit, Department of Statistical Sciences, University of Cape Town, Rondebosch, 7700 South Africa

van Noordwijk A.J.
Netherlands Institute of Ecology, P.O. box 40, NL 6666 ZG Heteren, The Netherlands

Walsh A.J.
National Parks and Wildlife Service, Wexford Wildfowl Refuge, North Slob, Wexford, Ireland

White G.C.
Dept. of Fishery and Wildlife Biology, Colorado State University, Fort Collins, CO 80523, U.S.A.

Wotawa M.A.
Colorado Cooperative Fish & Wildlife Research Unit, Colorado State University, Fort Collins, CO 80523, U.S.A.

Marked Individuals in the
Study of Bird Population
J.-D. Lebreton and P. M. North (eds.)

Instinctive Statistics

George M. Jolly

38, Silverknowes Terrace, Edinburgh, EH4 5HJ, U.K.

Summary. Examples are discussed to illustrate the intuitive or instinctive approach to statistics. The mathematician, Adolphe Quetelet's ideas in the early ninteenth century contributed to the development of biometry as a science. Similar thinking in the 1930's and 40's by the biologist, C.H.N. Jackson, led to important advances in mark-recapture methods for estimating animal populations. R.A. Fisher, P.H. Leslie and others made further improvements in this field. It is suggested that the importance of instinctive statistics should not be overlooked.

1. Introduction

At the Bird Census Symposium at Asilomar, California in 1980, I recall speculation regarding the advances likely to be made in the following ten years, particularly in mark-recapture methods as applied to ornithology. We foresaw a merging of recently developed theory to provide more realistic models for avian populations. Further tests would also be required to discriminate among models. These aims, it seems, have been achieved in large measure.

Some biosticians (if I may coin a word which gives more equable weighting to the two disciplines) may see the approaching decade as equally predictable. Before looking ahead, however, I thought this might be a suitable moment to reflect on the earlier intuitive thinking from which present mark-recapture theory arose.

2. A. QUETELET - an early influence in biometry.

Cormack's (1968) review refers to the earliest contributors of ideas from Francis Bacon, Laplace, Lincoln, Petersen, onwards but of necessity choice here must be restricted. In the early 1800's in the wake of a new interest in probability stimulated by Laplace, Poisson, Cauchy and others, the Belgian mathematician, Quetelet, realised the potential of their discoveries. Following the death of his father in 1803, young Adolphe Quetelet was encouraged by his mother in a wide range of pursuits including mathematics, and was soon attracting attention as a dramatist, poet, artist, and flautist of moderate ability. After obtaining a Doctorate for a work on conic sections, he became, at age 25, Professor of Mathematics at the Athenaeum in Brussels.

As a teacher Quetelet was an inspiration to his students. Meanwhile, he developed an interest in physics sufficiently to become astronomer of the Royal Observatory, and in the silence of the night sky under "the inflexible laws which regulate the march of worlds" his interest in measurement, hence statistics, evolved. From there he delved into the effect of social and other factors on the level of crime, ever searching for the causes of events. He was impressed by "the sad perspective of seeing the same crimes reproduced in the same order calling down the same punishments in the same proportions". Here was a social law comparable to a law in physics. Quetelet re-iterated these conceptions with such persuasion that the study of social conditions was soon elevated to the level of a scientific discipline.

By the 1830's Quetelet had acquired an international reputation through his numerous publications and foreign visits. Although his friend included Laplace, Poisson and Fourier, he sought not to advance theory as such, but rather to apply the new-found techniques to others fields of study. He worked to provide a reliable framework for life insurance, and in 1829 became active in the new Statistics Bureau of Holland, encouraging censuses to be conducted in both Holland and Belgium.

More than a century earlier, De Moivre had discovered the Normal distribution. This had been put to use by astronomers and physicists, also by the military in problems relating to artillery fire. Quetelet was fascinated. He found that data on the chest measurements of 5378 Scottish soldiers grave good agreement with the Normal curve. So also did the heights of 100,000 French conscripts, except for a congestion below the required height and a scarcity above it, indicating "that some 2000 have escaped service by reducing their height two or three centimetres". (Unfortunately, neither the technique of achieving this reduction nor the consequence of Quetelet's detection of it is recorded in the literature.)

However, the importance of Quetelet's work lies primarily in his method. The age of biometry had dawned. His studies on criminals led to their physicial measurements being used for identification purposes, and later as the basis of anthropology. Investigations such as that on time of blossoming of the lilac from such data as could be acquired over the years 1839-1844, from 16 European cities and New York State, foreshadowed the principles of regression and adjustement for non-orthogonality. (For more detail see Jolly and Dagnelie, 1967.)

3. C.H.N. JACKSON - a pioneer of mark-recapture

Whereas Quetelet was one of several leading contributors who helped to mould biometry, the intuitive efforts of Jackson were arguably the dominant factor that enable mark-recapture to become a viable quantitative method in the study of mobile populations. As an entomologist, Jackson's motivation was the study of the tsetse fly in East Africa.

In his first paper (1933), Jackson describes an attempt to estimate the number of tsese in a non-enclosed area by marking and releasing a fixed number daily until the proportion of marked individuals in the population reached a constant value. From the number of days required to achieve this he could calculate the tsetse death rate per day which in turn gave an estimate of the number of marked flies alive at any given time. Together with the proportion of the population marked, estimated from a random sample, the total number in the population could be estimated. The method, however, was abandoned on account of difficulty in determining the time at which the proportion of marked flies reached a constant value.

Jackson's subsequent papers (1937, 1939, 1940, 1944, 1948) use ingenious manipulation of ratios of numbers in the different recapture categories to estimate rates of gain and loss and the total number of flies in the population. In his trellis of recaptures each column refers to a specific date of release and each row to a date of capture. For the so-called "triple catch" situation where there are only two release and two recapture dates, giving three categories of recaptures, rates of gain and loss and population number are uniquely determined. Jackson's main formulae result from an extension of this idea on the assumption of a constant survival rate per unit of time (sampling at regular intervals), and are ratios of appropriate now summations (his "negative" method) or column summations (his "positive" method). The negative method is most suitable if releases are made on several occasions and recaptures are made in one or a few samplings. Conversely, the positive method would apply if releases were the result of a concentrated initial effort followed by sampling spread over time. Details of these procedures are given by Blower, Cook and Bishop (1981) and commented on by Cormack (1968).

4. R.A. FISHER - contributions to mark-recapture

Jackson attributes his estimate of population size to Fisher and its variance to W.L. Stevens.

Dowdeswell, Fisher and Ford (1940), and Fisher and Ford (1947) apply methods similar to Jackson's to butterflies and moths but using a new estimate for the assumed constant survival rate; they start with a trial value and proceed until an acceptably small difference is obtained between the means of the observed and expected number of days survived. This gives estimates for the number of marked individuals alive on each day, hence a daily population estimate on dividing by the observed proportion of marked to total recaptured on that day. The assumption of a constant survival rate is clearly acceptable for many insect populations over short study periods.

As Cormack (1968) points out, since a given recapture is included as often as that individual is released, Fisher's estimate of survival rate is not the most valid. The same comment, of course, applies to Jackson's method, as noted by Bailey (1951) who develops a maximum likelihood estimate for survival rate.

5. P.H. LESLIE and others - the last of the deterministic models

Leslie and Chitty (1951), recognising the inefficiency of the Jackson and Fisher approach, introduced a new method of grouping the data, classifying them by date of last previous capture. Leslie (1952) applied these results to the estimation of population numbers assuming survival either variable or constant over time. Variances were given. The problem was also discussed of estimation for a population subject to both death and immigration, an explicit solution for which was given by Jolly (1963).

It is perhaps interesting to reflect that the more realistic and more efficient stochastic models now in general use, because of their greater simplicity, tend to have an even greater intuitive appeal than their deterministic predecessors (see Seber, 1982).

6. Conclusions and Suggestions

We have glanced briefly at certain achievements of four original thinkers in the field of biometry over the space of a century and a half, starting not long after Laplace, in 1786, estimated the population of France with the help of a simple ratio of births to total population obtained from a sample of parishes.

With particular reference to mark-recapture methods, it is difficult not to envisage Jackson as the most influential contributor of his time. His persistent formulation and dissemination of ideas undoubtedly led to general recognition of the method's potential. In fact, Jackson's and Fisher's procedures remained popular with many biologists until the early 1980's in the absence of any published alternative appropriate to a constant survival rate assumption with immigration present. Hopefully, the existence of more powerful techniques will not dampen the instinctive approach. A flawless application of maximum likelihood could be based on a dubious assumption; might not Francis Bacon have agreed?

Considering the extraordinary degree of co-operation - direct and via publications - among an ever growing number of biosticians, and with such papers in mind as Lebreton et al. (1992), I feel it would be presumptuous to offer any suggestions for future work. The increasing flexibility of models and the possibility of combining techniques - mark - recapture, re-sighting, recovery and radio tagging - have been well utilized; the importance of specificity and parsimony is recognized; a versatile body of software now exists for model differentiation. The study of the dynamics of animal populations appears to be progressing confidently under its own momentum.

References

Bailey, N.T.J. (1951).- On estimating the size of mobile populations from capture-recapture data. Biometrika, 38: 293-306.

Blower, J.G., Cook, L.M. and Bishop, J.A. (1981).- Estimating the size of animal populations. George Allen and Unwin Ltd., London.

Cormack, R.M. (1968).- The statisties of capture-recapture methods. Oceanogr. Mar. Biol., An. Rev. 6: 455-506.

Dowdeswell, W.H., Fisher, R.A. and Ford, E.B. (1940).- The quantitative study of populations in the Lepidoptera. I *Polyommatus icarus (Rott.)*. Ann. Eugen., 10: 123-136.

Fisher, R.A. and Ford, E.B. (1947).- The spread of a gene in natural conditions in a colony of the moth *Panaxia dominula L.* Heredity, Lond., 1: 143-174.

Jackson, C.H.N. (1933).- On the true density of tsetse flies. J. Animal Ecol., 2: 204-209.

Jackson, C.H.N. (1937);- Some new methods in the study of Glossina morsitans. Proc. Zool. Soc., London 1936: 811-896.

Jackson, C.H.N. (1939).- The analysis of an animal population. J. Animal Ecol., 8: 238-246.

Jackson, C.H.N. (1940).- The analysis of tsetse fly population. I. Ann. Eugen. London., 10: 332-369.

Jackson, C.H.N. (1944).- The analysis of tsetse fly population. II. Ann. Eugen. London., 12: 176-205.

Jackson, C.H.N. (1948).- The analysis of tsetse fly population. III. Ann. Eugen. London., 14: 91-108.

Jolly, G.M. (1963).- estimate of population parameters from multiple recapture data with both death and dilution - a deterministic model. Biometrika, 50: 113-128.

Jolly, G.M. and Dagnelic, P. (1967).- Adolphe Quetelet, 1796-1874:153-179., in R.C. Olby (ed.) Early Nineteenth Century European Scientists. Pergamon Press Ltd, London.

Lebreton, J.-D., Burnham, K.P., Clobert, J. and Anderson, D.R. (1992).- Modeling survival and testing biological hypotheses using marked animals: an unified approach with case studies. Ecological Monographs, 62: 67-118.

Leslie, P.H. (1952).- The estimation of population parameters from data obtained by means of the capture-recapture method.II: the estimation of total numbers. Biometrika, 39: 363-388.

Leslie, P.H. and Chitty, D. (1951).-The estimation of population parameters from data obtained by means of the capture-recapture method. I: the maximum likelihood equations for estimating the death rate. Biometrika, 38: 269-292.

Seber, G.A.F (1982).- Estimation of animal abundance and related parameters. 2nd ed. Griffin, London.

Modelling complex effects in recoveries, live recaptures, and radio-tracking data

Marked Individuals in the
Study of Bird Population
J.-D. Lebreton and P. M. North (eds.)

Testing the Significance of Individual- and Cohort-Level Covariates in Animal Survival Studies

J. R. Skalski, A. Hoffmann, and S. G. Smith

Center for Quantitative Science, HR-20, University of Washington, Seattle, WA 98195, USA

Summary. Release-recapture and radiotelemetry models have been developed that permit the incorporation of individual- and cohort-level covariates based on the assumptions of proportional hazards in animal survival studies. When analyzing the relationship of cohort-level covariates and survival rates between populations or sampling periods within a population, likelihood ratio tests (LRT) proved highly unreliable in the presence of unaccounted-for variations in baseline survival rates. The greater the natural variation in survival, the more likely LRT falsely detected significant correlations with cohort-level covariates. Asymptotic F-tests based on Analysis of Deviance (ANODEV) provided near nominal tests of significance regardless of the degree of natural variation present. However, when analyzing the relationship of individual-level covariates and survival, LRT provide nominal tests of significance for the effects of measured traits despite the potential presence of unassignable sources of heterogeneity in individual survival probabilities.

1. Introduction

In recent years, there has been growing interest in describing the relationship between the survival rates of fish and wildlife and potential explanatory variables or covariates. These covariates range from environmental or man-induced covariates operating at the cohort level, to covariates describing the traits and exposure levels of individual animals to potential environmental stressors. Examples include testing pesticides and herbicides using field trials in avian and mammalian ecotoxicology; assessing the influence of phenotypic differences in entomology; determining the epidemiology of disease infestation in mollusca; characterizing the importance of social ranking in communal species; and evaluating the importance of individual traits on the fate of outmigrating salmon smolt. In all these cases, there is interest in determining both the biological and statistical significance of measured covariates on faunal survival rates.

Interpretation of the biological significance of an effect on survival is perhaps a more difficult task than the assessment of statistical significance. At a minimum, a realistic response model describing the relationship between covariates $\left(\underset{\sim}{x}\right)$ and shifts in survival probabilities [i.e., $\phi(x)$] is necessary. The survival model $\phi(x)$, must properly represent the response relationship if a reliable estimate of a cohort-wide survival rate (ϕ) is to be calculated where

$$\phi = \int_{-\infty}^{\infty} \phi(x) f(x) dx$$

and where $f(x)$ probability density function for the covariate in the cohort of tagged animals $(-\infty < x < \infty)$.

The various statistical models for tag analysis we present are generalizations of the release-recapture model of Cormack (1964) for survival estimation and the mark-recapture models of Seber (1965) and Jolly (1965). Our first generalization permits modeling capture probabilities and survival parameters in a proportional hazards framework. Sandland and Kirkwood (1981), Clobert (1981), and Clobert et al. (1985) were among the first to discuss special cases of the Cormack (1964) survival model by placing equality constraints among the parameters. Later, North and Morgan (1979), Clobert et al. (1988), Smith and Skalski (1989), Smith (1991), and Lebreton et al. (1992) generalized the ideas by reparameterizing the survival and/or capture parameters as functions of measured covariates. The second generalization we present extends the release-recapture models to the case of nonidentically distributed capture and survival probabilities among tag-individuals and leads to the use of multiple Bernoulli likelihood functions.

An important class of response models for survival relationships is the proportional hazards models (Cox 1972) originating in human epidemiology (Prentice 1973, Feigl and Zelen 1965). A particular proportional hazards formulation (Smith and Skalski 1989, Smith 1991) that is useful in analyzing tagging data assumes an exponential effect of a covariate on survival where

$$\phi(x) = \phi^{e^{\underset{\sim}{x}'\underset{\sim}{\beta}}} \qquad \text{(Eq. 1)}$$

and a multiplicative effect on an animal's life expectancy [i.e., $E\left(t_i \mid \underset{\sim}{x}\right)$] where

$$E\left(t_i \mid \underset{\sim}{x}\right) = \frac{E\left(t_i \mid \underset{\sim}{x} = 0\right)}{e^{\underset{\sim}{x}'\underset{\sim}{\beta}}} .$$

The ability to readily quantify effects on baseline survival and life expectancy distinguishes process-based models from other convenient parameterizations such as logit or probit. The recent publication by Lebreton et al. (1992)

emphasizes the use of a logistic parameterization for modeling survival rates but also discusses several other formulations. Choice of parameterization is analogous to the choice of a response model in classical statistical analyses. For instance, the logistic response implies an additive effect of a covariate on the log-odds [i.e., $\ln(p/(1-p))$] while Eq. (1) implies a multiplicative effect on life expectancy. A realistic parameterization is important if risk assessment based on the tagging results is planned.

The criteria for inclusion of a covariate in a survival model must include the standards of objectivity and reproducibility found in other fields of investigation. Typically, these criteria are partially fulfilled by statistical tests of significance based on the relationship between the magnitude of the measured trait and the variability among experimental units. Animal survival studies are no exception. However, the nature of the experimental unit in tests of significance depends on the type of covariate under investigation.

The following sections detail the structure of statistical tests of significance of individual- and cohort-level covariates based on a proportional hazards formulation. Analyses are illustrated using examples of bird tagging studies performed using program SURPH (SURvival with Proportional Hazards) in an X-View computing environment using a UNIX operating system on SUN® workstations. SURPH is capable of analyzing both release-recapture and radiotelemetry data for single and multi-release field studies.

2. Notation

The following notation and definitions are used in both the cohort-level model and the individual-level model. A cohort is defined in this paper as a group of tagged individuals comprising a release-recapture study at a site.

For both cohort and individual models, the following quantities are defined:

K = the number of sampling occasions.

ϕ = baseline survival in period 1 (for the cohort-level model this is the survival probability for individuals in the first cohort in period 1).

ρ_k = regression coefficient for period effects in period $k\,(k = 2, \ldots, K-2)$.

$\underset{\sim}{\tau}_k$ = regression coefficients for covariates such as treatment effects of interest in $k\,(k = 1, \ldots, K-2)$.

$\underset{\sim}{\beta}_k$ = general vector of regression coefficients $= (\rho_k, \tau_k)$ for period k.

p_{ik} = probability an individual from the *i*th cohort $(i = 2, \ldots, I - 1)$ is captured on the *k*th occasion $(k = 2, \ldots, K - 1)$, given it survived through the *(k-1)*st period. For a single cohort, this is just p_k.

λ_i = product of conditional survival in the last period $(K - 1)$ and capture on the last occasion k for all animals in the *i*th cohort $(i = 1, \ldots, I)$. For a single cohort, this is just λ.

For the cohort model, the following quantities are also defined:

N = the total number of cohorts in study.

a_{ik} = the number of previously marked animals caught on occasion k in the *i*th cohort $(i = 1, 2, \ldots, I; k = 2, 3, \ldots, K)$.

v_{ik} = the number of individuals marked on or prior to occasion k in the *i*th cohort that are known to be alive at the time of occasion $k + 1$ $(i = 1, 2, \ldots, I; k = 1, 2, \ldots, K - 1)$.

c_{ik} = the number of individuals in the *i*th cohort seen for the last time on occasion $k (i = 1, 2, \ldots, I; k = 1, 2, \ldots, K - 1)$.

χ_{jt_j} = probability an individual from the *i*th $(i = 2, \ldots, I - 1)$ cohort is not recaptured after the *k*th $(k = 1, \ldots, K - 1)$ occasion.

$\underset{\sim}{x}_{ik}$ = general vector of covariates for cohort i in period $k (i = 1, \ldots, I; k = 1, \ldots, K - 2)$.

For the individual-based model, the following quantities are also defined:

R = the total number of individuals in the study.

H = the number of possible capture histories.

$\underset{\sim}{x}_j$ = covariate vector of measured traits on the *j*th individual $(j = 1, \ldots, R)$.

$\underset{\sim}{t}_j$ = last occasion the *j*th individual is recaptured $(j = 1, \ldots, R)$.

χ_{jt_j} = probability the *j*th individual $(j = 1, \ldots, R)$ is last seen on occasion $t_j (t_j = 1, \ldots, K - 1)$.

$\underset{\sim}{(CH)}_j$ = vector of indicator variables for individual j $(j = 1, \ldots, R)$ that indicates its capture history (CH). This vector has elements $\{(ch)_{jh}; h = 1, \ldots, H\}$.

$I(jk)$ = indicator of whether the *j*th animal $(j = 1, \ldots, R)$ is recaptured on the *k*th occasion $(k = 2, \ldots, K)$.

ψ_{jh} = probability of the jth animal ($j = 1, \ldots, R$) has the hth capture history.

3. Modeling Capture Probabilities

With respect to our focus on survival rates, the capture probabilities in the tag-recapture models are little more than nuisance parameters. Still, it is important to choose a realistic yet parsimonious model for the capture probabilities. Using an overly general formulation for the capture rates will lead to lack of precision in modeling survival probabilities, while an overly restrictive model can bias the results for survival. Lebreton et al. (1992) propose preliminary modeling of capture probabilities using likelihood ratio tests in the content of a comprehensive model that includes the most general model possible for the survival probabilities. Because we have found instances where the LRT results for capture probabilities differ depending on the particular formulation for survival probabilities, we have devised methods of model selection for capture probabilities that are independent of the survival model.

Our approach to capture rate modeling is motivated by the method used by Manly and Parr (1968) to estimate capture probabilities. To estimate the probability of capture on sampling occasion i, the Manly-Parr model uses a subset of the data, namely those animals that were seen at least once before occasion i and at least once after occasion i. The capture probability is estimated as the proportion of this subset (hereafter called the "Manly-Parr subset") that were captured on occasion i itself.

For the cohort-level models, possible models for capture probabilities include: "P_{st}," in which a separate capture probability is assumed for each sampling occasion in each cohort; "P_s," in which each cohort has a unique capture probability that applies for all occasions; "P_t," in which the capture probabilities for all cohorts are equal on each occasion, but vary between occasions; and "P," in which a single capture probability is used for all occasions on all cohorts. Only P_t and P are relevant for single-population studies. An appropriate model is chosen from these and other candidates by arranging the Manly-Parr subset into a three-dimensional contingency table (cohort by period by captured/not captured) and performing a log-linear analysis (McCullagh and Nelder 1989). The approach is reminiscent of the goodness-of-fit tests (Tests 2 and 3) proposed by Burnham et al. (1987, pp. 71-77). To model capture probabilities as functions of both cohort and individual covariates, it is possible to perform logistic regression (McCullagh and Nelder 1989) analysis on the Manly-Parr subset. Logistic regression can also be used for model selection at the cohort level. This approach to capture model selection is incorporated in the SURPH program.

4. Analysis of Cohort-Level Covariates

Cohort-level covariates are defined as factors that assume the same level for all individuals in a release-recapture study at a specific location or during a sampling period. Meteorological or climatic factors, such as minimum temperatures or rainfall during a sampling period, are examples of cohort-level covariates. Another example is a treatment that is applied over the entire study site on which a cohort of tagged individuals reside. Cohort-level covariates are usually external to the individual animals. When using the SURPH model for cohort-level covariates, multiple cohorts serve as replicates. However, the model can be applied to single-cohort studies provided the level of the covariate changes from sampling period to sampling period.

4.1 Example

Before a new insecticide is licensed for use on cornfields, field trials must be performed to test for toxic effects of the compound on resident avian species. For the experiment, eight study sites are established in a corn growing area with starlings (*Sturnus vulgaris*) selected as the study species. Using a completely randomized design, four of the sites are assigned to receive insecticide application, and the other four sites serve as controls. The application of the compound occurs at the time of planting. The beginning of the release-recapture study of starlings coincides with planting. Sampling is conducted simultaneously on all sites at weekly intervals for a month, giving a total of four sampling occasions.

SURPH models covariate effects by reparameterizing the Cormack model (1964) survival probabilities in the form of Eq. (1). SURPH uses an offset modeling method whereby the survival probability in the first period in cohort #1 is defined to be ϕ and all other survival probabilities are defined relative to ϕ.

In the insecticide example, the covariates are indicator variables for the periods and the treatment. A separate treatment effect is estimated in each period (τ_k ;

$k = 1, 2, 3$), so one treatment indicator variable is required for each of the three periods. The four control populations and treatment populations serve as replicates in tests of effects.

4.2 Likelihood model

To define the SURPH likelihood, we make assumptions of independence and homogeneity of the survival probabilities of the birds within a period, permitting the modeling of the capture data as a multinomial sample. We also assume that the sampling on each site is independent from that on all other sites. Given these assumptions, the SURPH model has the likelihood (Smith 1991):

$$L\left(\phi, \underset{\sim}{\beta}, \underset{\sim}{p}, \underset{\sim}{\lambda} \,\middle|\, \left\{a_{ik}, v_{ik}, c_{ik}, \underset{\sim}{x}_{ik}\right\}\right) \propto$$

$$\prod_{i=1}^{N}\left(\left(\prod_{k=1}^{K-2}\left(\phi^{e^{\underset{\sim}{x}'_{ik}\underset{\sim}{\beta}_k}}\right)\left(p_{i,k+1}\right)^{a_{i,k+1}}\right.\right.$$

$$\left.\left.\cdot\left(1 - p_{i,k+1}\right)^{v_{ik} - a_{i,k+1}}\right)\lambda_i^{a_{iK}} \prod_{k=1}^{K-1} \chi_{ik}^{c_{ik}}\right) ,$$

(Eq. 2)

where $\underset{\sim}{\beta}_k = (\rho_k, \tau_k)$.

The objective of the experiment was to determine whether the insecticide application had adverse effects on the survival of starlings living near the treated cornfields. Symbolically, the hypotheses of interest are:

$$H_0 : \tau_k \leq 0 \quad \text{versus} \quad H_A : \tau_k > 0 \ . \qquad \text{(Eq. 3)}$$

A common approach to testing such hypotheses in the maximum likelihood framework is the likelihood ratio test (LRT). In this case, the model that includes the effect parameter τ_k is tested against the model that omits τ_k, giving a chi-squared test statistic with one degree-of-freedom.

4.3 Monte Carlo results for likelihood ratio test

Data sets were simulated under scenarios like that of the insecticide trial example. The simulations used completely randomized designs with $N = 2I$ populations, I of which received the treatment and I of which did not. Values of $I = 2, 4$, and 8 were used. Five $(K = 5)$ sampling occasions were simulated under the null hypothesis of no treatment effects.

First, data were generated by subjecting the individuals in each cohort to period-specific survival probabilities $\{\phi_k; k = 1, \ldots, K-1\}$ (without plot-to-plot variability) and occasion-specific capture probabilities $\{p_k; k = 2, \ldots, K\}$. Under the null hypothesis, survival probabilities were calculated as

$$\phi_{ik} = \phi + \delta_k \qquad \text{(Eq. 4)}$$

where

$\phi = 0.9$,

δ_k = adjustment to ϕ for period k $(k = 1, \ldots, K-1)$ $(\delta_k \sim U(-0.05, 0.05))$.

Under H_0, the distribution of the LRT for treatment effects is nominal [Fig. 1(a)]; i.e., the LRT results in nominal rejection rates. The period effect parameters (ρ_k) in the SURPH model account for the variability introduced by δ_k.

In a realistic situation, there is plot-to-plot variability in survival that is unrelated to treatments or periods. More realistic data were generated by subjecting each cohort to an extra plot-specific mortality factor. That is, survival was calculated under the null hypothesis as

$$\phi_{ik} = \phi + \delta_k + \epsilon_{ik} \qquad \text{(Eq. 5)}$$

where

ϵ_{ik} = extra mortality term for individuals in cohort i in period k $(\epsilon_{ik} \sim U(-0.1, 0.1))$ $(i = 1, \ldots, I_j; k = 1, \ldots, K-1)$,

and where ϕ and $\{\delta_k; k = 1, \ldots, K\}$ are defined above. In this more realistic scenario, the distribution of the LRT is biased; in other words, it rejects the null hypothesis of no-treatment effect far too often [Fig. 1(b)], when in fact there is no treatment effect.

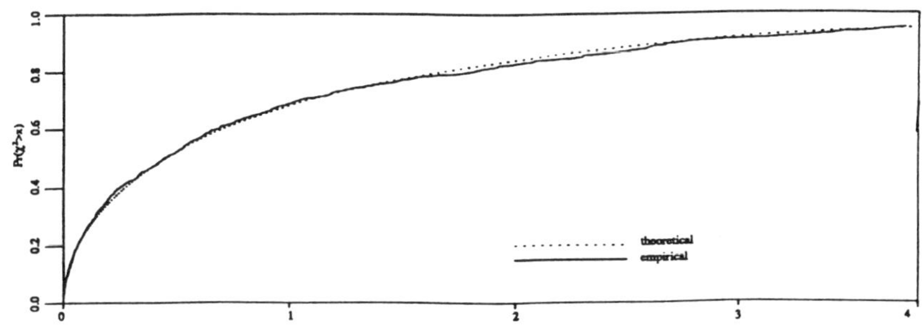

(a) Likelihood ratio test. Data generated without interpopulation variability.

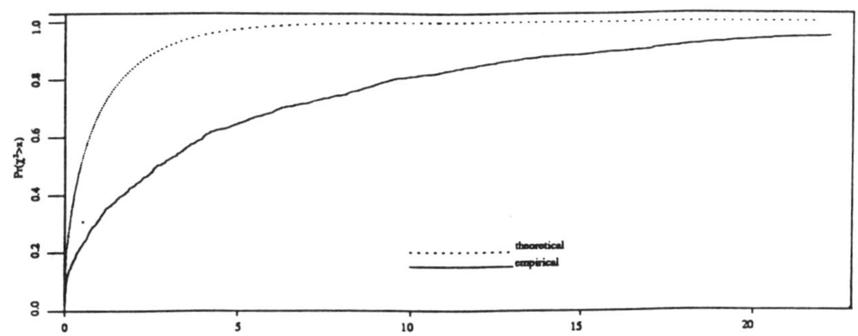

(b) Likelihood ratio test. Data generated with interpopulation variability.

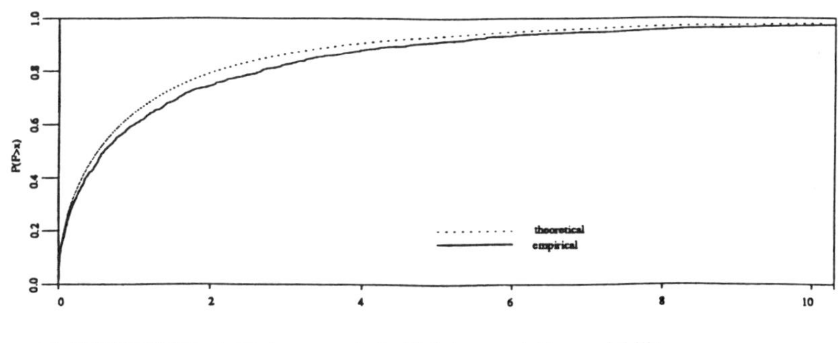

(c) ANODEV F-test. Data generated with interpopulation variability.

Fig. 1. Cumulative density plots for likelihood ratio test and ANODEV F-test under various simulation scenarios.

The LRT is based on sampling variability alone; it completely ignores the interreplicate variability between the cohorts (ϵ_{ik}). The ϵ_{ik} are analogous to the random error terms used in normal theory regression and analysis of variance. This error is the unexplainable variation always present in experiments of any kind, including wildlife investigations. As stated in Eq. (3), the LRT is intended to test the significance of the treatment parameter τ_k. However, the LRT chooses between two alternatives: (1) there is no variability between populations

$$\left(\phi_{ik} = \phi^{e^{\rho_k}}\right)$$

or (2) the total variability is explained by the treatment or covariate in question

$$\left(\phi_{ik} = \phi^{e^{\rho_k + \tau_k}}\right) .$$

Thus, data generated according to Eq. (4) conform to the LRT's null hypothesis. Consequently, the null model fits the data and the LRT is nominally distributed. However, when data are generated according to Eq. (5), the LRT's null model cannot fit adequately and the LRT breaks down. The simulation results show that when there is variability between tagged cohorts that is unrelated to the covariate, the LRT too often incorrectly attributes the extra variability to that covariate.

4.4 Analysis of deviance approach

In conjunction with the development of the SURPH model for cohort-level covariates, we have developed an approach to significance testing that accounts for interpopulation variability in a way that the LRT cannot (Smith 1991). Analysis of Deviance (ANODEV) is analogous to the familiar Analysis of Variance (ANOVA). Just as an ANOVA table partitions differences between models' sums of squares, an ANODEV table partitions differences between models' log-likelihoods (McCullagh and Nelder 1989). ANODEV tables give F-statistics for treatment effects that are interpreted like F-statistics from ANOVA.

The F-statistic from ANODEV tests effects relative to the total amount of variability in survival probabilities, accounting for both sampling variability and interpopulation variability. The F-statistic tests the hypotheses [Eq. (3)] in the context of an underlying model that includes interpopulation variability. The hypotheses under the ANODEV statistic can be expressed as:

$$H_O: \tau_k \leq 0 \quad (\text{i.e., } \phi_{ik} = \phi + \epsilon_{ik})$$
versus
$$H_A: \tau_k > 0 \quad \left(\text{i.e., } \phi_{ik} = \phi^{e^{\tau_k}} + \epsilon_{ik}\right) \qquad \text{(Eq. 6)}$$

Accounting for interpopulation variability in the ANODEV F-test results in nominal behavior of the test statistic [Fig. 1(c)]. The F-statistic is generally slightly liberal in the scenarios depicted, but not to the degree of the LRT. In simulations under a wider range of scenarios, the LRT was always more liberal in its rejection levels than the F-test. Similar results for LRT and ANODEV have been found for the case of exponential error (i.e., $\phi_{ik} = \phi^{e^{\epsilon_{ik}}}$ under H_O).

The ANODEV table for a generic multiple-population study is shown in Table 1. Analysis of Deviance begins by fitting a Cormack (1964) model independently to the data from each tagged cohort. Combining all the independent models gives the "Full" model of the data, analogous to the full model in ANOVA. The second ANODEV model is the "Period Means" model, which assumes a unique survival probability in each period but common across all populations. This is analogous to the "Grand Mean" model in ANOVA. The "Corrected Total" deviance is defined as twice the difference in the log-likelihoods of the Full and Period Means models

$$2\left\{\ln L\left(\hat{\underset{\sim}{\theta}}_F\right) - \ln L\left(\hat{\underset{\sim}{\theta}}_{PM}\right)\right\} .$$

The purpose of ANODEV is to determine how much of the Corrected Total deviance is "explained" by the covariates. The "Total Covariate" model contains all of the covariate effects in all of the periods. In the pesticide example, there is one treatment and three periods in which its effects can be estimated. Thus, there are three degrees of freedom for the Total Covariate deviance. The Total Covariate deviance is twice the difference between the log-likelihoods of the Total Covariate and Period Means models, and the Error deviance is the difference between the Corrected Total and the Total Covariate deviances. The Total Covariate deviance can be decomposed into the contribution of the covariates in each of the periods. The deviances attributed to the covariates in each period are sequential; that is, they depend on the order of other effects already in the model. The covariate effects are not decomposed into period contributions in Table 1. It is important to note the degrees-of-freedom for error are the same degrees-of-freedom obtained in an analysis of variance of the $\hat{\phi}_{ij}$ after pooling across periods. Wildlife investigations are not exempt from the general rules of error estimation found in any properly analyzed experiment.

The Total Covariate Deviance is equivalent to the likelihood ratio statistic, and the Mean Deviance Error (MDE) is an estimate of the degree of overdispersion in the data (interpopulation variability). Thus, the F-test statistic can be expressed as

$$F_{\mathrm{df,\,dfe}} = \frac{\mathrm{LRT/df}}{\mathrm{MDE}}$$

where df is the covariate degrees-of-freedom and dfe is the degrees-of-freedom for error. Lebreton et al. propose a similar test statistic based on a different overdispersion factor (Lebreton et al. 1992; p. 106) and outline a theoretical argument for the use of the statistic. The performance of this proposed statistic has not yet been explored using the type of Monte Carlo study reported here or any other method.

The Analysis of Deviance table for a contrived data set is shown in Table 2, including decomposition of the total treatment effect into period contributions. The MDE of 1.2353 indicates that interpopulation variability is present in the example, and must be accounted for when testing effect significance. The F-test for the treatment effects indicates that the effect is significant in the first period (P-value = 0.019) and not in the second (P = 0.135) or third (P = 0.496) periods, suggesting that the toxicity of the pesticide was acute and not persistent.

Fitting the model with period effects plus a treatment effect in only the first period (τ_1) gives estimates of the baseline survival probabilities of 0.911 (standard error 0.053) in the first period, 0.837 (s.e. 0.034) in the second period, and 0.967 (s.e. 0.039) in the third period. The estimate of the treatment effect parameter is $\hat{\tau}_1$ = 1.3821. Substituting $\hat{\tau}_1$ into the equation

$$\hat{\phi}^{e^{\hat{\tau}_1}}$$

gives 0.690 (s.e. 0.159) as the estimated survival probability in the first period in treated cohorts. Thus, the effect of the treatment in the first period was to reduce

Table 1. Analysis of Deviance Tables

Symbolic ANODEV Table. Covariate effects are not broken down by period.

Source	df[1]	Deviance[2]	Mean Deviance	F
Corrected Total	$(K-2)(I-1)$	$2\left\{\ln L\left(\hat{\underline{\theta}}_{F}\right)-\ln L\left(\hat{\underline{\theta}}_{PM}\right)\right\}$		
Total Covariate	$(K-2)r$	$2\left\{\ln L\left(\hat{\underline{\theta}}_{COV}\right)-\ln L\left(\hat{\underline{\theta}}_{PM}\right)\right\}$	$\frac{\mathrm{Dev(mod)}}{(K-2)r}$	$F=\frac{\mathrm{MD(Mod)}}{\mathrm{MD(Err)}}$
Error	$(K-2)(I-r-1)$	$2\left\{\ln L\left(\hat{\underline{\theta}}_{F}\right)-\ln L\left(\hat{\underline{\theta}}_{COV}\right)\right\}$	$\frac{\mathrm{Dev(Err)}}{(K-2)(I-r-1)}$	

1 K = number of sampling occasions; I = number of populations; r = number of covariates.

2 $\ln L\left(\hat{\underline{\theta}}_{F}\right), \ln L\left(\hat{\underline{\theta}}_{PM}\right), \ln L\left(\hat{\underline{\theta}}_{COV}\right)$ are the maximized log-likelihoods for the Full, Period Means, and Total Covariate models, respectively.

Table 2. Analysis of Deviance Table

Full ANODEV table for contrived tag-recapture data showing treatment effect decomposed by period.

Source	df	Deviance	Mean Deviance	F	P
Corrected Total	21	34.0178			
Covariate Total	3	11.7833	3.9278	3.1797	0.049
Treatment, Period 1	1	8.1622	8.1622	6.6077	0.019
Treatment, Per. 2 | Per. 1	1	3.0245	3.0245	2.4485	0.135
Treatment, Per. 3 | Pers. 1 & 2	1	0.5966	0.5966	0.4830	0.496
Error	18	22.2345	1.2353		

the survival probability of affected birds by 0.911 - 0.690 = 0.221. The $\hat{\tau}_1$ can be interpreted biologically by considering the hazard function (instantaneous rate of failure) and the expected lifespan. The proportional hazards assumption implies that hazard rate is estimated to be $e^{\hat{\tau}_1}$ = 3.98 times greater for the birds on the treated site than on the control sites. The implication for expected lifespan is that if the treatment effect remained constant at the same level as in the first period, then the birds on the treated site would be expected to live only $1/e^{\hat{\tau}_1}$ = 25.1% as long as control birds.

5. Analysis of Individual-Level Covariates

The ability to analyze the effects of individual-level covariates can also give valuable insights into the mechanisms driving a survival process. Individual-level covariates are defined as factors that may differ among the members of a population. Individual-level covariates are usually intrinsic to an individual such as age, sex, phenotype, genotype, social status, or experience. The model treats individual-level covariates as fixed. Some covariates such as sex and genotype are inherently fixed. Age can also be treated as fixed because it is equivalent to date of birth which is fixed. The values of other covariates such as weight and blood cholinesterase levels may vary over the course of the study. The model incorporates only the initial value of time-varying covariates. Incorporating updated values of a time-varying covariate presents difficulty in a mark-recapture analysis because the values can be updated only for those animals recaptured. Not knowing the covariate values for individuals that die and/or were not recaptured can seriously bias estimation of the survival function. This issue is not dealt with here.

The individual covariate model presented here is a likelihood model similar to the cohort model described above. Contrary to the performance in the cohort-level setting, LRT performs nominally in the individual setting, even in the presence of unascribed individual heterogeneity in survival.

5.1 Example

In an agricultural area, R sage grouse are banded with distinct leg bands. The purpose of the study is to quantify the relationship between amount of pesticide ingested, for which blood cholinesterase is a surrogate, and survival. The grouse are banded, measured, and released in the field. On three subsequent sampling occasions attempts are made to recapture the grouse. The sampling occasions are spaced one month apart. At the end of the study each grouse has a capture history describing the sampling occasions on which it was recaptured.

5.2 Likelihood model

There are eight possible capture histories in the grouse tagging study: (1111), (1110), (1101), (1011), (1100), (1010), (1001), and (1000). Each banded grouse represents a multiple Bernoulli random variable where the cells of the multiple Bernoulli are the eight potential capture histories. The probability the jth grouse has capture history (1010) is $\psi_{j(1010)}$. Thus, the likelihood function for the jth grouse is

$$L_j\left(\underset{\sim}{\psi}_j \mid \underset{\sim}{(CH)}_j\right) = \prod_{h=1}^{8} \psi_{jh}^{(ch)_{jh}} . \qquad \text{(Eq. 7)}$$

If the jth grouse's capture history is (1010), then Eq. (7) reduces to

$$L_j\left(\underset{\sim}{\psi}_j \mid \underset{\sim}{(CH)}_j\right) = \psi_{j(1010)} .$$

The capture history probabilities $\left(\underset{\sim}{\psi}_j\right)$ can be reparameterized according to Eq. (1). For a grouse with capture history (1010), its likelihood is

$$\psi_{j(1010)} = \phi^{e^{\underset{\sim}{x}_j'\underset{\sim}{\beta}_1}}(1-p_2)\phi^{e^{\underset{\sim}{x}_j'\underset{\sim}{\beta}_2}}p_3(1-\lambda) \quad .$$

Because the fate of each grouse is assumed to be independent from all other individuals, the overall likelihood model is the product

$$L\left(\underset{\sim}{\psi}\,\Big|\left\{\left(\underset{\sim}{CH}\right)_j\right\}\right) = L\left(S, \underset{\sim}{\beta}, \underset{\sim}{p}, \lambda \,\Big|\left\{\left(\underset{\sim}{CH}\right)_j\right\}\right)$$

$$\propto \prod_{j=1}^{R} L_j\left(S, \underset{\sim}{\beta}, \underset{\sim}{p}, \lambda \,\Big|\, \left(\underset{\sim}{CH}\right)_j\right)$$

$$L\left(\underset{\sim}{\psi}\,\Big|\left\{\left(\underset{\sim}{CH}\right)_j\right\}\right) = \prod_{j=1}^{R} \chi_{jt_j} \prod_{k=1}^{t_j} \phi^{e^{\underset{\sim}{x}_j'\underset{\sim}{\beta}_k}} p_{k+1}^{I(j,k+1)}(1-p_{k+1})^{(1-I(j,k+1))}$$

where

$$\chi_{jt_j} = \sum_{k=t_j+1}^{K-1}\left[\left(1-\phi^{e^{\underset{\sim}{x}_j'\underset{\sim}{\beta}_k}}\right)\prod_{l=t_j+1}^{k-1}\phi^{e^{\underset{\sim}{x}_j'\underset{\sim}{\beta}_l}}(1-p_l)\right]$$

$$+(1-\lambda)\prod_{k=t_j+1}^{K-1}\phi^{e^{\underset{\sim}{x}_j'\underset{\sim}{\beta}_k}}(1-p_k)$$

and

$$\underset{\sim}{\beta}_k = (\rho_k, \tau_k) \quad .$$

Likelihood ratio tests can be used to test between nested models. One hypothesis of interest is:

$$H_0 : \tau_1 = \tau_2 = 0$$

versus

$$H_A : \tau_1 \neq 0 \text{ or } \tau_2 \neq 0 \quad . \qquad \text{(Eq. 8)}$$

With this hypothesis, one tests whether or not a relationship exists between the covariate $\underset{\sim}{x}$ and survival in either period 1 or 2. The LRT statistic is

$$\chi_2^2 = 2\left(\ln L\left(\phi, \underset{\sim}{p}, \underset{\sim}{\rho}, \underset{\sim}{\tau}, \lambda \mid (CH)\right) - \ln L\left(\phi, \underset{\sim}{p}, \underset{\sim}{\rho}, \lambda \mid (CH)\right)\right)$$

and is distributed under the null hypothesis [Eq. (8)] as a chi-squared random variable with degrees of freedom equal to the number of parameters tested.

The failure of the LRT in the cohort-level model raises concern about the behavior of the LRT test statistic in the individual-level model. The close agreement between empirical and χ^2 cumulative distributions demonstrates the LRT performs nominally under H_o (Fig. 2) even in the presence of unaccounted-for variability at the individual level. The random variability was introduced in the individual survival probabilities according to the expression

$$\phi_{jk} = \phi^{e^{\rho_k}} + \epsilon_{jk}$$

where

$$\phi = 0.8,$$

$$\rho_2 = -0.75,$$

$$\rho_3 = 0.47,$$

and $\epsilon_{jk} \sim U(-0.05, 0.05).$

The capture probabilities used for the simulation were $p_1 = 0.9$, $p_2 = 0.6$, and $p_3 = 0.8$. Similar robustness of the LRT to unascribed error has been shown for uniform error added to the exponent of the survival model $\phi^{e^{\epsilon}}$. The nominal behavior of the LRTs in conjunction with the individual covariate model is in sharp contrast to the finding for the group covariate model. The reason the LRT is satisfactory for testing individual covariates appears to be because the unexplained variability between individuals is of the same form of variability (i.e., nonidentically distributed Bernoulli) already being modeled by the likelihood. However, in testing cohort-level covariates, the site-to-site variability is very different than the multinomial error being modeled. Consequently, the familiar LRT can be used to test for significance of individual covariates but not cohort-level covariates.

(a) $\phi(k)$

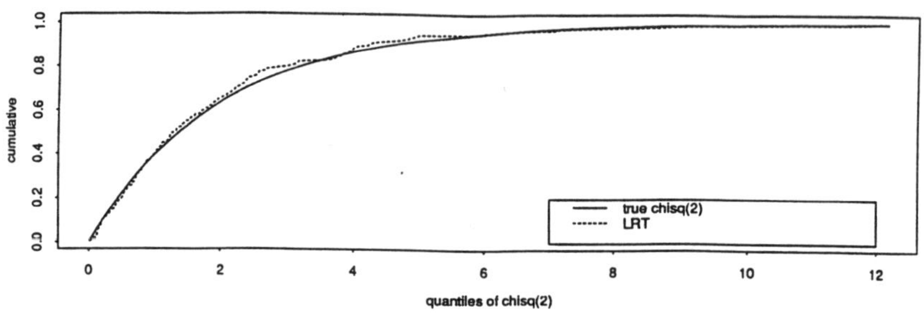

(b) $\phi(k) + U(j, k)$

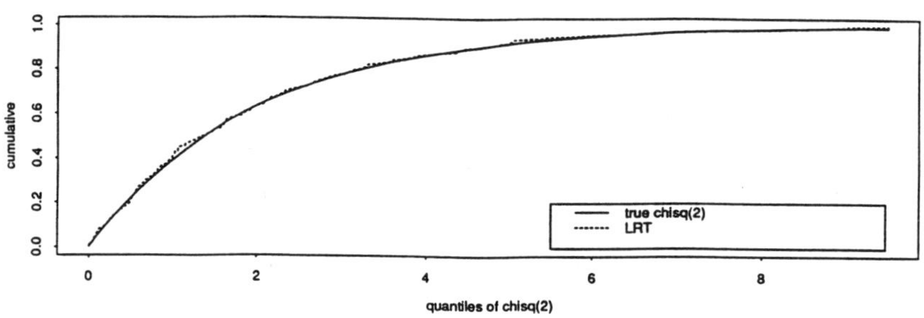

Fig. 2. **Simulation results on the distribution of the LRT statistic under (a) H_o with homogeneous survival of probabilities among individuals and (b) H_o with added random variability to individual survival probabilities. The simulation study consisted of 300 simulations. In each simulation 1000 grouse were banded and released.**

6. Discussion

The ability to assess effects of covariates on animal survival at both the cohort and individual levels provides ecologists with important research tools. Extrinsic factors, such as climatic variables, habitat conditions, design variables, and human-induced effects, all occur at the cohort level. For these factors, cohort-level tests are not only appropriate but also necessary.

Simulation studies confirm Fisher's (1947) statement that the proper error variance is that variance that incorporates all sources of variability contributing to the differences in response except the one being tested. In survival studies, as in any other model selection study, unaccounted for and extra-model variability will exist. Asymptotic F-tests based on analysis of deviance compare the total deviance with that portion attributable to the proposed survival model. In so doing, the asymptotic F-tests have nominal performance regardless of the degree of environmental variability in baseline survival rates.

Another suite of covariates that may affect survival are measurable at the level of the individual animal. Survival analysis at the individual level is crucial for understanding survival processes because ultimately mortality and selection occur at the level of the individual animal. These intrinsic traits include genotypic, phenotypic, behavioral, and social differences between animals. For these factors, LRT can be used to validly assess relationships between individual traits and survival using both release-recapture and radiotelemetry data. Analysis of individual-level covariates in animal survival studies opens up the possibilities of studying the interrelationships in population biology between genetics, behavior, and demographics as never before.

In certain types of investigations, options may exist to base survival studies at either the cohort or individual level. Choice of approach will then depend upon the organizational level that the covariate may operate and is measurable. Avian toxicology is an example where investigators may have a choice in study design and analysis. Currently, very large and expensive field trials are conducted to assess effects of pesticides and herbicides at the population or cohort level using replicate field plots. The natural variation in survival, logistics of such large studies, and study costs often work jointly to limit the statistical power of these field trials. Alternatively, individual birds at an exposure site can be captured, depressions in blood cholinesterase measured, and their fates monitored to test the relationship between organophosphate pesticide exposure and survival. The heterogeneity in exposure levels and survival probabilities between birds within a single population provides the basis to test survival relationships. Study costs under such circumstances can be greatly reduced and simultaneously provide greater resolution of survival relationships. Hence, the opportunity to study survival relationships using both individual- and cohort-level covariates permits investigators to seek creative study designs and meaningful explanatory variables to unravel the secrets of selective pressures driving the dynamics of wild populations.

Acknowledgements. This research was supported in part by the Bonneville Power Administration under contract DE-B179-90BP02341.

References

Burnham, K. P., Anderson, D. R., White, G. C., Brownie, C., and Pollock, K. (1987)..- Design and analysis methods for fish survival experiments based on release-recapture. Am. Fish. Soc. Monogr. No. 5. 437 pp.

Clobert, J. (1981).- Etude de la dynamique d'une population d'Etourneauz sansonnets (*Sturnus vulgaris*). These. Universite Catholique de Louvain. Louvain-la-Neuve, Belgique. 178 pp.

Clobert, J., Lebreton, J. D., Clobert-Gillet, M., and Coquillart, H. (1985).- The estimation of survival in bird populations by recapture or resighting of marked individuals. pp. 197-213 in Statistics in Ornithology, B. J. T. Morgan and P. M. North (Eds). Springer-Verlag, New York.

Clobert, J., Perrins, C. M., McCleery, R. H., and Gosler, A. G. (1988).- Survival rate in the Great Tit *Parus major* in relation to sex, age, and immigration status. J. Anim. Ecol., 57: 287-306.

Cormack, R. M. (1964).- Estimates of survival from the sighting of marked animals. Biometrika, 51: 429-438.

Cox, D. R. (1972).- Regression models and life tables (with discussion). J. Royal Statist. Soc. B., 34: 187-202.

Feigl, P., and Zelen, M. (1965).- Estimation of exponential survival probabilities with concomitant information. Biometrika, 21: 826-838.

Fisher, R. A. (1947).- The design of experiments. 4th Edition. Hafner, New York. 240 pp.

Jolly, G. M. (1965).- Explicit estimates from capture-recapture data with both death and immigration--stochastic model. Biometrika, 52: 225-247.

Lebreton, J. D., Burnham, K. P., Clobert, J., and Anderson, D. R. (1992).- Modeling survival and testing biological hypotheses using marked animals: a unified approach with case studies. Ecol. Monogr., 62: 67-118.

Manly, B. F. J., and Parr, M. J. (1968).- A new method of estimating population size, survivorship and birth rate from capture-recapture data. Trans. Soc. Brit. Ent., 18: 81-89.

McCullagh, P., and Nelder, J. A. (1989).- Generalized linear models. Chapman and Hall. New York. 511 pp.

North, P. M., and Morgan, B. J. T. (1979).- Modelling heron survival using weather data. Biometrics, 35: 667-681.

Prentice, R. L. (1973).- Exponential survival with censoring and explanatory variables. Biometrika, 60: 279-288.

Sandland, R. L., and G. P. Kirkwood. (1981).- Estimation of survival in marked populations with possibly dependent sighting probabilities. Biometrika, 68: 531-541.

Seber, G. A. F. (1965).- A note on the multiple recapture census. Biometrika, 52: 249-252.

Smith, S. G. and Skalski, J. R. (1989).- Statistical design and analysis of avian field trials in environmental toxicology. pp. 407-421 in Pesticides in Terrestrial and Aquatic Environments. Proceedings of a National Research Conference, D. L. Weigmann (Ed.). Virginia Water Research Center, Blacksburg, Virginia.

Smith, S. G. (1991).- Assessing hazards in wild populations using auxiliary variables in tag-release models. Ph.D. dissertation. Univ. Washington. 338 pp.

Marked Individuals in the
Study of Bird Population
J.-D. Lebreton and P.M. North (eds.)

Flexibility in Survival Analysis from Recapture Data: Handling Trap-Dependence

R. Pradel

Centre d'Ecologie Fonctionnelle et Evolutive, C.N.R.S., B.P.5051, 34 033 Montpellier CEDEX 1, France.

Summary: It is well known that survivorship and behaviour of animals can be affected by handling. The deleterious effect of capture can be minimized by taking special care during handling, but it is in general more difficult to avoid changes in behaviour; "trap-shy" animals will tend to avoid the traps while "trap-happy" animals will seek them, particularly when they are baited. General trap-dependence can be treated as age-dependence provided the data are suitably set out. However, trap-dependence effects are likely to be strongest immediately after a capture event. Thus models that consider a different capture probability whether an animal was caught on the previous occasion are particularly relevant. If we denote by m an immediate trap-effect (following Sandland and Kirkwood, 1981) the straightforward generalization of the Cormack-Jolly-Seber model (S_t,p_t) can be denoted as (S_t,p_{m*t}). I present a goodness-of-fit test for this model and a test of comparison with the Cormack-Jolly-Seber model as simple tests on 2*2 contingency tables.
The model (S_t,p_{m*t}) which allows different trap-effects in different years, such as trap-shyness alternating with trap-happiness cannot be used without caution to provide reliable parameter estimates due to identifiability problems. But it is most often sufficient to consider a model with an additive time and trap-effect on a convenient scale (S_t,p_{m+t}). The models (S_t,p_{m*t}), (S_t,p_{m+t}) and (S_t,p_m) can be useful not only when genuine trap-dependence is involved but also to account for other structural effects that mimic trap-dependence or can be alleviated through it, in particular general heterogeneity of capture probabilities among individuals.

1. Introduction

The Capture-Mark-Recapture (CMR) method, currently employed for the study of survivorship in natural animal populations, involves the handling of animals at least once (for mark setting purpose). Manipulations of animals may also take place at future observations. It has long been known that survivorship and behaviour of animals can be affected by handling. Fortunately, the deleterious effect of capture can be minimized by taking special care during handling. This is all the more fortunate because it is impossible to test for a handling effect on survival (Burnham et al., 1987 p.218 *et seq.*) unless, -but this is not very usual,- some control, not handled, animals are available; for example, Brownie and Robson (1983) considered the effect of marking by contrasting the future of newly marked animals with the future of previously marked animals resighted (hence not handled) at the same time. On the contrary, it is usually impossible to avoid changes in behaviour: "trap-shy" animals will tend to avoid the traps while "trap-happy" animals will seek them, particularly if they are baited. Ignoring this last phenomenon can lead to strongly biased estimates of survival. Such effects are probably not uncommon; It is thus important to be able to take them into account. I envisage here the context of Capture-Mark-Recapture experiments on open populations in which animals are handled when observed and I focus my attention on the aftermath of capture (the mark setting event is not treated differently from any other capture event).

From the modelling point of view, effects of capture implies that apparent survival (the complement to mortality *plus* definitive emigration) and recapture probabilities depend at any time on the previous capture history of an animal. Pollock (1975) gave the likelihood function for those cases when a shift in survival after a capture lasts at least as long as a shift in catchability. Yet, it is the reverse (the shift in catchability outlasting the shift in survival) that is likely to be more frequent in real situations. As a simpifying hypothesis one may reasonably consider that only the last event of capture matters. This amounts to neglect a possible enhancement of the trap-effect by repeated captures.

Under these hypotheses, I derive a very general approach to modelling trap-dependence. I then examine in more details the model which considers immediate trap-effect on capture (ITEC) and models derived by simplifying hypotheses (additivity of trap and time effects, no time effect). Goodness-of-fit test for ITEC and tests of comparison between models are provided. Winter trapping data for Teal in the Camargue are used to illustrate the method.

2. Trap-dependence limited to the previous capture event

When all that matters is the time elapsed since the last capture, there is no need to develop special models. Instead, it is possible to modify the data so as to resort to existing age and time dependent models. The trick is to make "time elapsed since the previous capture" appear in a new data set as "time elapsed since marking". To achieve this aim, each real animal is to be replaced by as many imaginary animals as there are events of capture. Each new imaginary animal is recaptured at most once, and, if so, immediately removed. At the same time, a new imaginary animal is supposed to be captured for the first time, marked and released (fig.1). This is repeated as long as the real animal is recaptured.

fig.1: from trap- to age-dependence. 0= no capture; 1= capture; 2= capture and removal.

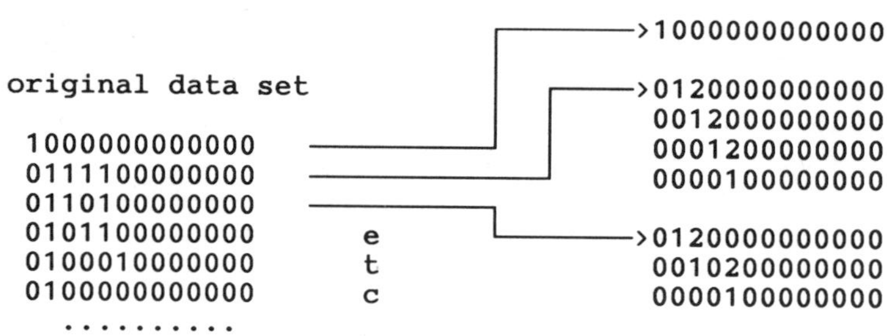

Now, trap-dependence in the old data set can be treated as age dependence in the new data set. It is thus possible to use the frame of an age and time dependent model (Pollock, 1981) allowing for losses at capture. When trap-effect is limited to capture or to apparent survival only, or is allowed to encompass a limited number of occasions, software are available for numerical fits (Pradel and Lebreton, 1991).

Yet, most general models are generally uselessly complex and, anyhow, suffer from important and pervasive identifiability problems, partly inherited from the age and time dependent models, partly due to the peculiarity of the transformed data.

3. Immediate trap-effect on capture

There have not been enough studies to evaluate the timescale over which trap-dependence operates. However trap-dependence effects are likely to be the strongest on the occasion immediately after capture. A good starting point is thus to assume that the capture effect does not last more than one occasion -and because a difference in survival can be avoided and is anyway not testable as noted in the introduction,- the straightforward generalization of the Cormack-Jolly-Seber model (CJS) would consider an immediate trap-effect on capture only (ITEC). Within the frame proposed by Wilkinson and Rogers (1973), as applied to Capture-Mark-Recapture models (Lebreton *et al.*, 1992), the notation for this model might be S_t, P_{m*t}. S represents apparent survival parameters, P capture parameters, m represents trap-effect as in Sandland and Kirkwood (1981), t, time effect. I will explain this model further by means of an example.

The data for this example have been kindly provided by La Tour du Valat biological station situated in the Camargue (southern France). Teal *Anas crecca* were trapped at the Saint-Seren marsh during many wintering seasons. I am here dealing with a restricted subset for the sake of simplicity. A more complete analysis (Pradel *et al.*, submitted) has been submitted for publication. In a first step, I tested the suitability of CJS using the goodness of fit test of Burnham *et al.* (1987).

This test is made up of 2 components: one compares the future of animals belonging to different cohorts (Test 2.C); The other compares the future of animals of different subcohorts inside a cohort (Test 3.S). Cohorts are defined as animals last released at the same time. Subcohorts contrast animals of the same cohort with different previous capture histories.

Table 1: Goodness of Fit Tests to CJS model (Burnham et al., 1987); notations after Burnham et al., 1987.

	Component	Chi-square	df	P-level
between subcohorts	TEST 3.S	21.5790	19	.3057
between cohorts	TEST 2.C	80.0731	20	.0000
overall		101.6521	39	.0000

No difference is expected between subcohorts under either CJS or ITEC. Between cohorts, no difference is expected under CJS; under ITEC, the proportion of animals belonging to cohort i which are recaptured at occasion i+1 is expected to be different from the corresponding proportion from any previous cohort. Indeed, their capture probability at time i+1 is expected to be higher in case of trap-happiness, lower in case of trap-shyness. The overall test for CJS (Tab. 1) is highly significant, mainly because of the high value of Test 2.C, thus leading to the rejection of CJS. Let us examine the raw data.

Table 2: Observed Recaptures: animals released at i, first recaptured at j. * indicate where ITEC is expected to differ from CJS. R(i) number of animals last released at occasion i.

i	R(i)	j=2	3	4	5	6	7	8	9	10	11	12	13
1	6	0*	1	1	0	0	0	0	0	0	0	0	1
2	13		2*	1	0	1	0	0	1	0	0	0	0
3	9			2*	1	0	1	0	0	0	0	0	0
4	34				9*	3	2	2	1	0	0	0	0
5	26					14*	0	0	0	1	0	1	0
6	95						40*	5	0	0	1	0	1
7	118							42*	5	2	2	0	0
8	128								37*	4	4	2	0
9	73									30*	0	0	3
10	54										15*	3	1
11	36											10*	1
12	24												4*

Test 2.C4

Test T.2C itself has one component for each occasion from 2 to 11. For instance, the contingency table for test 2.C4 is marked off in Table 2. Under ITEC, the cell number at the bottom left is not related to the other values. Under CJS, it is constrained by the proportionality of rows and columns. It is possible to isolate this cell in order to test CJS versus ITEC. This can be done in 2 steps: First, cohort 4 is set aside;

		5	6	7	8	9	10	11	12	13
	1	0	0	0	0	0	0	0	0	1
(I)	2	0	1	0	0	1	0	0	0	0
	3	1	0	1	0	0	0	0	0	0

	5	6	7	8	9	10	11	12	13
1 to 3 pooled	*1*	*1*	*1*	*0*	*1*	*0*	*0*	*0*	*1*
4	9*	3	2	2	1	0	0	0	0

Second, occasion 5 is set aside.

	5	>5	6	7	8	9	10	11	12	13
1 to 3 pooled	1	4	1	1	0	1	0	0	0	1
4	9*	8	3	2	2	1	0	0	0	0
	(III)^					(II)				

The same decomposition can be repeated for the other occasions. All the information to test CJS versus ITEC is contained in the 2x2 contingency tables number III. Only the contingency tables number I and II are needed to assess the goodness of fit of ITEC. These contingency tables would help, among other things, to detect a long term effect of trap-dependence. I propose to name the chi² tests associated to the contingency tables number III, Tests $2.CT_i$ where T is for Trap-effect and the chi² tests associated to the contingency tables number I and II together Tests $2.CL_i$ where L is for Long term effect. The practical calculation of these tests involves usually some pooling because of the sparseness of the data.

Table 3: Goodness of Fit Tests to CJS and to ITEC

	Component	Chi-square	df	P-level
GOF Test to ITEC	3+2.CL	38.8959	29	.1037
Test of CJS when ITEC fits	2.CT	62.7552	10	.0000
GOF Test to CJS	2+3	101.6521	39	.0000

ITEC is accepted for the current data (Tab.3).

As regard the identifiability of the parameters, it has long been known that CJS has two non-identifiable parameters at the last occasion whose product is identifiable. If K is the number of capture-recapture occasions, ITEC involves K-2 more capture parameters than CJS while the test of CJS versus ITEC has only K-3 degrees of freedom (contingency tables $2.CT_i$ exists for i ranging from 2 to K-2). As a consequence, ITEC has additional identifiability problems. Here again, like for the more general models considered in the previous section, the non identifiability is most probably pervasive and no parameter is separately identifiable (Burnham, pers. comm.).

4. Reduced parameters models based on ITEC

More assumptions are needed to make the parameters identifiable. ITEC allows for different trap-effect in different occasions. For example, it may happen that trap-happiness at one occasion comes after trap-shyness at the previous occasion. This is very uncommon in the real world. On the contrary, it seems quite sensible to imagine a constant trap-effect on a reasonable scale. (We will see below that the logit scale has theoretical advantages). Additive models (Pradel, 1988) have been used many times in other contexts (for instance, additive age and rainfall effects on apparent survival: Kanyamibwa *et al.*, 1990, Peach *et al.*, 1991; additive effect of sex on catchability: Pradel *et al.*, 1990). The model with additive trap- and time-effects on a logit scale (ITEC+) can be fitted numerically using program SURGE4 (Pradel and Lebreton, 1991) but we can also test ITEC+ versus ITEC from the T2.CT_i contingency tables.

Table 4: Cell probabilities in table T2.CT_i. n_i and r_i are the row marginals. p and p' denotes capture probabilities succeeding respectively to non capture and to capture events.

	occasion i+1	occ. i+2 to K-1
cohort 1 to i-1	$n_i*S_i*p_{i+1}$	$n_i*S_i*(1-p_{i+1})$
cohort i	$r_i*S_i*p'_{i+1}$	$r_i*S_i*(1-p'_{i+1})$

To do that, we may use G^2 test instead of the Pearson chi^2. What is needed to calculate each component of the G^2 for any T2.CT_i table is the log odds ratio of capture at the first occasion versus non capture (Tab.4). It can easily be shown to be exactly equal to the difference between capture probabilities taken on a logit scale: LOR=logit(p_{i+1})-logit(p'_{i+1}). Under ITEC+, LOR are equal across occasions; under CJS they are equal to 0. The calculations can be brought about under GLIM (Baker and Nelder, 1978); see Table 5 for details.

Table 5: Correspondence between log linear models and SURGE4 models: Data from tables 2.CT_i are fitted in GLIM using a Poisson error and a log link function. The effects considered are line effect (l), column effect (c) and time effect (t).

GLIM model	SURGE4 model	abbreviation
l*c*t	S_t, P_{t*m}	ITEC
l*c*t-l.c.t	S_t, P_{t+m}	ITEC+
(l+c)*t	S_t, P_t	CJS
A test would involve other tables than T2.CT	S_t, P_m	

There is perfect agreement between both methods (fig.2). The numerical fit further allows the adjustment of model S_t,P_m.

In the Teal example, the tests given in fig.2 show a very significant trap-effect that may, however, be considered constant on a logit scale (ITEC+ versus ITEC, $chi^2(9)=6.39$ $p=0.70$). To ignore the trap-effect would lead to a severe underestimation of the apparent survival probabilities (fig.3). For reasons of parsimony, model S_t,P_m may be preferred to ITEC+ (LRT, $chi^2(10)=15.97$, $p=0.10$)

Fig.2: Comparison of tests brought about under GLIM (generalized linear model context) and likelihood ratio tests computed from the difference in deviance provided by SURGE4 (numerical fit). G^2 values for the tests are given with the number of degrees of freedom between parentheses. NS = non significant at 5%; *** highly significant, $p<0.001$. See Table 5 for the model notation.

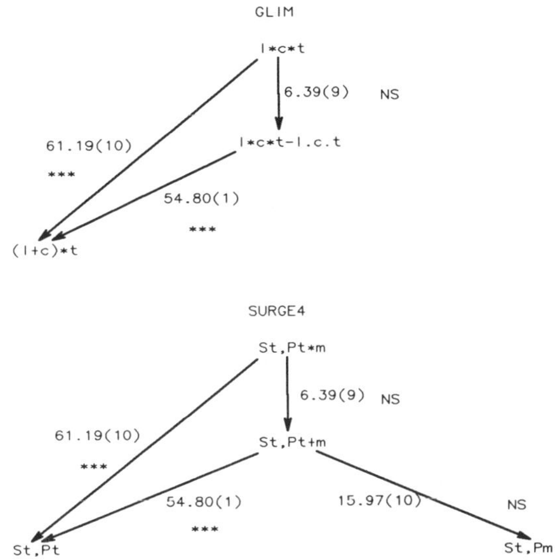

Fig.3: Estimates of apparent survival under a valid model correcting for trap-dependence (ITEC+ dashed line) and an invalid one (CJS plain line) exhibiting a strong negative bias.

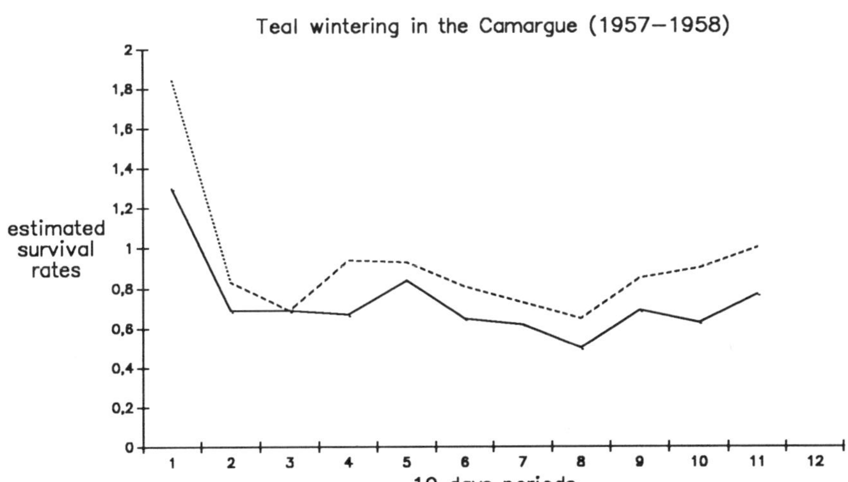

4. Conclusions

The example studied shows a very strong trap-effect on capture (P=0.11, P'=0.44 model S_t,P_m). This is probably to be related to the brevity of the interval between capture occasions: 10 days. One can wonder whether the trap-dependence effect can last longer and what could be done in that case. One can also wonder whether such effect can appear when capture occasions are yearly.

It should be possible to make up tests of a trap-effect ranging over 2 (or more) occasions by further splitting the T2.C contingency tables; but it is not as easy as before (2 or more rows and columns are affected) and the data are anyway generally too scarce to make the attempt worthwhile. The numerical approach is more hopeful and easier to implement. Just remember that, due to the particular presentation of the data we used, trap-effects have been replaced by "age"-effects. Thus, it is merely a question of, as a first step, fitting a 3 "age"-classes model, the first two "age"-classes ranging one interval each. It is even possible to smooth the effect of trap-dependence over time. Such models were not needed for the Teal.

We did find a trap-effect in an unpublished study of herring gulls where visits are yearly but in a study of breeding Blue Tits (Blondel *et al.*, 1992) where birds are captured in nest boxes, no trap-effect was apparent. Are long-lived species more susceptible to respond to being trapped? Another study dealing with Flamingos (Lebreton *et al.*, 1992) cautions us against hasty conclusions. Lebreton *et al.* found "trap-dependence" effects but this cannot be genuine trap-dependence because Flamingos are not physically recaptured. One must always keep in mind that models can have different interpretations. The conditions that led us to design the present model (shift in behaviour induced by capture) are not the only possible interpretation for a different probability of capture after a capture event. The same is expected to happen in the presence of heterogeneity in the proneness to capture. This could explain the Flamingos case because Flamingos are probably largely faithful to their place of previous nesting and some nests are more easily sighted than others. Thus, models considering trap-dependence might also be useful to treat, at least partially, other phenomena that mimic genuine trap-dependence. Among them are the important cases of random temporary migration to a less accessible zone (Sandland and Kirkwood, 1981) with, as an extreme, temporary emigration, and heterogeneity of capture rates between individuals as suspected in the Flamingos example.

Although it is important to recognize the trap-effect whenever it exists in order to avoid the bias on the survival estimates, general trap-dependence models have severe identifiability problems that seem to propagate from the last occasion back. Probably in many cases, limited trap-dependence can be a good alternative. Yet even ITEC has still unclear identifiability problems, most likely all pervasive (Burnham, pers. comm.). For the moment, ITEC+ is probably the best solution as long as it fits; like for CJS, only the last occasion parameters are not separately identifiable.

Acknowledgements. Comments on an earlier version by S. Baillie and R.M. Cormack helped me to improve the paper.

References

Baker, R.J. and Nelder, J.A. (1978).- The GLIM system, release 3, Generalized Linear Interactive Modelling. Numerical Algorithm Group, Oxford.

Blondel, J., Pradel, R. and Lebreton, J.D. (1992).- Low fecundity insular blue tits do not survive better as adults than high fecundity mainland ones Journal of Animal Ecology, 61: 205-213

Brownie, C. and Robson, D.S. (1983).- Estimation of time-specific survival rates from tag-resighting samples : a generalization of the Jolly-Seber model. Biometrics, 39 : 437-453.

Burnham, K.P., Anderson, D.R., White, G.C., Brownie, C. and Pollock, K.H. (1987).- Design and analysis methods for fish survival experiments based on release-recapture. American Fisheries Society Monograph n° 5, 437 p.

Kanyamibwa, S., Schierer, A., Pradel, R. and Lebreton, J.D. (1990).- Changes in adult survival in a western european population of White Stork (Ciconia ciconia). Ibis, 132 : 27-35

Lebreton, J.D., Burnham, K.P., Clobert, J. and Anderson, D.R. (1992).- Modelling survival and testing biological hypotheses using marked animals : a unified approach with case studies. Ecological Monographs, 62: 67-118.

Peach, W., Baillie, S. and Underhill, L. (1991).- Survival of british Sedge Warblers Acrocephalus schoenobaenus in relation to west African rainfall. Ibis, 133: 300-305.

Pollock, K.H. (1975).- A K-sample tag-recapture model allowing for unequal survival and catchability. Biometrika, 62: 577-583.

Pollock, K.H. (1981).- Capture-recapture models allowing for age-dependent survival and capture rates. Biometrics, 37 : 521-529.

Pradel, R. (1988).- Nouvelles approches statistiques en capture-recapture, un exemple : le test de parallélisme. Unpubl.report, USTL, Montpellier, France, 63 p.

Pradel, R. and Lebreton, J.D. (1991).- User's Manual for Program SURGE version 4.1 Mimeographed document. C.E.P.E./C.N.R.S. BP5051, 34033 Montpellier, CEDEX, France., 35pp.

Pradel, R., Clobert, J. and Lebreton, J.D. (1990).- Recent developments for the analysis of capture-recapture multiple data sets. An example concerning two Blue Tit populations. The Ring, 13, 193-204.

Pradel, R., Rioux, N., Tamisier, A. and Lebreton, J.D. (submitted).- Turnover of individual birds in their winter quarters: Example of teal in the Camargue.

Sandland, R.L. and Kirkwood, G.P. (1981).- Estimation of survival in marked populations with possibly dependent sighting probabilities. Biometrika, 68 : 531-541.

Wilkinson, G. N. and Rogers, C.E. (1973).- Symbolic description of factorial models for analysis of variance. Applied Statistics 22 : 392-399.

Marked Individuals in the
Study of Bird Population
J.-D. Lebreton and P. M. North (eds.)

The Flexibility of GLIM Analyses of Multiple Recapture or Resighting Data

R.M. Cormack

Department of Mathematical and Computational Sciences, University of St Andrews, St Andrews, Fife KY16 9SS, Scotland.

Summary. Birds are not inanimate objects blindly following laws of chance. To understand what we can safely learn from mark-recapture (CMR) records requires careful exploratory analysis of all the data. The paper advocates an approach to this when there are no continuous covariates measured on individuals, so that all the information is contained in the numbers of individuals having each of the (2^s-1) capture histories from s samples. Careful analysis is defined in terms of a set of loglinear models, from which an appropriate model is selected by examination of residuals as well as comparison of deviances, bearing in mind the biological interpretation of parameters and of residual patterns. Monte Carlo simulations are used to validate model-based inferences from the analysis.

The easy and flexible way in which a general interactive statistics software package can illuminate CMR data is demonstrated on an 8-year winter study of lazuli buntings (*Passerina amoena*). The biological background suggests that the population may contain both transients and "residents" and that trap dependence is unlikely. The analysis confirms these prior beliefs while indicating that other individual differences between birds remain, thus offering insights and interpretations different from those given by Burnham et al. (1987).

1. Introduction

The philosophy behind this approach to mark-recapture (CMR) data is to provide a framework within which the data can focus the biologist's attention on the behaviour of individual animals. In this sense it is the antithesis of an approach in which the first step is to formulate a precise probability model and identify minimal sufficient statistics for that model.

A model is a compromise between believable assumptions and uninterpretable complexity. Birds are not coloured balls dropped by controlled combinatorial mechanisms into a range of boxes. The biologist will have prior expectations as to how the particular species behaves, both when undisturbed and in response to the marking procedure, and should have the freedom to explore the data in the light of his expectations.

Such freedom carries with it the responsibility of understanding how a particular type of behaviour may be reflected in the available data. This is made easier by keeping the data in all analyses at the most disaggregated level, not changing the basis of each analysis by amalgamating different categories. In a CMR study this level is that of the capture history ω of an individual throughout the study, e.g. seen in samples 2, 4, 5 and 6, but not in samples 1, 3 or 7. In the absence of distinguishing information on individuals, the set of numbers n_ω of individuals observed to have each capture history ω forms the totality of information from the captures. Extra information on subpopulations, e.g. sexes , or on samples, e.g. their duration, can be incorporated in the framework. Data on individuals' quantitative covariates, such as weight, are better handled in the logistic framework developed recently by Alho (1990) and Huggins (1989).

2. Principles

The approach therefore (Cormack, 1985, 1989) is to describe the 2^s-1 counts n_ω from an s-sample study in terms of the complete set of main effects and interactions between s factors, representing the s samples, each at 2 levels - seen or not seen in that sample. Model formulation parallels the classical analysis of an agricultural experiment with s factors each either present or absent in any one treatment, or the log-linear analysis of a 2^s table of counts, expounded by Bishop, Fienberg and Holland (1975). Technically my approach differs from the latter in two respects:

first, in its parameterization, where the symmetry of representation of the two levels of each factor is inappropriate for a CMR study in which 'Seen' is known and 'Unseen' is unknown;

second, in its computer implementation, where the non-hierarchical nature of many CMR models means that they must be expressed in terms of regression variables rather than factors.

A set of (2^s-1) counts can be fitted exactly by the complete set of (2^s-1) regression variables. Some of these have specific biological interpretations. In the complete set:

1 variable M has as its coefficient an overall population 'size';

s variables P_i have as their coefficients functions of capture probabilities p_i;

(s-1) variables D_i have as their coefficients functions of survival rates ϕ_i and capture probabilities p_i with D_{s-1} identical to P_s;

(s-1) variables B_i have as their coefficients functions of immigration rates with B_1 identical to P_1;

(s-1) variables I_i have as their coefficients functions of dependence between successive samples, though I_1 is identical to B_2 and I_{s-1} to D_{s-2};

s variables T_i have as their coefficients functions of the numbers seen in one sample only, though T_1 is identical to D_1 and T_s to B_{s-1}.

Apart from the identities stated, these variables are linearly independent. The analysis explores, in the light of the biologist's expectations, which of these sets of variables are required in the model, and what other variables from the complete set or what combinations of them might plausibly be added to the set of regression variables. It then considers what behavioural patterns would be expected to reveal themselves in an excess or deficit in numbers of animals having particular capture histories. Inclusion of further terms in the model and patterns of residuals from standard models can be explored in order to gain information about the behaviour of the individuals making up the population. Any set of capture histories can easily be given weight zero and thus omitted from the data analysis, to explore how sensitive parameter estimates are to these particular observations.

Particularly with sparse data, asymptotic results should not be the only basis for trust in inferences. The behaviour of estimates, test statistics and residuals should be examined by generating Monte Carlo realizations of data from a proposed model, using parameter values estimated from the observed data. Cross-validation can also be practised, using a randomly selected subset of birds as a data set from which the model is selected, and validating the chosen model on the remaining subset of birds.

3. Technicalities

In the model formulation the only distributional assumption made is that the variance V of n_ω is a specified function of the expected number μ in that category, usually $V=\mu$, but one of the overdispersed forms $V=\phi\mu$ or $V=\mu+\phi\mu^2$ (see, e.g. Breslow, 1990) can be examined if clustering of birds within capture histories is suspected. In the absence of further assumptions, inference is based on quasi-likelihood (Wedderburn, 1974). With $V=\mu$, significance tests between nested models appeal to the asymptotic χ^2 distribution of deviance differences: there is accumulating evidence that this is a good approximation even for very sparse data, for which the residual deviance is not χ^2. Model selection follows from such tests and from residual patterns.

Different types of residual have been proposed as expressions of how different an observed count n_ω is from its fitted value μ (Pierce and Schafer, 1986; McCullagh and Nelder,1989, pp37-40). Choice among them remains a subject of debate. Here we use the deviance residuals,

$$[2\{n \ln(n/\mu)-n+\mu\}]^{1/2}\, \mathrm{sgn}(n-\mu)\ .$$

Their use is consistent with the use of the residual deviance as a measure of goodness-of-fit (and to provide an estimate of any overdispersion parameter ϕ) and the use of maximum likelihood or quasi-likelihood as our method of estimation. Although the estimate of ϕ from the residual mean deviance is inconsistent, it may have better properties in small samples (Nelder and Lee, 1992) than the alternative of Pearson's χ^2 divided by its degrees of freedom (McCullagh and Nelder, 1989; Breslow, 1990). The latter is consistent with the goodness-of-fit χ^2 statistic and estimation by minimum χ^2 , but not with maximum likelihood, which can result in negative increments between nested models (see e.g. Table 3). Some readers may doubt the wisdom of analysing such disaggregated and sparse data by quasi-likelihood (QL) without any distributional framework. However, the approaches are related:

1. If the assumption, commonly made for counts, of a Poisson distribution is added, QL estimates are maximum likelihood.

2. Survival estimates have the same asymptotic variances under the Poisson assumption as under the usual CMR multinomial assumption (Cormack and Jupp, 1991). Variances of estimates of population size obtained under the Poisson assumption need only a simple modification to convert them to multinomial variances.

3. Likelihood-ratio tests based on the disaggregated data are precisely the same as those based on efficiently aggregated data.

4. Transients

In many studies the possibility exists that the individuals in the study area at any time are a mixture of residents and transients, individuals passing through the area whether on account of migration, dispersal, hunting, territory-seeking, or some other cause. Transients will swell the numbers seen on only one occasion. If this possibility cannot be ruled out on biological grounds then the numbers seen only once should be compared with the corresponding numbers predicted by the model fitted to the rest of the data omitting these seen once - an extended jackknife.

There are two ways of achieving this within the log-linear parameterisation. One is by including in the model all interactions of order (s-1), the variables T_i, since each such interaction fits exactly one of the capture histories having a single capture. The other is by editing the weight vector to be zero for the units representing individuals seen only once: this does not imply that no individuals are observed with these capture histories. Analysis of this model will then give fitted values for the number of residents that would be predicted to be seen only once. The transient population at each sample period is then estimated by subtracting this estimate from the numbers observed only at that sample. Since no information is available on the survival rate of transient birds, estimates of survival pertain to the resident population. The only bias incurred by weighting out particular observations from the analysis is that incurred in any process of estimating parameters in a model selected from the same data : the fact that some residents will by chance be seen only once still remains in the model.

5. Example - Lazuli Buntings (*Passerina amoena*)

One of the examples given by Burnham et al (1987 : pp 348-361) to illustrate the use of programs RELEASE and SURVIV analyses an 8-year study by Allen W Stokes of lazuli buntings wintering in Logan, Utah.

5.1.Analysis in Burnham et al. (1987)

This analysis goes through the following stages.

1) Provisional acceptance of the standard Jolly-Seber model. Tests 2.C.i look at those birds known to be alive in Years i and (i+j) and ask if being captured in Year i influences the chance of next being captured in Year (i+j). Small numbers often restrict the feasibility of the test to j=1, recapture in the sample immediately after first marking. The Pearson χ_1^2 goodness-of-fit statistics are 1.74, 1.68, 0.32, 2.29 and 0.21 for i=2 to 6 respectively - an overall χ_5^2 of 6.25.

2) Tests 3.Sm.i look at birds caught in Year i and known to be alive in Year (i+1) and ask if those first caught in Year i have the same chance of being seen in Year (i+1) as those seen earlier. Burnham et al note that only one of these tests (i=5) makes a

major contribution ($\chi_1^2 = 6.68$) to the overall χ_5^2 of 11.25. In fact, all years suggest a shortfall in the numbers seen in the year after first marking, even among those known to remain in the population - the χ_1^2 from the sums of the corresponding cells of the five 2x2 tables is 9.18.

Small numbers often restrict the comparisons to the two groups - newly marked at i, and previously marked. In this form this is a test for temporary emigration after first marking.

3) Tests 3.SR.i all show an overwhelming rejection of the hypothesis that individuals have the same probability of being seen again after first capture as after any subsequent recapture. Burnham et al. (p.351) conclude "that the banded population consists of a mixture of birds that are probably never seen again after banding and birds that are commonly recaptured".

Their analysis proceeds by partitioning the data into two groups or arrays: "new captures and previous captures". "The birds that return and are recaptured fit the Jolly-Seber model fairly well." Jolly-Seber estimates of survival ϕ_i and capture probability p_i , separately for both arrays, are given in columns 2 and 3 of Table 4. Three models then examine if certain parameters can be regarded as taking common values:

A : survival ϕ and capture probability p constant over years, p differing between groups;

B : ϕ constant, the same for both groups: p_i year-specific and differing between groups;

C : ϕ_i year-specific but the same for both groups; p_i as in B.

Model A is rejected by a likelihood-ratio test ($\chi_{10}^2 = 57.70$) in comparison with Model B . Model B is not rejected against Model C by such a test ($\chi_4^2 = 2.08$) but is rejected by a goodness-of-fit χ_9^2 of 23.0. Comparison of the two χ^2 values suggests that Model C would also be rejected. Estimates of ϕ_i and p_i under Model C are given in columns 4 and 5 of Table 4.

The biological interpretation of Model C, which seems to be accepted by Burnham et al. (1987), is as follows. All birds seen in that area in one winter have the same probability of being there again in any subsequent winter. However their probability of being seen remains low until they are seen for a second time, after which they become more visible in subsequent winters. I find this inherently implausible.

5.2. Analysis using GLIM

GLIM (Payne, 1986) is a widely available general statistical package for the regression analysis of linear or generalized linear models for univariate data. The complete set of GLIM instructions to analyse these data is given in Table 1. Two external files are called. The first, **lazuli.dat,** contains the data in standard order (Table 2). The second, **recap.glm**, a slightly modified version of that listed in Cormack (1985), contains GLIM subfiles **cr3** to **cr8** to construct the regression variables required for s=3 to s=8 samples, and a general subfile **cr1** comprising macros **estimate** and **rplot** which are called later when estimates of the biological parameters or examination of residuals are required.

Table 1 : Complete GLIM input for sequence of analyses discussed. The input required by or the output obtained from any instruction is summarised in the Table indicated in the final column.

GLIM input	Interpretation	Refer Table
`$units 255   $data r   $`	!	
`$dinput 21  $`	! Prompt for data file	
`lazuli.dat`	! Table 2 gives data	2
`$input 22 cr1 cr8  $`	! Prompt for macros file: 8 years	
`recap.glm`		
`$fit p1+p2+p3+p4+p5+p6+p7+p8  $`	! MODEL M1 : Closed population M_t	3
`$fit +d1+d2+d3+d4+d5+d6+b2+b3+b4+b5+b6+b7  $`	! MODEL M2 : Jolly-Seber	3
`$use estimate  $`	! gives estimates as Table 12 in B	4
`$use rplot  $`	! residuals vs. capture frequency f	5
`$fit +pbd-p2-p3-p4-p5-p6-p7  $`	! MODEL M3 : constant capture prob.	3
`$fit -pbd+p2+p3+p4+p5+p6+p7+i2+i3+i4+i5+i6  $`	! MODEL M4 : open with dependence	3
`$fit -i2-i3-i4-i5-i6+t2+t3+t4+t5+t6+t7  $`	! MODEL M5 : J-S plus transients	3
`$use estimate  $`	! gives J-S estimates for residents	4
`$use rplot  $`	! residuals for residents	5
`$fit +h  $`	! M6 - test for residual heterogeneity	3
`$edit 1 w  0     $fit  .  : -h  $`	! remove f_8 : M7 and M8 - redo test,JS	3
`$use estimate  $`	! estimates when f_8 omitted	4
`$fit +i2+i3+i4+i5+i6  $`	! M9 - tests any dependence now	3
`$cal pp3=p2+p4+p5+p6+p7  $`	! vector for constant prob except p_3	
`$fit -i2--i3-i4-i5-i6-p2-p4-p5-p6-p7+pp3  $`	! MODEL M10 : J-S residents, no f_8, constant capture prob except p_3	3
`$use estimate  $`	! estimates from this final model	4
`$use rplot  $`	! deviance residuals from final model	5
`$dis r  $`	! Pearson residuals from model	
`$dis v  $`	! Covariance matrix of estimates	
`$stop   $`		

Table 2: Counts n_ω of Lazuli buntings, in standard order by row of capture history ω, starting with ω = (11111111), (21111111), (12111111), (22111111), (11211111) , (21211111),... where the ith number is 1 for a history including being seen in the ith sample, 2 if not seen.

```
1 0 0 1 1 1 0   2 0 0 0 0 0 0 0    11 0 0 0 0 0 0 0    0  0 0 0   0   0   0    0   12
0 0 0 0 0 0 0   0 0 0 0 0 0 0 0     1 0 0 0 0 0 0 0    0  0 0 0   0   0   0    0   46
0 0 0 1 0 0 0   0 0 0 0 0 0 0 0     0 0 0 0 0 0 0 0    1  0 0 0   0   0   0    0    7
0 0 0 0 0 0 0   0 0 0 0 0 0 0 0     1 0 0 0 0 0 0 0    1  0 0 0   0   0   1    0  100
0 0 0 0 0 0 0   4 0 0 0 0 0 0 0     5 0 1 0 1 0 0 0    0  0 0 0   0   0   0    0   27
0 0 0 0 0 0 0   0 0 0 0 0 0 0 0    12 0 0 0 0 0 0 0    3  0 0 0   0   0   0    0  194
0 0 1 0 0 1 0   3 0 0 0 1 0 2 1    22 0 0 0 0 0 1 0    2  0 0 1   0   0   1    0  210
2 1 0 2 3 1 0  16 0 0 0 2 0 2 0   203 2 4 0 3 0 6 0  198  2 6 2  54  20 339  132
```

My prior expectations were that birth and death will occur between samples, that for a small passerine the danger of negative estimates of them will be slight, that definition of a population of birds resident in the same area over a number of winters will be difficult, so that heterogeneity is likely, perhaps explainable by a mixture of subpopulations - residents and transients. Temporary absence in the year after first marking was thought to be unlikely. Analysing data at a distance I do not know if a measure of sampling effort each winter is available.

The first **$fit** instruction tries the standard Schnabel model for a closed population, called M1 in Table 3. GLIM responds with the residual deviance, given for each model in that table. The next **$fit** adds death and birth parameters to form the Jolly-Seber model M2 : the huge improvement is visible in the difference in deviance. Since this is the basic model for an open population a table of parameter estimates (survival and capture probability, given in Column 1 of Table 4, and population and immigrant numbers, not discussed here) can be obtained by **$use estimate** and a summary of residuals by **$use rplot**. A complete listing of all 255 Pearson residuals would be given by **$dis r** after any model fit, but is so overwhelming that it should be used sparingly as a final exploration: **rplot** is only one of many possible ways of examining the residuals. It plots them against capture frequency, and tabulates the average of the residuals for those seen once, twice, etc. Before investigating the heterogeneity revealed by the excess of birds seen all 8 times or only once (Table 5), the next two **$fit** statements ask whether or not constant capture probability is acceptable (M3 : answer 'No'- the deviance difference is large) and, returning to unequal capture probabilities, how strong the evidence is for dependence between successive samples (M4 : some evidence but not insistent). The latter test is comparable to Burnham et al's 2.C.i tests. Next (cf their 3.SR.i tests) the dependence parameters are removed and factors t, which give a perfect fit for the numbers seen once only, are added - a huge reduction in deviance (M5). M5 can alternatively be fitted by weighting out birds seen once by the instructions **$cal w= w*%ne(pc,7) $fit .** instead of adding the t parameters to the model. This has the advantage that **$dis r** gives the numbers predicted by the rest of the data to be seen once only :

	18.4	81.3	55.2	75.4	25.3	8.5	50.9	9.1
compared with values	100.0	183.5	195.6	194.6	185.6	47.3	331.0	132.0

predicted by the Jolly-Seber model fitted to the complete data. Again **$use estimate** gives estimates for the resident population. After each fit of a submodel of Jolly-Seber type, **$dis e** would display the GLIM estimates, allowing a check that birth and death are not negative. Also, after any **$fit** those who prefer χ^2 to deviance as their measure of goodness-of-fit can obtain it by **$look %x2 $** . In this, capture histories are not pooled in any way: the χ^2 differences, including one negative one, are shown in Table 3.

Use of **rplot** after this model suggests that some heterogeneity remains even after transients have been removed. If birds in a closed population differ in their intrinsic catchability, too many will be seen once or twice or on most occasions, and too few an intermediate number of times. This suggests formally testing for heterogeneity by adding to the model a term which is quadratic in the frequency of capture: **$fit +h $** (M6) gives a highly significant reduction in the deviance. The most extreme residual

Table 3 : Deviances (DV) and degrees of freedom (d.f.) from the sequence of models fitted. DVdiff and χ^2 diff give the changes in deviance and in Pearson's χ^2 between nested models,with d.f. given in the intervening column. The model with which comparison is made is given in the final column.

	Model	d.f.	Deviance	DVdiff	d.f.	χ^2 diff	Model comparison
M1	Closed	246	854.2				
M2	Jolly-Seber	234	219.41	634.79	12	3136410	with Model 1
M3	J-S : constant p	239	233.85	-14.44	-5	+158	with Model 2
M4	Open with dependence	229	213.28	6.13	5	264	with Model 2
M5	J-S plus transients	228	125.18	94.23	6	3416.5	with Model 2
M6	Added heterogeneity	227	115.80	9.38	1	24.9	with Model 5
M7	with f_8 removed	226	115.52	0.28	1	1.3	with Model 6
M8	Transients and no f_8	227	120.82	4.36	1	15.3	with Model 5
M9	Added dependence	222	114.86	5.97	5	33.1	with Model 8
M10	Constant p except p_3	231	122.30	-1.48	-4	-0.7	with Model 8
	further heterogeneity	230	117.71	4.59	1	36.0	with Model 10

Table 4: Estimates of survival ϕ and capture probability p first from the overall Jolly-Seber model, then for the various subcategories defined by Burnham et al. (1987), finally for GLIM models M5, M8 and M10. Suffices AM and GM refer to arithmetic mean and geometric mean of survival estimates.

	J-S	Burnham et al				GLIM		
		Newly marked	After re-release	Model C New	Model C Old	Without transients	No f_1 or f_8	No f_1 or f_8 Constant p
ϕ_1	0.488	0.575						
ϕ_2	0.223	0.270	0.454	0.328		0.485	0.470	0.455
ϕ_3	0.551	0.544	0.591	0.544		0.674	0.660	0.652
ϕ_4	0.324	0.255	0.720	0.412		0.729	0.723	0.702
ϕ_5	0.408	0.494	0.408	0.419		0.518	0.507	0.514
ϕ_6	0.333	0.296	0.508	0.401		0.556	0.551	0.607
ϕ_{AM}	0.388	0.406	0.536	0.421		0.592	0.582	0.586
ϕ_{GM}	0.372	0.383	0.525	0.415		0.585	0.575	0.579
p_2	0.378	0.321				0.687	0.667	0.637
p_3	0.230	0.137	0.497	0.102	0.538	0.382	0.364	0.364
p_4	0.374	0.200	0.900	0.198	0.893	0.633	0.621	0.637
p_5	0.472	0.428	0.621	0.271	0.744	0.606	0.594	0.637
p_6	0.451	0.284	1.000	0.297	1.000	0.612	0.605	0.637
p_7	0.619	0.513	0.835	0.400	0.883	0.714	0.708	0.637

Table 5 : Means of deviance residuals for each capture history, categorised according to frepuency of capture f, under each of Models M2, M5 and M10.

f	M2	M5	M10
8	3.67	2.03	
7	0.37	-0.03	-0.00
6	0.11	-0.18	-0.16
5	0.13	-0.17	-0.16
4	0.03	-0.26	-0.24
3	-0.23	-0.38	-0.36
2	-0.69	0.22	0.19
1	0.58		

was caused by the one bird seen on all eight occasions. It can be given zero weight in the analysis by **$edit 1 w 0 $**. Models (M7) with, and (M8) without, further heterogeneity are examined by **$fit . : -h $** . Unfortunately, although removal of that individual significantly reduces the deviance (χ_1^2=4.36) from M5, other heterogeneity remains (χ_1^2=5.30, comparing M7 with M8). Since no estimates are obtainable from M6 or M7, M8 might be regarded as acceptable with reservations, and estimates obtained from it. The question is then whether or not, for this model and reduced data set, there is any evidence of dependence between samples (M9) - the χ_5^2 of 5.97 suggests some, but not too serious.

The estimates suggest that sampling effort may have been constant in all years except the third. This would need to be confirmed by the experimenter: if so, such a model (M10) can be fitted by the final **$fit** statement in Table 1. The addition to the deviance (χ_4^2 =1.48) is acceptable. Adopting this as the new provisional model **$use estimate $** and examine the complete pattern of residuals by **$use rplot $dis r $** the latter providing Pearson residuals.

To check on evidence for temporary emigration after first marking, though a priori this seems less likely than in a study with samples days or weeks apart, all capture histories can be examined whose records in the three years from first sighting are Seen-Unseen-Seen. In standard order, capture histories starting

121+++.... are Units 3, 11, 19, 27, ... Those starting
2121++.... are Units 6, 22, 38, 54, ... Those starting
22121+.... are Units 12, 44, 76, 108, ... and so on.

The pattern of suffices doubling makes it easy to pick out all categories indicating such temporary emigration. In them, the sum of observed counts are

4, 10, 3, 3, 13, 7 while for M10 the fitted values are
2.8, 10.4, 2.1, 4.8, 9.1, 7.3 showing no consistent suggestion

of temporary emigration.

The sparseness of the data results in a residual deviance of 122.3 from M10 which is clearly not a χ^2 with 231 df. To assess its significance, sets of Poisson observations are simulated with means given by fitted values for n_ω under M10. Different models (e.g. M8 and M10) can then be fitted to the simulated data sets.

The residual deviances from 200 runs of M10 have mean 100.3, SD 16.1, 90% percentile 122.4 and 95% percentile 129.5. The observed 122.3 is 21st highest of the 200 runs.

The geometric mean ϕ_{GM} of the survival rates for residents was 0.572 with SE 0.039 (0.044 when M8 was fitted).

The deviance differences between M8 and M10 have mean 4.0, SD 2.67, 90% percentile 7.71 and 95% percentile 9.18. Since the data were generated under the more restrictive M10, this should asymptotically be a χ_4^2 with mean 4.0, SD 2.82, 90% percentile 7.78, 95% percentile 9.49 - confirmation that the distribution remains a good approximation for deviance differences even for data with this degree of sparseness. The observed 1.48 is 33rd smallest of the 200 runs.

Similar assessments can be made of differences from other models, of specific residual patterns (e.g. those indicative of temporary emigration or behavioural heterogeneity), of the behaviour of χ^2 differences - the list is limited only by the experimenter's interest.

Conclusions about this data set can be summarised as follows. The dominant feature is the excess of individuals seen only once, provisionally interpreted as a mixture of two subpopulations, called 'residents' and 'transients'. Some evidence of heterogeneity remains among the residents, the most obvious, but not the only, aspect of this being the one individual seen every year. Behavioural disturbance in response to capture is slight. An estimate of annual average survival for residents of 0.572, with 95% CI of (0.49,0.65), is unlikely to be grossly in error. Some reduction in its variance, but no change in its value, has been achieved by assuming capture probability constant for 5 out of the 6 intermediate years. More precision would be achieved if this assumption were extended to the final sample, though the data can provide no information on the validity of this.

6. Discussion

With many data sets, the type of analysis used in the above example will lead to the same general biological conclusions as the type of analysis possible within the range of specialised computer packages now available for CMR and ring-recovery data, impressively discussed and illustrated by Lebreton et al. (1992) and references therein. There are several combinations of data type and model which can be analysed by one or other of these packages but not, yet or possibly ever, by GLIM. One such is a model with constant annual survival. However, nothing is gained by such a model in estimating or understanding survival: the precision of the estimates of average (geometric or arithmetic mean) survival will be almost identical. Moreover, absolutely constant annual survival is not a believable null hypothesis. Although a constant capture probability is equally unbelievable, its assumption can increase the precision of survival estimates, at the cost of increasing bias if the assumption is false.

There is a fundamental difference in approach between specialist CMR packages and the type of analysis illustrated here, whether implemented in GLIM or other general software: what takes priority - the model or the data? In the specialised packages the model does. Residuals are never mentioned. The computer revolution has completely changed the statistician's approach to analysing regression models - we look at residuals, influence, data plots and transformations: a statistical analysis is aimed at extracting information from the data. GLIM makes this attitude available to CMR analysts. The framework encourages investigation of residuals. Lebreton et al. claim that GLIM is "useful only to statisticians". The irony is that by using a specialised package the analyst is forced to think in the way of the creator of that package: only by using a general package analysing full capture histories will the biologist find flexibility really to examine the data.

7. Extensions

Further examples of data anlyses can be found in Cormack (1985,1989).

Other analyses and models which fit readily into the log-linear framework, and which have not been reported previously include:

the use of score tests (Pregibon, 1982) to aid selection between nested models;

radiotelemetry data - CMR models with capture probabilities p_i=1;

CMR with death on capture;

the models (s_i, f_i) , (s, f_i) and (s,f) given by Brownie et al. (1985) for ring-recovery studies.

In the second of these the GLIM **$offset** facility is used. In the third, the additional data on deaths on capture are input in a particular order after the complete data. The extension to ring-recovery models requires a different standard order for input and a different set of regression variables. Programming details are available from the author.

Acknowledgement. Financial assistance was provided by The Wellcome Trust

References

Alho, J.M. (1990). Logistic regression in capture-recapture models. *Biometrics* **46**, 623-635.

Bishop, Y.M.M., Fienberg, S.E. and Holland, P.W. (1975). *Discrete Multivariate Analysis : Theory and Practice.* Cambridge, Massachusetts: The Massachusetts Institute of Technology Press.

Breslow, N.E. (1990). Tests of hypotheses in overdispersed Poisson regression and other quasi-likelihood models. *Journal of the American Statistical Association* **85**, 565-571.

Brownie, C., Anderson, D.R. , Burnham, K.P. and Robson, D.R. (1978, 1985). *Statistical inference from band-recovery data - a handbook.* U.S. Fish and Wildlife Service Resource Publication 1st ed. No. 131, 2nd ed. No. 156.

Burnham K.P., Anderson, D.R., White, G.C., Brownie, C. and Pollock, K.H. (1987). *Design and analysis methods for fish survival experiments based on release-recapture.* Bethesda, Maryland: American Fisheries Society.

Cormack, R.M. (1985). Examples of the use of GLIM to analyse capture-recapture studies. In *Lecture Notes in Statistics, no.29. Statistics in Ornithology* , B.J.T. Morgan and P.M. North (eds.), 243-273. New York: Springer-Verlag.

Cormack, R.M. (1989). Log-linear models for capture-recapture. *Biometrics* **45**, 395-413.

Cormack, R.M. and Jupp, P.E. (1991). Inference for Poisson and multinomial models for capture-recapture experiments. *Biometrika* **78**, 911-916.

Huggins, R.M. (1989). On the statistical analysis of capture experiments. *Biometrika* **76**, 133-140.

Lebreton, J-D., Burnham,K.P., Clobert, J. and Anderson, D.R. (1992). Modeling survival and testing biological hypotheses using marked animals: a unified approach with case studies. *Ecological Monographs* **62**, 67-118.

McCullagh, P. and Nelder, J.A. (1989). *Generalized Linear Models,* 2nd.edition. London: Chapman and Hall.

Nelder, J.A. and Lee, Y. (1992). Likelihood, quasi-likelihood and pseudo-likelihood : some comparisons. *Journal of the Royal Statistical Society, Series B* **54**, 273-284.

Payne, C.D. (1987). *The GLIM System, Release 3.77.* Oxford: Numerical Algorithms Group.

Pierce, D.A. and Schafer, D.W. (1986). Residuals in generalized linear models. *Journal of the American Statistical Association* **81**, 977-998.

Pregibon, D.A. (1982). Score tests in GLIM with applications. In *Lecture Notes in Statistics, no.14. GLIM 82 : Proceedings of the International Conference on Generalized Linear Models,* R.Gilchrist (ed.), 87-97. New York: Springer-Verlag.

Wedderburn, R.W.M. (1974). Quasilikelihood functions, generalized linear models and the Gauss-Newton method. *Biometrika* **61**, 439-447.

Marked Individuals in the
Study of Bird Population
J.-D. Lebreton and P. M. North (eds.)

Estimating Survival of Radio-Tagged Birds

Christine M. Bunck[1] and Kenneth H. Pollock[2]

[1]*U.S. Fish & Wildlife Service, Patuxent Wildlife Research Center, Laurel, MD 20708, and*
[2]*Department of Statistics, North Carolina State University, Raleigh, NC 27695-8203*

Summary. Parametric and nonparametric methods for estimating survival of radio-tagged birds are described. The general assumptions of these methods are reviewed. An estimate based on the assumption of constant survival throughout the period is emphasized in the overview of parametric methods. Two nonparametric methods, the Kaplan-Meier estimate of the survival function and the log rank test, are explained in detail. The link between these nonparametric methods and traditional capture-recapture models is discussed along with considerations in designing studies that use telemetry techniques to estimate survival.

1. Introduction

Radio-tagging and the subsequent tracking of birds are useful techniques for studying survival, movement and habitat use. After capture, tagging and release, searches for radio-tagged birds are made at regular and frequent intervals. The nearly continuous monitoring provides data on survival that are different than data modelled by traditional capture-recapture (Burnham et al., 1987; Pollock et al., 1990; Lebreton et al., 1992) or ring-recovery methods (Brownie et al., 1985). Data from radio-tagged birds contain confirmed mortalities that can be used directly in estimating survival rates, as well as incomplete data records that result when radio-tagged birds disappear during the study and their fates (dead or alive) are unknown.

Methods for estimating the survival rate of radio-tagged birds have largely been adapted from other applications. One common parametric method treats time as discrete units and is an extension of an estimate for nest survival that was proposed by Mayfield (1961). Nonparametric methods have been taken from medical applications (Cox and Oakes, 1984) and modified to accommodate the unique problems that arise in the study of highly mobile animals, such as birds (Pollock et al., 1989a; Pollock et al., 1989b).

Here we first introduce some terminology and discuss the assumptions of methods applied to survival data from radio-tagged birds. Next we provide a brief overview of parametric approaches and a more detailed summary of nonparametric methods. Lastly, we discuss several issues related to the design of studies and explain the connection between the nonparametric methods and methods based on open capture-recapture models.

2. Terminology and Assumptions

After birds are radio-tagged, the objective of the field effort is to monitor these radio-tagged birds from release to death or to the end of the study. If the death of the bird is not observed during the study, then the record of survival for that bird is said to be right censored (a term taken from the medical literature; see Cox and Oakes, 1984 or Lawless, 1982). This data record reflects the knowledge that the bird was alive from its entry into the study to the time that it was last observed; however, its status after that time is uncertain. Such censoring can occur because a bird emigrates from the study area or loses its transmitter or because its transmitter fails. Birds alive at the end of the study are also considered right censored because their deaths have not been observed during the study period.

The methods described here share five general assumptions.

1. The sample is representative of the population to be studied. This assumption is essential for most studies and is common to all methods. A capture technique that obtains a random sample from the study population is required for satisfying this assumption.

2. Each bird's fate is independent of any other in the sample. This assumption is also common to all the methods described here and to other methods for estimating survival rates, such as those based on capture-recapture and ring-recovery models. It is a critical assumption for deriving estimates.

3. Survival is not influenced by radio-tagging. If this assumption is not satisfied, then the estimates will be biased. Transmitter design and the attachment method must be evaluated and tested before beginning a study of survival with telemetry techniques.

4. Censoring is independent of a bird's fate. This assumption is important for demonstrating that estimates are unbiased. It would be violated if the transmitters of radio-tagged birds are routinely destroyed when the birds are caught and killed by predators or consumed by scavengers and the investigator unknowingly records such deaths as censors. To avoid this type of bias, investigators need to routinely and frequently monitor radio-tagged birds.

5. Survival time is known exactly. This assumption is also important in deriving the properties of estimates and tests, but simulation studies have demonstrated that it can be relaxed (Heisey and Fuller, 1985).

3. Parametric Methods

Parametric methods are based on the five general assumptions described previously and on additional assumptions about characteristics of the data. The additional assumptions can be simple, such as assuming that survival is constant throughout the study period, but most often completely characterize the data by selecting a specific form for the probability distribution. Parametric methods can model time to death in discrete time units or on a continuous scale.

3.1. Discrete Models

A common parametric approach is a generalization of an estimate of nest survival that was proposed by Mayfield (1961) and was applied to data from radio-tagged cottontail rabbits (Trent and Rongstad, 1974). To derive this estimate, the probability of survival is assumed to be constant throughout a period. Time is viewed in discrete units and each time unit or trial is independent of the next. With these additional assumptions about the structure of the data, maximum likelihood (ML) theory can be applied to derive an estimate of survival per time unit in the following way. Let p be the probability of surviving a unit of time and c be the probability of censoring occurring in a unit of time. These two probabilities are assumed to be independent (see assumption 4 above). Let $X=1$ if the animal dies on day t during a period of J days and $X=0$ otherwise; let $Y=1$ if the animal is alive at the end of J days and $Y=0$ otherwise; and let T be the time at which the animal was censored or found dead during the period of J days, or be J if the animal is alive at the end of the period. Then the joint probability of (X, Y, T) is

$$f(x,y,t) = \begin{cases} (p(1-c))^{t-1}\, c & 0,0,t=1,\ldots,J \\ (p(1-c))^{t-1}(1-p)\,(1-c) & 1,0,t=1,\ldots,J \\ (p(1-c))^{t} & 0,1,J. \end{cases}$$

Based on this joint probability distribution, the likelihood for a sample of n animals can be written and solved to obtain the following ML estimate for p

$$\hat{p} = 1 - \frac{\sum_k x_k}{\sum_k t_k - (n - \sum_k y_k - \sum_k x_k)} \qquad for\ k=1,2,\ldots,n.$$

The numerator of the previous expression is the number of deaths and the denominator is total exposure time minus the number of days that "ended" with censoring. The asymptotic variance is

$$var(\hat{p}) = \frac{\hat{p}^2 \ (1-\hat{p})^2}{(1-\hat{p})^2 \ (\sum_k t_k + \sum_k y_k - n) \ + \hat{p}^2 \sum_k x_k} \quad for\ k = 1,2,\ldots,n.$$

Asymptotic results for ML estimates can be used to construct confidence intervals and tests that are based on the Gaussian distribution.

Although much of the published work on the ML estimate of p has focused on studies of nest survival and on the case with no censoring (Johnson, 1979; Hensler and Nichols, 1981; Bart and Robson, 1982; Hensler, 1985), the results generally apply to studies of radio-tagged birds and to the case when censoring is present. In particular, simulation results (Bart and Robson, 1982) indicated that a square root transformation of $\hat{p}$ improved its distributional properties.

Heisey and Fuller (1985) recognized that survival rates are not likely to be constant over long periods of time and provided a generalization of this approach. They set up intervals within which the assumption of constant survival is reasonable, computed an estimate for each interval assuming constant survival within it, and used the product of these statistics to estimate survival for the entire period. They also developed estimates of cause-specific mortality and constructed likelihood ratio tests for factors with more than two levels.

A second discrete model that has been applied to survival data from radio-tagged animals is the multinomial model (White, 1983). Estimation and testing associated with this model is comprehensively treated in White and Garrot (1990).

3.2. Continuous Models

These parametric models view the time scale for survival as a continuous measure and characterize the structure of the data with a specific probability distribution. Some probability distributions commonly used in medical applications for modelling survival (or time to death) are the exponential, Wiebull, and lognormal (Gross and Clark, 1975; Lee, 1980; Cox and Oakes, 1984). These models have not been used for estimating survival of radio-tagged birds largely because it is difficult to support the choice of a specific probability model. Hazard functions of free-ranging birds are often time-dependent and therefore not smooth enough to be modelled by continuous probability distributions. Random processes such as weather can introduce peaks or shifts on the real time scale in hazard functions of free-ranging birds.

4. Nonparametric Methods

Nonparametric methods are not completely free of assumptions but are based on fewer and less specific assumptions than parametric methods. The nonparametric methods described here are based on the five general assumptions described in Section 2 and require no futher assumptions or characterization of the data.

4.1. Estimate of the Survival Function

Nonparametric estimates of survival have been widely used for analyzing data from telemetry studies because they avoid the problem of choosing a specific probability model. The Kaplan-Meier or product limit estimate (for overview see Lee, 1980 or Cox and Oakes, 1984) is frequently used to estimate the survival function ($S[t]$), the probability that a bird survives at least t units of time. The Kaplan-Meier estimate can be obtained by considering the number at risk and the number of deaths at each time that a death is recorded. Let a_1, $a_2, \ldots, a_g$ be the times at which deaths were recorded; $r_1, r_2, \ldots, r_g$ be the numbers of birds at risk or the risk set at these times; and $d_1, d_2, \ldots, d_g$ be the numbers of deaths observed at these same times. Then the Kaplan-Meier estimate of the survival function at time t is

$$\hat{S}(t) = \prod_{j|a_j<t} (1 - d_j/r_j) .$$

This estimate is simply the product of a series of conditional survival probabilities. The asymptotic variance, attributed to Greenwood (Cox and Oakes, 1984), is estimated by

$$var[\hat{S}(t)] = [\hat{S}(t)]^2 \sum_{j|a_j<t} \frac{d_j}{r_j(r_j-d_j)}.$$

An approximate 95% confidence interval for the survival function at time t_0 can be based on the standard Gaussian distribution and is expressed as

$$\hat{S}(t_0) \pm 1.96 \, [var \, \hat{S}(t_0)]^{1/2}.$$

Computations for the Kaplan-Meier estimate are illustrated with data taken from a study of female black ducks wintering in the coastal marshes of New Jersey, USA (Pollock et. al, 1989b). Birds were all captured, radio-tagged, and released by 15 December 1983 and tracked through 15 February 1984.

Calculations of the Kaplan-Meier estimate for the hatch-year birds are shown in Table 1. The number at risk (r_j) is decreased when deaths (d_j) are recorded and when birds were censored (e.g., disappeared from the study area).

Table 1. Kaplan-Meier estimate of survival ($\hat{S}[t]$) for hatch-year female black ducks wintering in the coastal marshes of New Jersey, USA. Data are from Pollock et al. (1989b).

j	Days in Study	r_j	d_j	Number censored	$\hat{S}(t)$
1	6	31	0	1	1.0000
2	7	30	1	0	0.9667
3	14	29	0	1	0.9667
4	22	28	1	0	0.9321
5	26	27	2	0	0.8631
6	27	25	1	0	0.8257
7	29	24	1	0	0.7941
8	32	23	1	0	0.7596
9	34	22	2	0	0.6905
10	37	20	1	0	0.6560
11	40	19	1	0	0.6214
12	44	18	1	0	0.5869
13	49	17	0	1	0.5869
14	56	16	0	2	0.5869
15	57	14	0	1	0.5869
16	58	13	0	1	0.5869
17	63	12	0	12	0.5869

Estimated survival ($\hat{S}[t]$) changes only when a death is recorded. Although the tracking and relocating of the ducks was conducted daily, computations are required only when a death is recorded.

The most convenient way to present these results is graphically, as in Figure 1. When a death is recorded, a drop in the step function representing estimated survival occurs. No change in height occurs at censored values, but the risk set decreases as shown in Table 1. The pattern of survival throughout the study period is described by the estimates of $S(t)$ and their confidence intervals. The estimate

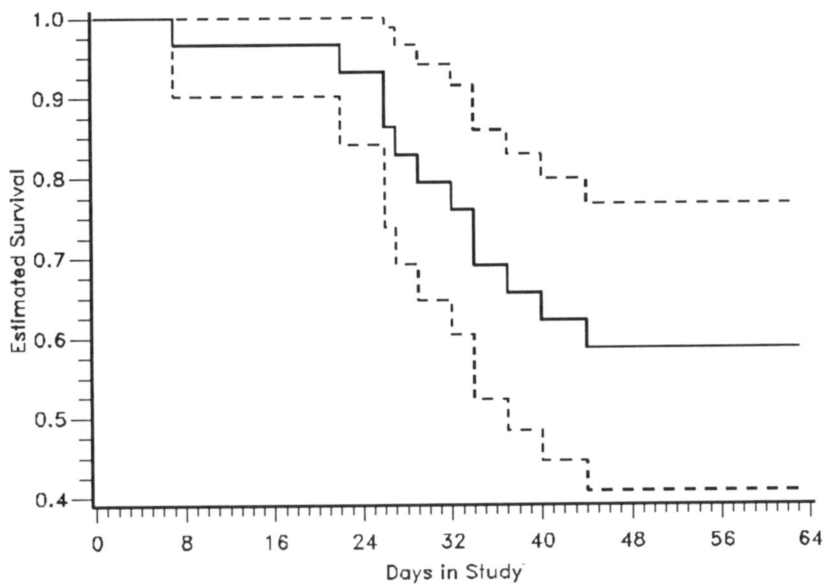

Figure 1. Estimated survival (solid line) of hatch-year female black ducks wintering in the coastal marshes of New Jersey, USA. Dashed lines indicate the 95% confidence intervals for the estimates.

of $S(t)$ at the end of the study and its confidence interval are often reported as a single summary of overall survival during the study period.

Typically r_j, the number of birds at risk, decreases as deaths and censors occur, but additions of newly tagged birds can be used to increase the risk set. Such increases in the risk set can be accommodated by the Kaplan-Meier estimate of the survival function (Pollock et al., 1989a). This staggered entry of birds into the risk set can result from the length of time that is required to capture, tag and release birds. Entry also can be staggered by design to reduce the variance of the conditional estimates of survival that follow the entry point of the newly tagged birds. Survival estimates and, in particular, their variance from times preceding the entry point will not be affected. In all cases, the newly tagged birds are assumed to have the same underlying survival as the previously tagged birds that are still alive.

4.2. Two-Sample Tests

Linear rank statistics provide a framework for comparing survival functions among groups. If there is no censoring, familiar nonparametric tests can be applied, such as the Wilcoxon rank sum test for comparing two samples and the Kruskal-Wallis test for comparing k ($k>2$) samples (Hollander and Wolfe, 1973). When censoring is present, adjustments to these tests are required. Many adjustments for censoring have been developed for each of the linear rank statistics (Lee, 1980).

The log rank test is a two-sample test that easily can accommodate censoring and commonly has been applied to data from radio-tagged birds. Let $a_1, a_2, \ldots, a_k$ be the observed death times when the two samples are combined. Let d_{1j} and d_{2j} denote the number of deaths in sample 1 and 2 at time a_j, respectively; and r_{1j} and r_{2j} denote the number at risk in sample 1 and 2 at time a_j, respectively. Let d_j denote the total number of deaths recorded $(d_{1j} + d_{2j})$ and r_j denote the total number at risk $(r_{1j} + r_{2j})$ for both samples at time a_j. At each of the k times when deaths occur, the data can be represented as a 2 x 2 contingency table with the expected value (E) and variance of d_{2j} in the jth table given as

$$E(d_{2j}) = d_j r_{2j} / r_j$$

and

$$var(d_{2j}) = d_j r_{1j} r_{2j} (r_j - d_j) / r_j^2 (r_j - 1) .$$

Assuming conditional independence and asymptotic normality of the d's, a test statistic that has an approximate chi-squared distribution with 1 degree of freedom can be obtained by combining the results from the k contingency tables. This test statistic,

$$X^2 = \frac{\left[\sum_j d_{2j} - \sum_j E(d_{2j})\right]^2}{\sum_j var(d_{2j})} \qquad for\ j = 1, 2, \ldots, k,$$

evaluates the hypothesis that the survival functions of the two groups do not differ.

The log rank test directly generalizes to a study with a staggered entry or multiple releases of birds into the sample (Pollock et al., 1989a). The risk sets (r_{ij}) can decrease as deaths or censors occur and increase as newly tagged birds are added to each sample. Other linear rank tests can be adjusted for censoring, but their scoring functions cannot easily accommodate additions after the time origin. Again newly tagged birds are assumed to have the same underlying survival as the previously tagged birds in each sample. Pollock et al. (1989a) present detailed calculations for this test statistic when all birds are marked at the start of the study and when birds are added to the risk set throughout the study period.

5. Discussion

In closing, we briefly discuss several issues that must be carefully considered when designing a telemetry study focused on estimating survival. We also describe the robustness of the nonparametric methods when relocation probabilities are less than one and develop a link to open capture-recapture models. Lastly we suggest some topics for future research.

5.1. Definition of Time Origin and Scale

Choosing a time origin and scale for measuring survival requires careful thought and consideration. In many medical applications, each subject can have a unique date of entry. Survival is measured from that entry date and on a scale that represents time in the study or time since diagnosis or time on treatment. Subjects are generally considered equivalent at the time that they enter the study and begin treatment irrespective of the actual calendar date of entry.

For some ecological studies, a biological time origin (e.g., hatching, fledging, or beginning of treatment) can be defined for each member of the sample and survival would be recorded as time since that event. For other studies, no biological event can provide a common origin for all individuals. In these studies, calendar time is used as the scale. Often calendar time is the appropriate scale, because important mortality factors can change dramatically along that scale (e.g., opening and closing of hunting seasons or severe winter weather).

5.2 Definition of Study Area

Careful definition of the study area, particularly as it relates to the relocation effort and the unit of the time scale, is required at the beginning of a telemetry study. The time unit (e.g., week or day) for summarizing survival should be defined by the time that is required to search the study area for every radio-tagged bird, such that the relocation probability is relatively constant for each individual in the sample. If all the radio-tagged birds on the study area do not have the same daily relocation probability, then a day should not be used as the time unit for computing survival rates.

Data obtained from searches outside the study area or after the end of the study should not be used to update the status of birds that were censored. This updating can bias estimates because it is more likely to happen for birds that are alive at the end of the study than for those that have died during the study but were not found. Therefore, censoring will no longer be independent of the bird's fate. Also there is some interpretation problem in using data from a bird that may have emigrated from the area early in the study period and resided elsewhere when estimating the survival of birds that remained at the study site.

5.3. Homogeneity of Survival

Implicit in the parametric and nonparametric methods described above is an assumption of homogeneity of survival among individuals. Basically, all animals are assumed to have the same chance or probability of surviving. For parametric methods based on continuous probability models, this assumption is evident because all individuals are described by the same probability distribution. For the other methods, this assumption is evident because a conditional binomial approach is used to obtain the estimated variances. If survival rates are heterogeneous, the point estimates are valid and describe an "averaged" survival rate among individuals, but the estimated variances described above are no longer appropriate.

5.4. Benefits of Staggered Entry or Multiple Releases

Two study designs were described for nonparametric methods, one with a single release of radio-tagged birds and the second with multiple releases of radio-tagged birds. Some recent work (William A. Link, personal communication) establishes the situations in which each of these designs would be optimal. A design with multiple releases will more precisely estimate conditional probabilities than one with a single release of radio-tagged birds. If the total number of marked birds is the same for both designs, then a design with multiple releases will more precisely estimate unconditional probabilities (e.g., survival at the end of the study) only if survival is low and if more than half the radio-tagged birds die during the study. The greatest benefit of multiple releases is that the total number of radio-tagged birds in the study would be increased because the logistics of a single, short period of capture, tagging and release often limits the total number of birds that could be included in a study.

5.5. Relocation or Capture Probability

Methods to estimate survival of radio-tagged birds differ from those used to estimate survival in capture-recapture studies in two ways. First, these methods use confirmed deaths in the estimation process; open capture-recapture models do not. Also these methods do not explicitly model relocation or capture probabilities; open capture-recapture models include parameters for capture probabilities. These methods do not specifically model relocation probability because the relocation probability is implicitly assumed to be one.

A Monte Carlo study was conducted to evaluate the robustness of the Kaplan-Meier estimate of $S(t)$ to this implicit assumption about relocation probability. The performance of the traditional form of the estimate was compared with a modification that did not assume relocation probability was always one. The modified estimator is an extension of the staggered-entry setting and includes birds in the risk set only at the times that they were actually relocated. The traditional estimate was positively biased and the bias increased as the relocation probability decreased. The mean square errors of the two estimates were similar at high relocation probabilities but differed greatly at low relocation probabilities as shown in Figure 2.

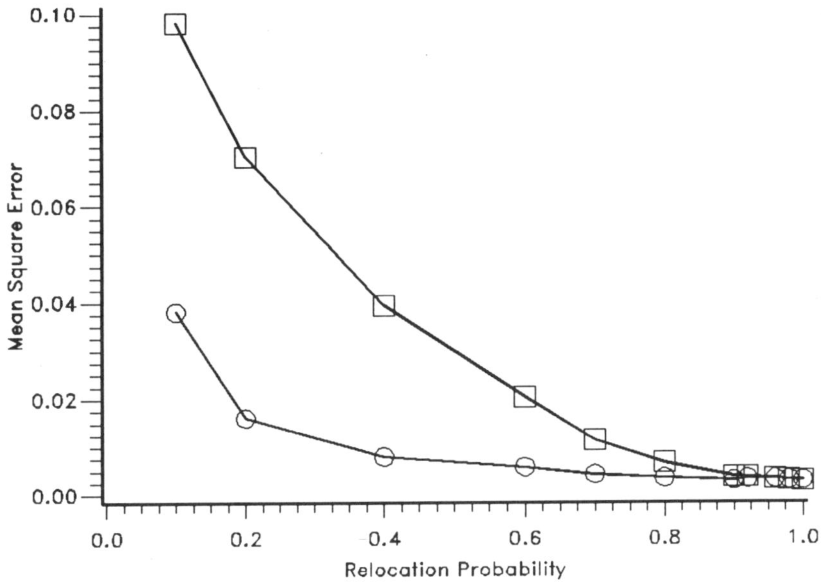

Figure 2. Mean square error of Kaplan-Meier estimate of $S(t)$ in its traditional form (□) and in a modified form that accommodates relocation probabilities that are less than one (○).

For the traditional form of the Kaplan-Meier estimate, conditional survival is described by the ratio of the number alive at time $i+1$ and the number alive at time i. The number alive at time $i+1$ is computed by subtracting the number of confirmed deaths between times i and $i+1$ from the number of radio-tagged birds alive at time i. The number of deaths between times i and $i+1$ is known only if the relocation probability is one. If it is less than one, then the number of observed deaths is only an estimate of the true number of deaths and becomes a random quantity. Estimates of survival from open capture-recapture model have a similar form. The losses between times i and $i+1$ due to death or emigration are estimated and these estimates are used to compute the survival rate.

A general framework that encompasses the nonparametric models (such as the Kaplan-Meier estimate) and open capture-recapture models can be developed for analyzing data from radio-tagged birds. Relocation or capture probabilities, the number of radio-tagged birds that are alive, and the number of those that have died would be estimated and then used to estimate apparent emigration rate and actual survival rate. We are working on the formal development of such a framework.

5.6. Number of Parameters

Future research also should focus on approaches for reducing the number of parameters estimated in the nonparametric methods. Parametric models have many fewer parameters but are unattractive because hazard functions of bird populations are not easily characterized by smooth functions. The Kaplan-Meier estimate of the survival function does no smoothing. A conditional probability for survival is estimated for every interval between successive deaths; this process results in a model with many parameters. Complete smoothing of these probabilities, or averaging them, leads to a model with constant survival that is equivalent to characterizing survival with an exponential distribution. Some intermediate amount of smoothing, such as localized smoothing or a weighted average of neighboring values, may achieve a more appropriate tradeoff between an inefficient, overparameterized model and a highly efficient, oversimplified model for estimating survival of radio-tagged birds.

5.7. Other Topics

Here we have just briefly discussed a few statistical tools that can be applied to survival data from radio-tagged birds. Many other topics such as the use of covariates and competing risks are relevant to survival data from radio-tagged birds. The general references in this paper will provide an introduction to these and related topics.

Acknowledgements. We thank G.C. White, R.E. Kenward, M.H. Wilson, and an anonymous reviewer for contributing insightful suggestions and comments on an earlier draft; C.-L. Chen for assisting with the simulations; and C.M. Hagen for her patience in preparing the final copy of the manuscript.

References

Bart, J. and Robson, D.S. (1982).- Estimating survivorship when the subjects are visited periodically. Ecology, 63:1078-1090.

Brownie, C., Anderson, D.R., Burnham, K.P. and Robson, D.S. (1985).- Statistical inference from band recovery data - a handbook. 2nd edition. US Fish and Wildlife Service Resource Publication 156.

Burnham, K.P., Anderson, D.R, White, G.C., Brownie, C. and Pollock, K.H. (1987).- Design and analysis methods for fish survival experiments based on release-recapture. American Fisheries Society Monograph, 5.

Cox, D.R. and Oakes, D. (1984).- Analysis of survival data. Chapman and Hall, New York.

Gross, A.J. and Clark, V.A. (1975).- Survival distributions: reliability applications in the biomedical sciences. John Wiley & Sons, New York.

Heisey, D.M. and Fuller, T.K. (1985).- Evaluation of survival and cause-specific mortality rates using telemetry data. Journal of Wildlife Management, 49:668-674.

Hensler, G.L. (1985).- Estimation and comparison of functions of daily nest survival probabilities using the Mayfield method. pp 289-301 *in* Statistics in Ornithology, B.J.T. Morgan and P.M. North (Eds). Springer-Verlag, New York.

Hensler, G.L. and Nichols, J.D. (1981).- The Mayfield method of estimating nesting success: a model, estimators and simulation results. Wilson Bulletin, 93:42-53.

Hollander, M. and Wolfe, D.A. (1973).- Nonparametric statistical methods. John Wiley & Sons, New York.

Johnson, D.H. (1979).- Estimating nest success: the Mayfield method and an alternative. Auk, 96:651-661.

Lawless, J.F. (1982).- Statistical models and methods for lifetime data. John Wiley & Sons, New York.

Lebreton, J.-D., Burnham, K.P., Clobert, J. and Anderson, D.R. (1992).- Modeling survival and testing biological hypotheses using marked animals: a unified approach with case studies. Ecological Monographs, 62:67-118.

Lee, E.T. (1980).- Statistical methods for survival data analysis. Lifetime Learning Publications, Belmont, California.

Mayfield, H.F. (1961).- Nesting success calculated from exposure. Wilson Bulletin, 73:255-261.

Pollock, K.H., Nichols, J.D., Brownie, C. and Hines, J.D. (1990).- Statistical inference for capture-recapture experiments. Wildlife Monograph, 107.

Pollock, K.H., Winterstein, S.R., Bunck, C.M. and Curtis, P.D. (1989a).- Survival analysis in telemetry studies: the staggered entry design. Journal of Wildlife Management, 53:7-15.

Pollock, K.H., Winterstein, S.R. and Conroy, M.J. (1989b).- Estimation and analysis of survival distributions for radio-tagged animals. Biometrics, 45:99-109.

Trent, T.T. and Rongstad, O.J. (1974).- Home range and survival of cottontail rabbits in southwestern Wisconsin. Journal of Wildlife Management, 38:459-472.

White, G.C. (1983).- Numerical estimation of survival rates from band-recovery and biotelemetry data. Journal of Wildlife Management, 47:716-728.

White, G.C. and Garrot, R.M. (1990).- Analysis of wildlife radio-tracking data. Academic Press, New York.

Marked Individuals in the
Study of Bird Population
J.-D. Lebreton and P. M. North (eds.)

Extensive Parameterisation of Survival Models for Recovery Data Analysis

J.Rinne[1], H.Lokki[2] and P.Saurola[3]

[1]*Department of Meteorology, University of Helsinki, Hallituskatu 11-13, SF-00100 Helsinki, Finland*

[2]*Department of Computer Science, University of Helsinki, Teollisuuskatu 23, SF-00510 Helsinki, Finland*

[3]*Ringing Centre, Zoological Museum, University of Helsinki, Pohjoinen Rautatiekatu 13, SF-00100 Helsinki, Finland*

Summary: Unless properly selected, constraints in life table modelling may produce linear dependence between cells causing the model to be non-identifiable. In our case study, based on 30-year Tawny Owl (Strix aluco) nestling recovery data, we found a unique solution with the following model (no external factors or other covariates were included):
-survival rates were allowed to vary with calendar year (i) for the first three age-classes, but not for older age-classes ($S_{1i}, S_{2i}, S_{3i}, S_4, S_5, S_6, \ldots$).
- also the recovery rate was allowed to vary with calendar year (f_i). In addition, a separate first year recovery rate was applied.
The large number of parameters made the model identifiable. In some cases the year-dependent values justified the assumption of one general value for the parameter. The number of parameters was then reduced by computing weighted means of the year-dependent values. The technique so developed gave robust and reasonable values for survival and recovery rates, without making use of any covariates. However, the study of mutual correlations indicates difficulties in the estimation of the first year survival and recovery rates. Modifications of the model (e.g. application of a covariate) did not solve this problem.

1. Introduction

One can attempt to estimate survival and recovery rates from the ringing recoveries collected into so called life tables. A probability model describing the life table is constructed. The first problem encountered is that there are too many parameters to be estimated. Traditionally this is avoided by assuming that some of the parameters are equal. The second problem is that such assumptions tend to increase linear dependence between the cells of the model. The model becomes non-identifiable. Therefore additional constraints are needed. Such a model was studied in Rinne, Lokki and Saurola (1990) and will be used here for comparison.

Morgan and Freeman (1989) showed that, paradoxically, by adding more parameters to the model, the model is made identifiable. In this paper we will first discuss the linear dependence within a model. We also investigate various models using real data to demonstrate how, by adding more parameters, the linear dependence can be reduced to a minimum.

Our aim is not to validate the models in a statistical sense. Rather we study the mathematical problems met in the model construction. We study models at the upper limit of parameter numbers and thereby illustrate how the functional dependence within the model can be reduced. The notation used in Brownie et al. (1985; N,f,S) will be applied.

2. The traditional approach

First we will discuss problems encountered with the traditional approach in determining survival and recovery rates. The following discussion is meant for biologists that are unfamiliar with the problem, therefore, we repeat basic features that can be found in several references (e.g. Cormack 1970, Lakhani and Newton 1983).

Let the number of ringed birds in 1991 be N_{91}. From these $N_{91}S_1$ survive over the first winter to the nesting season in 1992, $N_{91}(1-S_1)$ do not. If f_1 is the recovery rate during that winter, then $f_1N_{91}(1-S_1)$ birds will be found during the winter in question. From those that survived the first winter, $f_2N_{91}S_1(1-S_2)$ will be found during the next winter. This gives us two known numbers of recoveries, theoretical values of which are $f_1N_{91}(1-S_1)$ and $f_2N_{91}S_1(1-S_2)$. From these we want to determine f_1, f_2, S_1 and S_2. This is not possible because there are four parameters to be estimated from two known numbers. When we add one more observation and ringing year, the following life table results.

N_{91}	$f_{1,91}N_{91}(1-S_{1,91})$	$f_{2,91}N_{91}S_{1,91}(1-S_{2,91})$	$f_{3,91}N_{91}S_{1,91}S_{2,91}(1-S_{3,91})$
N_{92}		$f_{1,92}\,N_{92}(1-S_{1,92})$	$f_{2,92}N_{92}\,S_{1,92}(1-S_{2,92})$
N_{93}			$f_{1,93}\,N_{93}(1-S_{1,93})$

or symbolically { $[f_{1i}],[f_{2i}],[f_{3i}],[S_{1i}],[S_{2i}],[S_{3i}]$ } with the subscript i refering to the year: 91, 92 or 93.

The cell values in the table are theoretical ones. In real conditions there will be additional noise. There are now 6 known values and twelve unknown parameters. These are $f_{1,91}$, $f_{2,91}$, $f_{3,91}$, $f_{1,92}$, $f_{2,92}$, $f_{1,93}$, $S_{1,91}$, $S_{2,91}$, $S_{3,91}$, $S_{1,92}$, $S_{2,92}$, $S_{1,93}$. If there are no additional observations, then the only way to continue is to assume additional relationships between these parameters which we will refer to the *initial* parameters. It is reasonable to assume that the survival rates of birds of the same age are similar among different cohorts, i.e. $S_{1,91}=S_{1,92}=S_{1,93}=S_1$ and $S_{2,91}=S_{2,92}=S_2$. Similarly, $f_{1,91}=f_{1,92}=f_{1,93}=f_1$ and $f_{2,91}=f_{2,92}=f_2$.

Now there are 6 known values and 6 parameters to be determined (although $f_{3,91}$ and $S_{3,91}$ are not separately identifiable). The principle of parsimony requires that the number of parameters should be made smaller. One has to make additional assumptions. A natural choice is to make a distinction between young and old birds. Then we have

N_{91}	$f_1N_{91}(1-S_1)$	$fN_{91}S_1(1-S)$	$fN_{91}S_1S(1-S)$
N_{92}		$f_1N_{92}(1-S_1)$	$fN_{92}S_1\,(1-S)$
N_{93}			$f_1N_{93}(1-S_1)$

or symbolically { f_1, f, S_1, S }. Now there are four unknowns and six observations.

The unknowns are determined so as to give the best fit to the observations. In the normal case such a solution exists. If there are enough data, the confidence limits become acceptable.

Thus the problem is solved but only at the cost of additional assumptions. One cannot be sure that the separation of birds into young and old with regard to recovery and survival rates is valid. In any case we have at least some kind of solution and we can perhaps be satisfied. A solution has been found.

3. The linear dependence

The last statement of the previous paragraph is wrong. There is no unique solution. By removing parameters we have added linear dependence into our life table. This is best seen by transforming the data to be relative values, i.e. by dividing the numbers in each cohort with the corresponding number of birds ringed. The theoretical model of the life table for four unknowns and six observations given in relative numbers becomes

1	$f_1\,(1\text{-}S_1)$	$f\,S_1(1\text{-}S)$	$f\,S_1\,S(1\text{-}S)$
1		$f_1\,(1\text{-}S_1)$	$f\,S_1(1\text{-}S)$
1			$f_1(1\text{-}S_1)$

The three terms on the longest diagonal are identical. They represent three measurements of the same variable, $f_1(1\text{-}S_1)$. From these one cannot determine separately f_1 and S_1. Note here the difference between the cells and the parameters. For a given value of $f_1(1\text{-}S_1)$, the parameters f_1 and S_1 are functionally dependent. However, the cell probabilities are identical, excluding possible noise, and are thus linearly dependent. We have four unknowns but only three independent measurements, $f_1(1\text{-}S_1)$, $fS_1(1\text{-}S)$ and $fS_1S(1\text{-}S)$. The problem is thus non-identifiable. There are solutions but many more than one.

In the original table the cells were linearly independent. Because there were too many parameters, we decreased their number with additional assumptions. However, simultaneously, without knowing it, we added linear dependence to the theoretical model. The initial problem, too many parameters, remained. By adding questionable assumptions we only made the model worse.

It turns out that if there are enough ringing and recovery years, then the survival rates can be made estimable by adding enough assumptions (Rinne et al. 1990). For instance, one can assume one general recovery rate, f, and survival rates for each age, S_1, S_2, S_3 and so on. Often there are not enough data and still more assumptions are needed. The usual approach is to assume that the survival rates of very old birds are equal, for instance, $S_7=S_8=S_9$. This assumption is made because it seems reasonable to expect this in nature. However, the assumption is really unfair. One cannot guarantee that in the actual data the survival rates of older birds are equal. Instead, it would be fair to give certain parameters suitable values. For instance, in our illustrative life table one could decide that $f=0.12$. Then there will be three unknowns and three independent cells. With such a choice, one states openly that a value has been taken from "the hat" and there is no attempt to mislead.

4. Changing back to the initial parameters

Our experience suggests that the idea presented by Morgan and Freeman (1989), allowing first year survival to be calendar year dependent, can be extended further to allow the survival rates of birds of higher age groups to depend also upon the calendar year. So one can assume age dependency in recovery rates without making the model non-identifiable.

We first assume one additional cohort

$$\begin{array}{lllll}
1 & f_1(1-S_1) & fS_1(1-S) & fS_1S(1-S) & fS_1SS(1-S) \\
1 & & f_1(1-S_1) & fS_1(1-S) & fS_1S(1-S) \\
1 & & & f_1(1-S_1) & fS_1(1-S) \\
1 & & & & f_1(1-S_1)
\end{array}$$

This is of the type $\{ f_1, f, S_1, S \}$. Now there are four unknowns and four independent observations. The difference between their numbers equals zero. That difference is taken as an indicator of the degrees of freedom. We replace the second year survivals by the initial parameters and get

$$\begin{array}{lllll}
1 & f_1(1-S_1) & fS_1(1-S_{2,91}) & fS_1S_{2,91}(1-S) & fS_1S_{2,91}S(1-S) \\
1 & & f_1(1-S_1) & fS_1(1-S_{2,92}) & fS_1S_{2,92}(1-S) \\
1 & & & f_1(1-S_1) & fS_1(1-S_{2,93}) \\
1 & & & & f_1(1-S_1)
\end{array}$$

or symbolically $\{ f_1, f, S_1, [S_{2i}], S \}$.

There are seven unknowns and seven independent cells. The degrees of freedom remain the same, but now we can estimate more parameters with less assumptions. This is the result of trying to make the cells independent. Could it be done better?

Yes, we can replace the first year survival rates by the initial parameters (i.e. we introduce the calendar year dependence for first year survival, cf. Morgan and Freeman 1989) resulting in

$$\begin{array}{lllll}
1 & f_1(1-S_{1,91}) & fS_{1,91}(1-S) & fS_{1,91}\,S(1-S) & fS_{1,91}SS(1-S) \\
1 & & f_1(1-S_{1,92}) & fS_{1,92}(1-S) & fS_{1,92}\,S(1-S) \\
1 & & & f_1(1-S_{1,93}) & fS_{1,93}(1-S) \\
1 & & & & f_1(1-S_{1,94})
\end{array}$$

or symbolically $\{ f_1, f, [S_{1i}], S \}$.

All ten cells are linearly independent. Because there are seven unknowns, there are three degrees of freedom. We thus have more parameters, less assumptions about them, and more degrees of freedom. This is, in the sense of degrees of freedom, the best model so far. The reason for that is the removal of S_1. Because it affects all the cells, it is an effective means of producing independence.

Instead of changing S_1, we might try to change recovery rates for older birds as follows

1	$f_1(1-S_1)$	$f_2S_1(1-S)$	$f_3S_1S(1-S)$	$f_4S_1SS(1-S)$
1		$f_1(1-S_1)$	$f_3S_1(1-S)$	$f_4S_1S(1-S)$
1			$f_1(1-S_1)$	$f_4S_1(1-S)$
1				$f_1(1-S_1)$

The numbers of unknowns, independent cells and degrees of freedom are 6, 7 and 1, respectively. The model is thus the second best. The added recovery rates affect all cells except those of young birds.

With larger life tables one can add more initial parameters. The best choices, for a change back to the initial parameters, are obviously S_1, S_2 and f. They each affect nearly all or all the cells. Another very effective approach is to increase the independence in the different directions of the table by simultaneously making S_1 year-dependent and f year- plus age-dependent.

The yearly variation of the recovery rates of youngest birds can be interesting. However, making them year-dependent doesn't help to solve the linear dependence problem because the number of degrees of freedom is not increased.

So far all the cells are taken to have comparable values. Yet those consisting of younger birds include more recoveries whereas some cells concerning older birds may be empty. The result is that it can be difficult to use the initial parameters or both the recovery rate and survival rate of first year birds. The latter affects the whole life table. However, if the observations are concentrated on the first two years of the life table, then we can add 4+4 parameters into a life table of the previous illustrative case when the main data consist of 7 cells. Thus, in reality, the degrees of freedom has been decreased.

Removing the last cell (i.e. the latest cohort) helps if both f_1 and S_1 are replaced by the initial parameters. Then one data cell and two unknowns are removed, resulting in another additional degree of freedom.

In conclusion, the linear independence within the model can be increased by coming back to the initial parameters. This is most effective when applied to the rates of the younger birds as most of the cells are then affected. Changing survival rates is more effective than changing recovery rates. A combination of both can be best. Only a limited number of parameters can be changed back. That limit is determined by the number of data rich cells.

5. Illustrative case studies

5.1. Material and methods

During 1961-1990, 19 903 Tawny Owls were ringed as nestlings and 2 267 of them were later reported dead. The longevity record for the Finnish Tawny Owls is 17 years, which means that for the first 13 cohorts, our data set is 'complete'. For more details about the data, see Rinne et al. (1990). The material (as updated until 1990) will be applied in the case studies, where we test how far back to the initial parameters one can go. Thus the aim here is not to make a final survival analysis of the Finnish Tawny Owls but to study model variations.

The computations are based on the program SURVIV (White 1983).

In the case studies we apply models where there are survival rates for single cells (of type S_{ij}). With such a model, new technical difficulties may arise. Cases where the data indicates $S_{ij}=0$ (no more recoveries) or $S_{ij}=1$ (no changes in the number of recoveries) should be identified beforehand. In such cases, it is often better to fix the value of that survival rate to 0 or 1 as the data indicate. Otherwise the solution may be difficult to compute or the model can become non-identifiable. This practice is related to that applied in our 'baseline solutions' (Rinne et al. 1990). Further, depending on the model, it can be advantageous to eliminate the last short cohorts.This holds true especially when a parameter appears only in one model cell.

In complicated models, the ability to converge upon a solution may depend on the choice of the initial guess. Therefore, one may need to try several such guesses. These are best found by first studying less complex models. In the most sensitive case we have met, using identical models, one converged, the other one didn't. The only difference was that the parameters for the program SURVIV were given in a different order.

5.2. Models

We now apply the previous suggestions concerning linear dependence and investigate different models with the Tawny Owl data. The results are compared with those in Rinne et al. (1990). There the model applied was of the type $\{f,S_1,S_2,S_3,\ldots,S_{16}\}$. Thus S_{17} was set to zero ('baseline solution'). Because the data here have been updated, we have rerun all those computations and present updated results. We refer to this model as model 1.

In model 1 there are 17 parameters. We first change back to the initial parameters for survival rates. The test model 2 will be $\{f,[S_{1i}],[S_{2i}],[S_{3i}],S_4,S_5,\ldots,S_{16}\}$. This model uses 101 parameters, a significant extension of the 17 parameter model! The solution was found, though warnings about the dominance of rounding errors and the small number of significant digits resulted.

Model 2 was at the limits of identifiability. We nevertheless added one more parameter and applied separate recovery rates for the youngest as opposed to all other birds and so constructed model 3: $\{f_1,f,[S_{1i}],[S_{2i}],[S_{3i}],S_4,S_5,\ldots,S_{16}\}$. The model became easier to identify and no problems were encountered, although the results indicate that applying one global recovery rate, i.e. model 2, would otherwise have been justified. The recovery rates found for younger and older birds were similar: 0.1213 and 0.1303, respectively. Accordingly, the maximum log-likelihood was practically the same as in model 2. In this case a correct additional assumption of equal recovery rates of first year and older birds cause a change from model 3 to model 2 giving similar results but making the problem less identifiable! This is a good illustration of the effect of dependence between cells.

We continued by adding more recovery rates. For each winter we applied a separate recovery rate for older birds. This resulted in model 4 or $\{f_1,[f_i],[S_{1i}], [S_{2i}],[S_{3i}],S_4,S_5,\ldots,S_{16}\}$. We now have 30 recovery rates instead of one. In comparison to the 17 parameters in model 1, we now have 130 parameters. In spite of this enormous addition of parameters, the model was identifiable.

5.3. Additional models for studying the first year rates

Model 1 could not be extended by taking a separate parameter for the recovery rate of young birds; the solution was no longer unique. Instead, the model was changed to $\{w;f_1,f,S_1,S_2,S_3,S\}$ applying recovery rates for young and old birds separately. These could be determined because a covariate, w for the snow depth, was added to the model (see Rinne et al. 1990). This modified model will be referred as to model 1a.

In model 4 the recovery rates of the older birds are year-dependent. One can assume that a similar yearly variation also affects the younger birds. We therefore modelled f_{1i} equal to $g*f_i$, where g is a parameter to be estimated and f_i is the recovery rate of older birds during the actual winter (i). Model 4a became $\{g; [f_i], [S_{1i}],[S_{2i}],[S_{3i}],S_4,S_5,\ldots,S_{16}\}$.

5.4. Results

We first summarize the basic models as follows

Model 1	$\{f,S_1,S_2,S_3,\ldots,S_{16}\}$
Model 2	$\{f,[S_{1i}],[S_{2i}],[S_{3i}],S_4,S_5,\ldots,S_{16}\}$
Model 3	$\{f_1,f,[S_{1i}],[S_{2i}],[S_{3i}],S_4,S_5,\ldots,S_{16}\}$
Model 4	$\{f_1,[f_i],[S_{1i}],[S_{2i}],[S_{3i}],S_4,S_5,\ldots,S_{16}\}$

The survival and recovery rates found with these models were as follows

	f_1	$f_{>1}$	S_1	S_2	S_3	$S_{>3}$
Model 1	0.121	-	0.435	0.602	0.677	0.732
Model 2	0.125	-	0.450	0.612	0.691	0.722
Model 3	0.121	0.130	0.433	0.611	0.690	0.722
Model 4	0.124	0.148	0.447	0.589	0.648	0.752

In the models 2-4 general survival rates such as S_1, S_2 and S_3 were not used as parameters. The values given are arithmetic means of survival rates [S_{1i}] (some weighted means for model 4 will be discussed later). The general survival rates of oldest birds ($S_{>3}$) were computed from a model in which we assumed additionally $S_4=S_5=S_6=...=S_{16}$. This can be seen as a complex computation of a mean.

The maximum log-likelihoods were -667.082, -514.333, -514.282, -480.695 in models 1-4, respectively, showing an increase from model 1 to model 4 as expected in a set of nested models. Most of the confidence ranges of the survival and recovery rates have remained reasonable though the number of parameters has increased considerably.

Recovery and survival rates estimated with model 4 are shown in Figs. 1-4. For each parameter, excluding S_1, there exists a certain value which falls within nearly all confidence ranges. Therefore, we computed, instead of arithmetic means, the more accurate weighted means for model 4 applying the formulae given in the Appendix. These are 0.091, 0.450, 0.616 and 0.689, respectively, for f_1, S_1, S_2 and S_3. For f_1, the weighted mean differs clearly from the directly computed mean (0.124). The weighted mean more readily conforms to the one determined graphically in Fig. 1 (about or less than 0.10).

The estimated standard deviations for f_1, S_1, S_2 and S_3 became 0.010, 0.039, 0.012 and 0.012, respectively. The correlations over time between the estimates seemed to be very small (Fig. 5). The dependence between the parameters tended to be weak, except in the following two cases: 1) the correlations between the first year survival and recovery rates (Fig. 6) varied around 0.5, 2) similar distribution of positive correlations was found for correlations between [f_i] and [S_{2i}].

For the additional models we found

	f_1	$f_{>1}$	S_1	S_2	S_3	$S_{>3}$
Model 1a						
-Rinne et al.	0.094	0.241	0.226	0.611	0.665	0.733
-updated data	0.242	0.075	0.707	0.600	0.668	0.725
Model 4a	0.081	0.363	0.156	0.592	0.683	0.730

For model 1a the results are given for both data in Rinne et al. (1990) and the present updated data. The log-likelihoods for these models and 4a were -561.448, -632.233 and -474.435, respectively.

The survival rates of older birds are similar in models 1-4 and also in modified models 1a and 4a. In contrast, the recovery rates and the first year survival rates vary widely in these models. A remarkable change has been caused in model 1a by a small change in data. The correlations between the first year survival and recovery rates were very strong in models 1a and 4a. Hence these models are close to non-identifiability. The attempts in section 5.3 to find better models for first year rates have thus been unsuccessful.

In order to study the case of model 1a, we constructed a similar version of model 4 by applying the snow depth as a covariate. To our surprise, no changes were found.

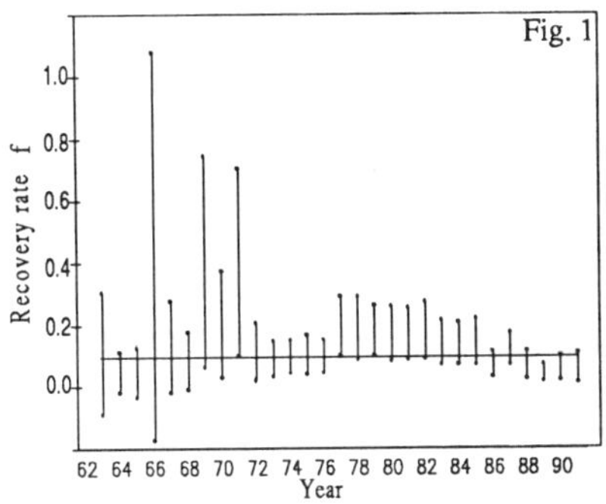

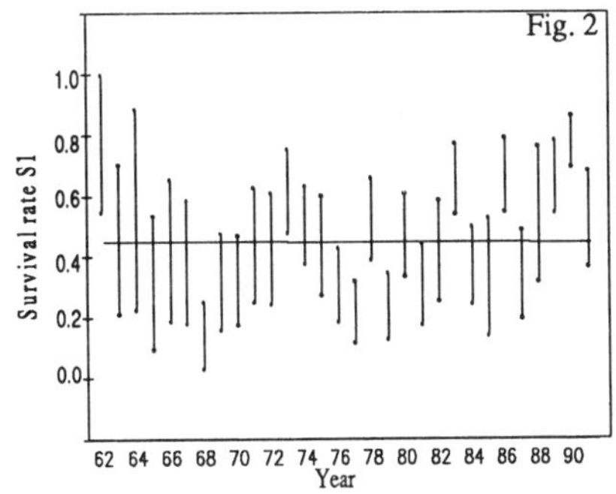

Figs. 1-2. The year dependent recovery (f; Fig. 1) and survival (S_1; Fig 2) rates as estimated by model 4. The 95 % confidence intervals are shown. The horizontal lines have been drawn to give the best graphical fit within the confidence limits.

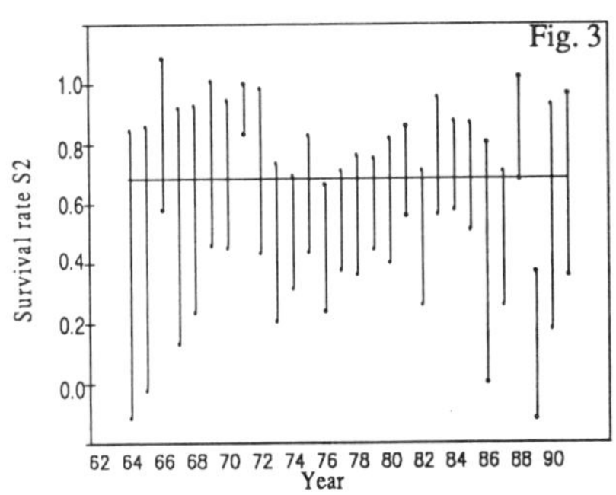

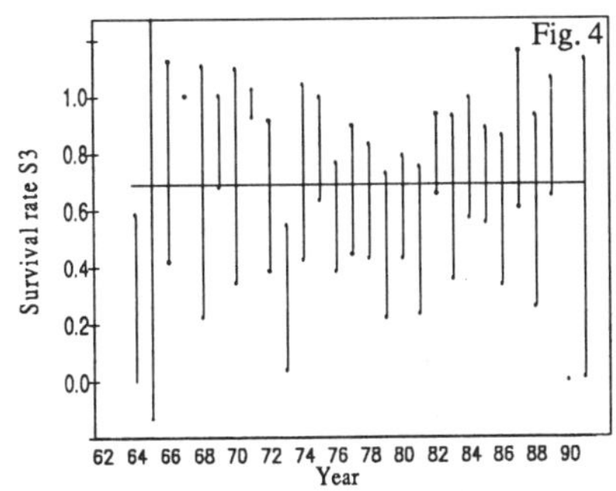

Figs. 3-4. As in Fig. 2 but for the second year survival (S_2; Fig. 3) and third year survival rates (S_3; Fig. 4).

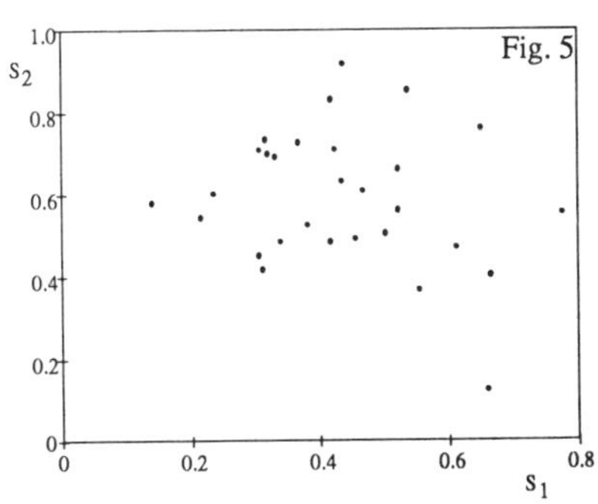

Fig. 5. Second year survival rates S_2 versus first year survival rates S_1 as estimated by model 4.

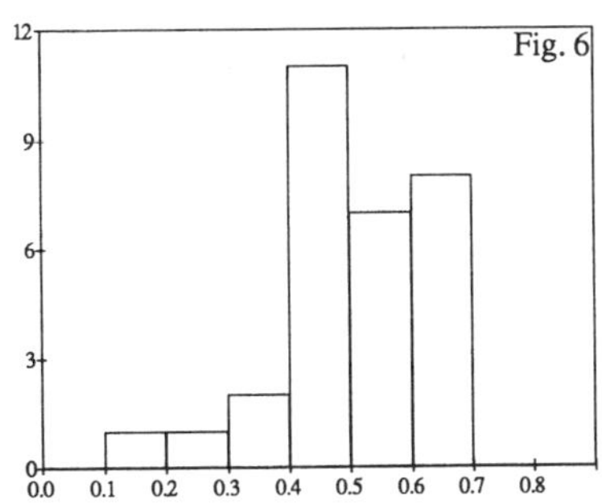

Fig. 6. Histogram of the frequencies of the correlations (given by SURVIV) in model 4 between the first year recovery rate f_1 and the yearly first year survival rates S_{1i}.

The covariate neither improved nor worsened the model. However, snow depth should be a reasonable covariate. When the yearly snow depth values were compared with the first year survival rates in model 4, reasonable dependence was found.

6. Discussion

The survival and recovery rates are very similar in models 1-4. Their differences give some insight into the effect of the different formulations of the model. The small variation reflects some robustness in the results.

The first year rates are exceptions. The correlations were high in model 4 (Fig. 6). Neither the application of a reasonable covariate (model 1a) nor the construction of a link between young and old birds (model 4a) aided in the estimation of the recovery and survival rates of the youngest birds.

Model 4 indicated some dependence between the second year survival rate and the recovery rates of older birds. The results reflect this only in the latter parameter whereas the estimates of S_2 (and survival rates of older ages) are rather similar in all models.

We conclude that it was possible to find robust estimates for survival rates of older birds. The first year rates could be accepted only with caution.

Between models 2 and 3, there seemed to be no differences in the results and in the log-likelihood. However, the model with more parameters was easier to identify. The normal method of approach requires that model 2 is chosen due to the principle of parsimony. Thus one should select a model that is at the limit of identifiability. This is not the principle applied in this paper. We select a model where the functional dependence is as weak as possible. As a consequence, the number of the parameters increases significantly. However, the results are given as mean values. For example in models 2-4 we gave values of S_1, S_2 and S_3, though such parameters initially were not present in the models. This approach is supported by Figs. 3-4 which show that only one representative value for S_2 and S_3 can be found. Further, for the older (>3 years) birds we computed one general value. This was accomplished by applying a model in which the survival rates of older birds were assumed to be equal. Formally this is a new model. The use of weighted means was justified by the results of the main model indicating that the survival rates of older birds can be assumed to be equal. This assumption was not made for biological reasons and so, from our point of view, a new model was not developed.

In the approach presented in this paper one first studies the problem of identifiability and tries to determine as many parameters as possible. The next step is to estimate statistical parameters such as confidence limits or correlations. In this aspect there is a need for further developments.

Acknowledgements. The valuable comments made by three referees, among them B.J.T. Morgan and S. Freeman, and by the editors J.-D. Lebreton and P.M.North, are gratefully acknowledged. C. Taylor kindly and effectively checked the language.

References

Brownie, C., Anderson, D.R., Burnham, K.P. and Robson, D.S. (1985).- Statistical inference from band recovery data - a handbook. Resource Publication No. 156. Washington D.C.: United States Department of the Interior, Fish and Wildlife Service.

Cormack, R.M. (1970).- Statistical appendix. In 'Mortality and population change of Dominican Gulls in Wellington, New Zealand' by R.A. Fordham.- Journal of Animal Ecology, 39: 13-27.

Lakhani, K.H. and Newton, I. (1983).- Estimating age-specific bird survival rates from ring recoveries - can it be done? Journal of Animal Ecology, 52: 83-91.

Morgan, B.J.T. and Freeman, S.N. (1989).- A model with first-year variation for ring-recovery data. Biometrics, 45: 1087-1101.

Rinne, J., Lokki, H. and Saurola, P. (1990).- Survival estimates of nestling recoveries: forbidden fruits of ringing? The Ring, 13: 255-270.

White, G.C. (1983).- Numerical estimation of survival rates from band recovery and biotelemetry data Journal of Wildlife Management, 47: 716-728.

Appendix. Formulae for weighted mean values of parameters.

In some cases the parameters are year-dependent resulting in a number of values. For instance, instead of one single value there may be 30 different values for first year survival rates. To characterize these, we computed mean values by applying the following *ad hoc* formulae. An estimated mean value was computed as follows

$$\bar{S}_2=\Sigma\ w_k S_{2k} / \Sigma\ w_k$$

where index k runs over all years and the weights are given by $w_k=1/s^2(S_{2k})$. Here $s(S_{2k})$ is the standard deviation of S_{2k}, as given by program SURVIV (cases with exceptionally small s were excluded). The symbol of a mean, the overbar, will be omitted in the main text. Estimates of covariances were computed in the following manner

$$\hat{cov}(\bar{S}_2, \bar{S}_3)=\Sigma\Sigma\ w_k\ w_l\ \hat{cov}(S_{2k}, S_{3l}) / \Sigma\Sigma\ w_k\ w_l\ ,$$

where indices k and l run over years and $\hat{cov}(S_{2k}, S_{3l})$'s are the covariances given by SURVIV. Standard deviations were then determined in the usual manner.

Marked Individuals in the
Study of Bird Population
J.-D. Lebreton and P.M. North (eds.)

Merging Recoveries and Recaptures to Estimate Survival Probabilites

T.B. Oatley and L.G. Underhill

Avian Demography Unit, Department of Statistical Sciences, University of Cape Town, Rondebosch, 7700 South Africa

Summary. The need for techniques for estimating and comparing survival rates from relatively small data sets of both recoveries and recaptures is considered. Methods based on the proportional hazards regression model and the kernel method are suggested. The methods are illustrated using recovery and recapture data from South Africa for two starlings, the introduced European Starling *Sturnus vulgaris* and the indigenous Pied Starling *Spreo bicolor*.

1. Introduction

The accumulation of recovery records is widely regarded as the objective of most ringing schemes, and all other activities tend to be subordinate to this ultimate objective. Mead (1984) considered that at least 500 recoveries for a species were needed before an analysis of survival could be attempted, and that once 10 000 recoveries had accumulated it was possible to examine both regional and annual differences in survival rates. Such criteria for individual species are attainable within the context of large national ringing schemes, as in Britain or in North America, or when the recoveries for a species are pooled across national schemes, as is achieved by EURING. But they are unattainable, in the foreseeable future, by most ringing schemes in Africa, Asia, Australasia and South America, in which the total number of ringers within the distribution of even a relatively widespread species is small, and the average ecological awareness of the public is by-and-large low, so that reporting rates are low in comparison with, say, Europe and North America. These two factors, small numbers of birds ringed for each species and low reporting rates of recoveries of these birds, combine to reduce the number of recoveries available for analysis.

There are therefore relatively few good data to enable comparisons between the survival rates of passerines, based on studies in the intensively worked temperate regions of Europe and North America, with those elsewhere (see, for example, Karr *et al.*, 1990, Rowley and Russell, 1991). This has implications for the science of ecology as a whole, and for conservation in particular. Much of the theory of ecology has been based on the north temperate experience, and general principles of population dynamics have been based on studies in this climate zone. This theory would be, at least, enriched (if not challenged) by extensive and comparative studies from other zones. Conservation management strategies for a species depend on adequate knowledge of population demographic variables, and the temptation to use "standard" temperate survival rates is great,

Table 1. Distribution of ringing effort with SAFRING rings over 824 species ringed up to June 1991.

Number of individuals ringed	Number of species
1–9	177
10–99	254
100–999	245
1 000–9 999	125
10 000–99 999	22
>100 000	1

Table 2. Number of birds ringed for species with more than 20 000 individuals ringed with SAFRING rings up to June 1991.

Species	Number of individuals ringed	Percentage recovery rate
European Swallow *Hirundo rustica*	161 310	0,42
Redbilled Quelea *Quelea quelea*	96 402	0,25
Cape Gannet *Morus capensis*	91 855	1,25
Yellowbilled Duck *Anas undulata*	58 478	1,66
Cattle Egret *Bubulcus ibis*	47 817	0,80
Masked Weaver *Ploceus velatus*	46 014	0,94
Laughing Dove *Streptopelia senegalensis*	40 354	1,48
Red Bishop *Euplectes orix*	36 862	0,31
South African Cliff Swallow *Hirundo spilodera*	36 740	0,24
Cape Sparrow *Passer melanurus*	36 301	0,67
Jackass Penguin *Spheniscus demersus*	30 379	2,05
Redknobbed Coot *Fulica cristata*	25 962	3,07
Curlew Sandpiper *Calidris ferruginea*	23 834	0,41

because of the lack of definitive studies for the species in question.

In southern Africa, the South African Bird Ringing Unit (SAFRING) provides ringing scheme services to South Africa, Namibia, Zimbabwe, Botswana, Moçambique, Malaŵi and Mauritius (Underhill *et al.*, 1991). In this entire subcontinent, a total of 104 ringers were active in the 1989/90 ringing year, and ringed 37 425 birds (Oatley, 1991). Between 1948, when bird ringing in southern Africa commenced, and June 1991, approximately 1,3 million birds of 824 species were ringed with SAFRING rings (Table 1). The only species with more than 100 000 individuals ringed in southern Africa is the European Swallow *Hirundo rustica*; for a further 12 species, between 20 000 and 100 000 individuals have been ringed (Table 2). The SAFRING recovery file contained approximately 23 000 records in June 1991, having increased by an average of 462 records per year during the 1980s. The file includes details of 364 species recovered with SAFRING rings (Table 3). This is less than half the number of species ringed. For only 19 species are there 200 or more recoveries (Table 4). The overall SAFRING recovery rate is 1.01%, less than half the recovery rate of 2.12% of the British ringing scheme (Mead and Clark, 1991). For all except six species (Table 5), the number of recoveries is still too small for meaningful analyses of survival probabilities, according to Mead's (1984) criterion (500 recoveries).

This depressing analysis is not unique to southern Africa. The problem is

Table 3. Numbers of birds with SAFRING rings recovered up to June 1991.

Number of recoveries	Number of species
0	460
1–9	243
10–99	93
100–999	27
>1 000	1

Table 4. Species with more than 200 recoveries of birds with SAFRING rings up to June 1991.

Species	No. of recoveries
Cape Gannet *Morus capensis*	1 150
Yellowbilled Duck *Anas undulata*	973
Redknobbed Coot *Fulica cristata*	798
European Swallow *Hirundo rustica*	684
Jackass Penguin *Spheniscus demersus*	623
Laughing Dove *Streptopelia senegalensis*	599
Cattle Egret *Bubulcus ibis*	431
Egyptian Goose *Alopochen aegyptiacus*	392
Masked Weaver *Ploceus velatus*	381
South African Shelduck *Tadorna cana*	364
Redbilled Teal *Anas erythroryncha*	335
Whitebreasted Cormorant *Phalacrocorax carbo*	332
Hartlaub's Gull *Larus hartlaubii*	321
Swift Tern *Sterna bergii*	262
Sacred Ibis *Threskiornis aethiopicus*	252
Cape Cormorant *Phalacrocorax capensis*	251
Cape Sparrow *Passer melanurus*	245
Redbilled Quelea *Quelea quelea*	237
Cape Vulture *Gyps coprotheres*	235

probably worse elsewhere, because the ringing scheme in southern Africa is one of the oldest and largest outside Europe and North America. Even within the Euro-American context, North (1990) observed that the development of sophisticated models for the analysis of recovery data had progressed far ahead of the volumes of data available, citing studies that had not been able to take full advantage of models in which parameters can vary by time, age, place, etc. It seems that the comment "more general models were tried, but there were too few (recovery) data to support them" (North, 1990) is, indeed, a worldwide problem.

However, many ringers in southern Africa, as elsewhere, have operated at single sites for extensive periods, generating large numbers of recaptures of birds ringed in previous years. The extreme values from such recapture studies produce impressive (by north temperate standards) longevity observations for individual birds, especially if this is considered in relation to the small numbers of birds ringed of each species in southern Africa, as compared to Britain (Table 5). The prime example of such a study is Hanmer (1981, 1983, 1985, 1987, 1989), whose study was in a small area of indigenous woodland at Nchalo (16 16 S; 34 55 E), Malaŵi, from March 1974 to July 1989, or 15 years. Up to mid-1984, Hanmer had ringed *c.* 19 000 birds at Nchalo; recaptures proved that 579 (3%) of these reached an age of over five years, and 36 individuals of 17 spe-

Table 5. Oldest known age at recapture of selected passerines in southern Africa and age at recovery of six British passerines, in relation to number of birds ringed.

Species	Age (years)	Total ringed[1]	Reference
Southern Africa			
White-throated Swallow *Hirundo albigularis*	>10	2 380	Earlé, 1987
Yellow-bellied Bulbul *Chlorocichla flaviventris*	12	392	Vernon, 1986
Terrestrial Bulbul *Phyllastrephus terrestris*	>12	811	Hanmer, 1989
Chorister Robin *Cossypha dichroa*	>25	211	Oatley, 1982
Starred Robin *Pogonocichla stellata*	>13.5	753	Oatley, 1982
Cape Reed Warbler *Acrocephalus gracilirostris*	>10.5	3 108	Hanmer, 1989
Yellow-breasted Apalis *Apalis flaviventris*	>11.5	226	"
Red-faced Cisticola *Cisticola erythrops*	>8	425	"
Black Flycatcher *Melaenornis pammelaina*	>8.5	191	Ginn, 1982
Paradise Flycatcher *Terpsiphone viridis*	>10.5	1 068	Hanmer, 1989
Brubru Shrike *Nilaus [brubru]*	>8	77	"
Collared Sunbird *Anthreptes collaris*	>11	630	"
White-bellied Sunbird *Nectarinea talatala*	>16	2 125	A.J. Tree, unpubl.
Brown-throated Weaver *Ploceus xanthopterus*	13	2 736	Hanmer, 1989
Bully Canary *Serinus sulphurata*	>8.5	1 959	"
Britain			
Robin *Erithacus rubecula*	>8.5	531 588	Mead and Clark, 1991
Garden Warbler *Sylvia borin*	>8.0	85 122	"
Willow Warlber *Phylloscopus trochilus*	>9.5	774 218	"
Great Tit *Parus major*	13.9	777 074	"
Chaffinch *Fringilla coelebs*	12.0	652 134	"
Starling *Sturnus vulgaris*	>22	1 092 865	Mead and Hudson, 1986

[1] To June 1991 for southern Africa (SAFRING unpubl.), and to December 1990 for Britain (Mead and Clark, 1991).

cies reached ages above 10 years (Hanmer, 1989). A Brown-throated Weaver *Ploceus xanthropterus* reached 13 years (Table 5), and a further six birds of this species reached 12 years. There can be little doubt that if her 15 year study had continued into the 1990s, these longevity records would have been extended, as each of her successive papers, at two year intervals, broke the records set in the previous papers. Similarly, what is probably the world's oldest record of a wild passerine is a recapture; a Chorister Robin *Cossypha dichroa* marked on 27 May 1956 was retrapped (for the first time) on 14 April 1981 only 300 m from the ringing site (Table 5, Oatley, 1982). This bird was at least six months old when ringed, so was at least 25 years old when recaptured. To prove that this was not a freak, another Chorister Robin, ringed as an adult on 25 December 1955 was last recaptured on 14 April 1971, when at least 16 years old (Oatley, 1982). The European Starling *Sturnus vulgaris* of at least 22 years (Table 5) needs to be seen in the context of over one million birds ringed and 35 898 recoveries of British-ringed starlings (Mead and Clark, 1991).

But can the full set of recaptures be used to supplement the meagre recovery data in the estimation of survival rates? In fact, recoveries and recaptures tend to complement each other; in southern Africa, species, especially small passerines, with low recovery rates frequently have higher recapture rates and *vice versa*

(Oatley, 1990, Fig. 2).

Retrap data have been submitted by ringers to SAFRING on a voluntary basis since 1982. By June 1989, the SAFRING recapture file contained 13 441 records of 269 species. During the 1988/89 ringing year, five times as many recaptures as recoveries were processed (SAFRING unpubl.).

Thus the need for statistical methods that extract the information contained in both recoveries and recaptures is an important one for several reasons:

1. Suitable data sets requiring this kind of approach exist.
2. No large data sets of recovery data only or recapture data only are likely to be available for a long time for many species.
3. Survival rates (and hence population demography) in the biogeographical regions where recovery data sets are smallest may be very different from the areas with adequate, in terms of Mead's (1984) criterion, data sets.

2. Methods for merging recoveries and recaptures

In this paper we consider potential methods for deriving estimates of survival probabilities from merged sets of recovery and recapture data. The proposed methods make the assumption that the recovery rate does not vary with age or time, and further development of the methods will be required to overcome these restrictions.

2.1. Modifications of Cormack-Jolly-Seber and Brownie methods

Buckland (1980, 1982) extended the Cormack-Jolly-Seber capture-recapture model so that recoveries of birds from the marked population could be incorporated into the analysis. Buckland and Baillie (1987) suggested the possibility of developing this method to incorporate recoveries from a national ringing scheme along with recaptures from smaller mark-recapture programmes such as the British Trust for Ornithology's Constant Effort Sites scheme. The models of Brownie *et al.* (1985) provided an extensive set of models for estimating survival and recovery rates for ring recovery data when the number of birds ringed in each year is known. Mardekian and McDonald (1981) extended these methods to take account of recaptures in the marked population. However, all these methods and their extensions are designed primarily for study of a single population, and require full knowledge of the numbers marked and recaptured in each sample. They are therefore unsuitable for the type of recovery-recapture data considered here, mostly generated as a result of small numbers of birds ringed at many sites over long periods of time. These methods are not considered further.

2.2. Proportional hazards regression model

A description of the proportional hazards regression model, developed by Cox (1972), is given Bradley *et al.* (1989), who, in a study of the survival of Short-tailed Shearwaters *Puffinus tenuirostris*, were amongst the first to apply this model to the estimation of avian survival probabilities, although they did not use it in its full generality. This model is an extension of the Kaplan-Meier empirical survival curves (Kaplan and Meier, 1958) and compares survival probabilities in a medical context when groups of patients are subjected to different treatments. The data consist of the lengths of time between treatment and death (these are recoveries, in ringing parlance!) and of lengths of time that patients were known to survive before they were "lost to follow-up" (final recaptures).

Besides the fact that both recoveries and recaptures can be included in the analysis of survival, the advantages of this model are that the so-called "baseline hazard function" can be either parametric or empirical, explanatory variables can be incorporated, and it is a generalized linear model (Nelder and Wedderburn, 1972, McCullagh and Nelder, 1989). An empirical baseline hazard function enables the analyst to avoid the stifling restrictions imposed by a parametric survival function (such as the exponential or Weibull distributions). The baseline hazard function may be thought of as the death rate function when there are no explanatory variables. The explanatory variables are, as in other generalized linear models, incorporated into the model via a "link" function, and modify the baseline hazard function through multiplicative terms that adjust it to take account of the explanatory variables. The explanatory variables may be continuous (for example, winter temperatures) or factor variables (for example, sex).

Because it is a generalized linear model, its statistical properties are relatively well known and understood, and it can be fitted with standard statistical packages, such as GLIM (Baker and Nelder, 1978) and GENSTAT (Genstat 5 Committee, 1987). However, the assumption that reporting rate is constant with respect to age and time is currently implicit in this model. The development of generalizations to this method that overcome this limitation would make application of the proportional hazards regression model to the estimation of avian survival rates more useful.

2.3. Kernel method

The parametric approach that appears to have been almost exclusively adopted for the modelling of survival probabilities of birds effectively assumes that the observed lifetimes fit a known parametric family of probability density functions. Nonparametric approaches, which allow less rigid assumptions about the data to be made, have not been tried. Silverman (1986) reviews these nonparametric approaches, focusing on kernel methods. The chief advantage of such methods is that they allow the data to speak for themselves, and this may highlight aspects of the data which may be hidden if the data are constrained to fit one member of some parametric family. Bivariate kernel methods have been applied to the problem of determining home ranges of animals using radio-tracking data (Button, 1986, Worton, 1989), but appear not to have been applied to the simpler univariate problem of estimating survival density functions.

The kernel estimator of a probability density function can be thought of as the sum of 'bumps' centred on the observations. The shape of the bumps is called the kernel function, and the area of individual bumps is arranged so that the total area under the curve defined by the sum of the bumps is equal to one, so that this final curve is a probability density function. Usually, each observation is given equal weight, so that if there are n observations, the area under the curve for each bump is $\frac{1}{n}$. All bumps can have the same shape, or the shapes can vary according to the frequency with which data points occur in a region. The choice of kernel function is important. There are two components to this choice: the shape of the kernel function, and the "width" of the kernel. The estimated density function inherits many of the mathematical properties of the chosen kernel, in particular they share continuity and differentiability properties (Silverman, 1986). The width of the kernel function determines the amount of smoothing that takes place. If too wide a kernel function is chosen, the fine structure of the density function is blurred, and, conversely, if it is too narrow, spurious fine structure is emphasized. Silverman (1986) describes guidelines for choosing the width (or window) for the kernel function, but recommends that comparing the results of several different widths generally provides greater insight into the data. In the example below, the kernel estimator, as defined by Silverman (1986), is applied to two data sets of purely recovery data. We used the Normal distribution for the kernel function, centred on the age of survival of each observation, and with standard deviation equal to the square root of the age of survival, in order to provide greater smoothing in those areas where the data were sparse. In other words, in the terminology of Silverman (1986), we used an adaptive kernel procedure.

How can the kernel method be modified to take account of recaptures as well as recoveries? The problem of extending the kernel method to handle censored data appears to have received no formal attention from statisticians, and the following should be regarded as a first approach in this direction, likely to be superceded by improved methods.

Clearly, we cannot follow the approach of erecting a bump at the point of each recapture, as this does not represent the time of death of the bird. The information contained in the recapture is that the bird died at some later time. This suggests erecting a bump to the right of each recapture. The key issue is deciding the shape for this bump. One obvious choice is to use the shape of the kernel density function obtained from the recoveries alone, but scaled down so that, if there are m recaptures, the area of each bump is $\frac{1}{m}$. These bumps should no longer be centred over the observations, but aligned so that the zero point is at the observation, and the entire bump lies to the right of it. The sum of these bumps then provides a density function for survival derived from the recaptures. An overall density function, derived from both recoveries and recaptures, is obtained by "mixing" the density functions derived from the recoveries and the recaptures; the logical choice of mixing parameter is in proportion to the two sources of data. This choice of kernel function for the recaptures is sensible because there is little information in recaptures made soon after ringing, and it is only with this choice that early recaptures have negligible effect on the shape of the final density function.

Mathematically, suppose we have observations on n recoveries made at times t_1, t_2,..., t_n, and m recaptures made at times u_1, u_2, ..., u_m. Suppose the kernel function is $K(x)$, and is centred on the zero. Then the kernel probability density function $f_1(x)$ at any time $x \geq 0$, estimated from the recoveries, is

$$f_1(x) = \frac{1}{n} \sum_{i=1}^{n} K(x - t_i), \tag{1}$$

and is zero for negative values of x. The equivalent $f_2(x)$ from the recaptures is

$$f_2(x) = \frac{1}{m}\sum_{i=1}^{m} f_1(x - u_i). \tag{2}$$

Then the overall probability density function, $f_3(x)$, is found by mixing $f_1(x)$ and $f_2(x)$ to obtain

$$f_3(x) = \frac{n}{n+m} f_1(x) + \frac{m}{n+m} f_2(x) \tag{3}$$

Alternative approaches are also possible. One suggestion is to iterate, so that the kernel estimated is used in the first iteration (as above), and then the new kernel used in the second iteration, etc. In other words, in the second and subsequent iterations, $f_3(x - u_i)$ computed from (3) is substituted in place of $f_1(x - u_i)$ in (2). However, the process does not converge satisfactorily, as on each iteration the density function is shifted to the right.

Another possibility is to replace $f_1(x - u_i)$ in (2) by separate conditional probability density functions $f_i(x|u_i)$, computed as follows for each of the m recaptures:

$$\begin{aligned} f_i(x|u_i) &= f_1(x) \Big/ \int_{u_i}^{\infty} f_1(x)\,dx \qquad && x \geq u_i \\ &= 0 && x < u_i. \end{aligned} \tag{4}$$

This is unsatisfactory because it results in discontinuities in $f_2(x)$ at each recapture, which are carried forward into $f_3(x)$ by equation (3).

In the example, the approach using equations (1), (2) and (3) has been used. Point estimates of percentiles of these density functions may be estimated in the usual way, and confidence intervals estimated by bootstrapping. This has been done for the median, quartiles and the 90th percentile for $f_1(x)$ and $f_3(x)$.

3. Example

To demonstrate what can be achieved using the proportional hazards model and the kernel method, we compare the SAFRING data sets for two species of starling (Sturnidae). The European Starling, a species introduced into South Africa from Europe in 1899 (Maclean, 1985), has been the target of considerable ringing effort (4 258 ringed and 103 recoveries), making it one of the eight passerines for which there are more than 100 recoveries in the SAFRING data bank. In addition, there are 28 recaptures. The indigenous Pied Starling *Spreo bicolor* occurs in the cooler parts of South Africa and is sympatric with the European Starling in some areas, but favours a rural rather than urban environment (Maclean, 1985). The data set is smaller, with 663 ringed, 12 recoveries and 22 recaptures.

Our attention was drawn to the European Starling because, despite the relatively large (for SAFRING) number of recoveries, no individuals had reached seven years of age, an age attained by two of the 12 Pied Starling recoveries. This suggests that despite its present environment, the European Starling

Table 6. Medians and quartiles of the survival times for European and Pied Starlings, derived from the empirical survivor functions of Cox (1972), using both recoveries and recaptures.

	European Starling	Pied Starling
Lower quartile	7	24
Median	15	37
Upper quartile	28	64

Table 7. Medians, quartiles and 90th percentile of the survival times for European and Pied Starlings, derived from the kernel method, for recoveries (n) only, and for both recoveries and recaptures (m). 95% confidence intervals were derived from 1000 bootstrap samples.

	European Starling		Pied Starling	
	recoveries	recoveries and recaptures	recoveries	recoveries and recaptures
Sample size	n=103	n=103, m=28	n=12	n=12, m=22
Lower quartile	6 (5–8)	7 (5–9)	10 (7–23)	21 (13–31)
Median	12 (10–16)	14 (11–18)	23 (12–48)	37 (26–62)
Upper quartile	24 (20–30)	27 (23–33)	48 (25–77)	65 (39–87)
90th percentile	40 (32–49)	43 (35–51)	77 (32–85)	89 (55–104)

still exhibits a typically north-temperate survival rate. We would expect an indigenous relative to enjoy a superior survival rate, more in keeping with other southern African passerines.

From the Kaplan-Meier empirical survival curves, the median survival time for the European Starling is 15 months, compared to 37 months for the Pied Starling (Table 6). The generalized Wilcoxon or Breslow test rejected the null hypothesis of equal survival curves for the two species ($P < 0.0005$). Since this test showed that there were significant differences between the survival times, Cox's proportional hazards model was used to model the survival times for the two species. Let

$$h(t,i) = h_0(t) \exp \beta_i$$

be the hazard, or instantaneous death rate for species i, $i = 1,2$, where $h_0(t)$ is the non-parametric baseline hazard. The proportional hazards model indicated that the instantaneous death rate for the Pied Starling was only 24% of that of the European Starling. The difference was significant ($P < 0.001$).

The kernel density functions for recoveries only and for both recoveries and recaptures were similar for European Starlings (Fig. 1), suggesting median life times of 12 and 14 months respectively (Table 7), and close to the value of

Figure 1. The kernel estimates of the probability density functions (PDF) of European Starlings computed (A) from 103 recoveries only and (B) from 103 recoveries and 28 recaptures

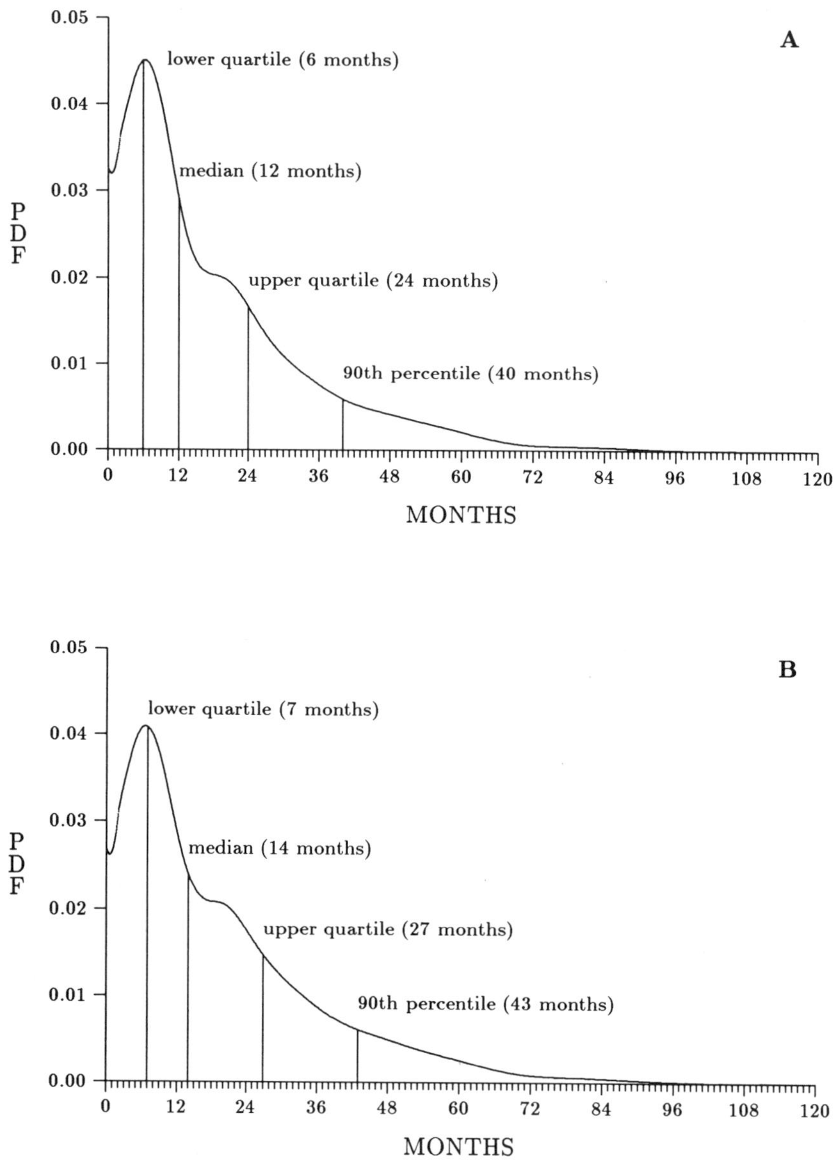

Figure 2. The kernel estimates of the probability density functions (PDF) of Pied Starlings computed (A) from 12 recoveries only and (B) from 12 recoveries and 22 recaptures

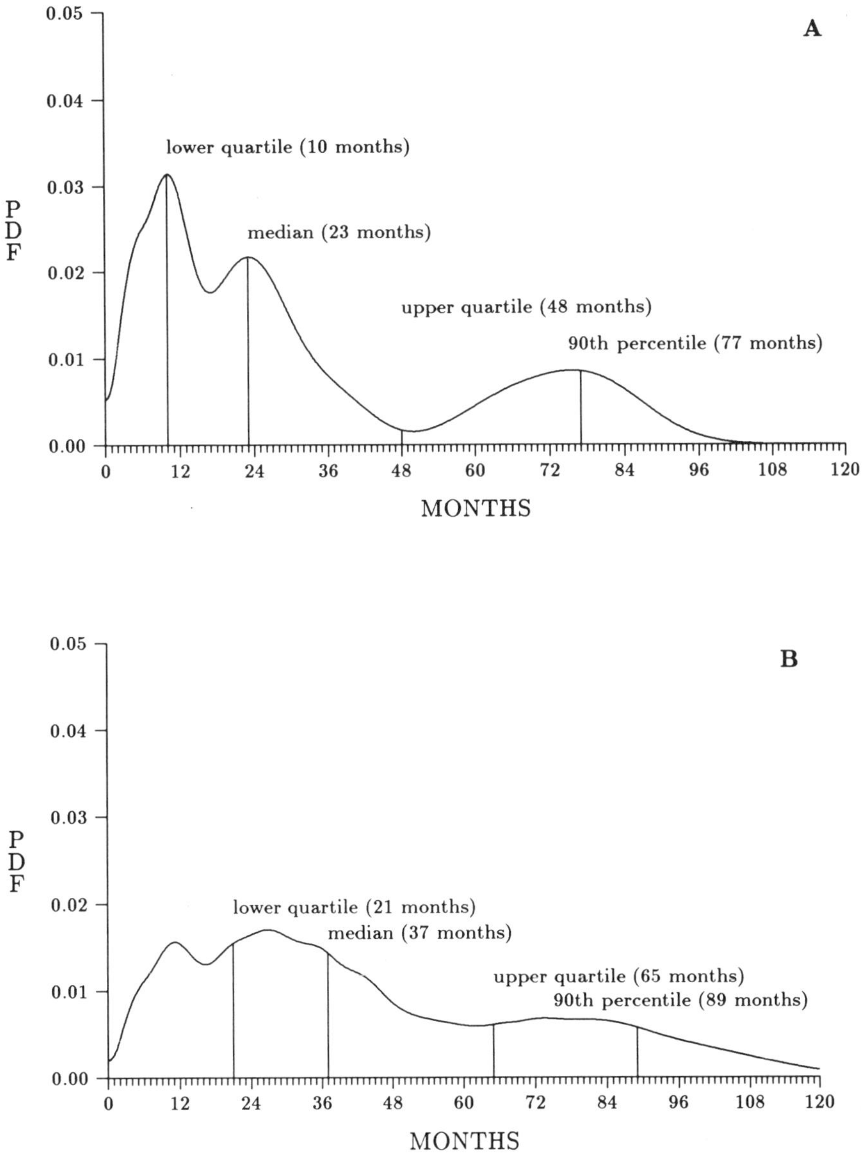

15 months obtained from the Kaplan-Meier survival curve. However, for the Pied Starling, due to the small sample size, these two curves differed considerably in shape (Fig. 2), and the median survival times were estimated as 23 months from recoveries only and 37 months from both recoveries and recaptures (Table 7). The latter value is the same as that obtained from the Kaplan-Meier survival curve.

4. Discussion

Mardekian and McDonald (1981) hinted at the major philosophical difficulty of combining recovery and recapture data, the problem that the recovery and the recapture rates differ. In fact, the underlying processes that generate recoveries and recaptures are very different. The recovery process is more or less continuous. In some studies, such as those of Hanmer (1981, 1983, 1985, 1987, 1989), recaptures only occur on the discrete occasions when birds are being ringed. Thus, if a ringing site is only operated at infrequent intervals, the intervals between ringing and recapture at that site can only take on a few specific values. In particular, if a site is operated for a few years and ringing there then discontinued, the recapture periods will be truncated. In the extreme situation, ringing may take place at an isolated site only once, in which case there are no retraps (unless the species is subject to movement or migration). In other studies (for example, studies of Barnacle Geese *Branta leucopsis* by Ebbinge *et al.* (1991)), birds with individually marked colour rings recognisable at a distance are searched for by a large team of observers, and resightings are made independently of ringing operations. The recapture process is then comparable to the recovery process, in the sense of being continuous, but may be more intense in that many more resightings are made than recoveries.

However, the concept of a reporting rate, or even a year specific reporting rate (denoted f_i by Brownie *et al.* 1985), is open to question. Apart from game species, where the reporting rate is simply the proportion of shot birds with rings which are reported, the reporting rate is an amalgam of different reporting rates related to different causes of death. Thus the probability of a bird killed by a cat (EURING circumstance code 61) or shot (code 01) being reported is far larger than for birds taken by a reptile, amphibian or fish (code 68) (EURING, 1979). Not only are causes of death likely to vary through time, so are the reporting rates associated with different causes. For example, the reporting rate associated with hunted birds is thought to have decreased in countries in which the hunting lobby is in conflict with conservationists. A further problem is that reporting rate may vary seasonally. This is particularly likely for migrants, where the reporting rate may depend on the part of the range in which it dies.

When considered in this light, the philosophical difficulty of combining recoveries and recaptures is no worse than the widely adopted approach of combining recoveries from a variety of circumstances, and treating the data set as a representative sample of life times from which an attempt must be made to estimate survival probabilities. As presented here, both the proportional hazards model and the kernel method assume that the reporting rate is constant over both time and age. With the small sample sizes envisaged when these methods are applied, there is insufficient information to estimate variations in reporting rate by any method, and this assumption needs to be clearly stated.

However, using the kernel method with large samples, may in fact show patterns in the variation of survival rates that were previously unsuspected. Preliminary analyses on large data sets of recoveries (i.e. using only (1)) revealed seasonal patterns in the kernel density that appear to be more likely to be due to variations in mortality than variations in reporting rate (unpubl. data).

In equation (3) it is assumed that each of the n recoveries and m recaptures has equal "weight". Alternative weightings could be used, but would need to be justified. If, for example, it could be shown that the recapture process generates samples at twice the rate of the recovery process, so that each recovery is "worth" two recaptures, (3) could be modified to take this into account:

$$f_3^\star(x) = \frac{2n}{2n+m} f_1(x) + \frac{m}{2n+m} f_2(x). \tag{5}$$

However, such weightings could easily be reduced to "fudge factors", and editors and referees should not be too easily convinced.

For the proportional hazards model, one is frequently interested in comparisons between two or more data sets. If it is reasonable to assume that all data sets have been subjected to the same variation in reporting rates, then this is absorbed into the baseline hazard function, and the comparisons remain valid. We would argue that this is the case with the European/Pied Starling comparison made in the example.

There are two other important differences between the use of the proportional hazards model to estimate and compare survival rates in a human medical context and in the context of wild bird populations. Firstly, in ringing studies, most individuals are "lost to follow up" immediately upon release. A further assumption with this method is therefore that the birds are randomly assigned to the three categories: recovered, recaptured and never seen after first being ringed and released. Secondly, implicit in the proportional hazards model is the assumption that individuals are "followed-up" (recaptured) on each occasion until they are lost to follow up. In other words, the probability of recapture on each occasion until the bird disappears is equal to one. Perhaps, therefore, the most useful area of application of the proportional hazards model will be in radio-telemetry studies, where almost every bird is either followed until death, or is censored by transmitter failure, emigration or the end of the observation period.

Acknowledgements. We acknowledge support from the University of Cape Town Research Committee, the Department of Environment Affairs, South Africa, (TBO), the Southern African Nature Foundation (TBO), the Foundation for Research Development (LGU) and the Schweizerische Vogelwarte (LGU). Professor J.M. Juritz assisted with the proportional hazards regression model.

References

Baker, R.J. and Nelder, J.A. (1978).- The GLIM System 3 Release Manual. Numerical Algorithms Group, Oxford.

Bradley, J.S., Wooller, R.D., Skira, I.J. and Serventy, D.L. (1989).- Age-dependent survival of breeding Short-tailed Shearwaters *Puffinus tenuirostris*. J. Anim. Ecol. 58: 175–188.

Brownie, C., Anderson, D.R., Burnham, K.P. and Robson, D.S. (1985).- Statistical Inference from Band Recovery Data. A Handbook. 2nd edition. United States Department of the Interior, Fish and Wildlife Service, Washington, Resource Publication 156.

Buckland, S.T. (1980).- A modified analysis of the Jolly-Seber capture-recapture model. Biometrics 36: 419–435.

Buckland, S.T. (1982).- A mark-recapture survival analysis. J. Anim. Ecol. 51: 833–847.

Buckland, S.T. and Baillie, S.R. (1987).- Estimating bird survival rates from organised mist-netting programmes. Acta Orn. 23: 89–100.

Button, P. (1986).- The estimation of home range. Unpublished B.Sc.(Hons) thesis, Department of Mathematical Statistics, University of Cape Town.

Cox, D.R. (1972).- Regression models and life tables. J. Roy. Statist. Soc. B. 34: 187–220.

Earlé, R.A. (1987).- Ringing and recovery details of four southern African swallow species. Safring News 16: 67–72.

Ebbinge, B.S., van Biezen, J.B. & van der Voet, H. 1991. Estimation of annual survival rates of Barnacle Geese *Branta leucopsis* using multiple resightings of marked individuals. Ardea 79: 73–112.

EURING (1979).- Code Manual: New EURING. Tring: EURING.

Genstat 5 Committee (1987).- Genstat 5 Reference Manual. Oxford University Press, Oxford.

Ginn, P. (1982).- Letter to editor. Safring News 11: 72–73

Hanmer, D.B. (1981).- Longevity from retraps. Safring News 10: 12–22.

Hanmer, D.B. (1983).- Longevity from retraps continued. Safring News 12: 56–63.

Hanmer, D.B. (1985).- Malaŵi longevity again. Safring News 14: 51–60.

Hanmer, D.B. (1987).- Getting older in Malaŵi. Safring News 16: 73–78.

Hanmer, D.B. (1989).- The end of an era — final longevity figures for Nchalo. Safring News 18: 19–30.

Kaplan, E.L. and Meier, P. (1958).- Nonparametric estimation from incomplete observations. J. Am. Statist. Assoc. 53: 457–481.

Karr, J.R., Nichols, J.D., Klimkiewicz, M.K., and Brawn, J.D. (1990).- Survival rates of birds of tropical and temperate forests: will the dogma surive? Am. Nat. 136: 277–291.

Maclean, G.L. (1985).- Roberts' Birds of Southern Africa. John Voelcker Bird Book Fund, Cape Town.

Mardekian, S.Z. and McDonald, L. (1981).- Simultaneous analysis of band-recovery and live-recapture data. J. Wildl. Management 45: 484–488.

Mead, C.J. (1984).- Rings and the ringers: after the 75th year. Ring. and Migr. 5: 14–19.

Mead, C.J. and Clark, J.A. (1991).- Report on bird ringing for Britain and Ireland for 1990. Ring. and Migr. 12: 139–175.

Mead, C.J. and Hudson, R. (1986).- Report on bird-ringing for 1985. Ring. and Migr. 7: 139–188.

North, P.M. (1990).- Analysis of avian ring recovery and live recapture data: where have we come from Wageningen to Sempach? Ring 13: 11–22.

Oatley, T.B. (1982).- The oldest passerine. Safring News 11: 21.

Oatley, T.B. (1990).- SAFRING statistics for the 1988–89 ringing year. Safring News 19: 27–35.

Oatley, T.B. (1991).- The 1989-1990 ringing year. Safring News 20: 42–52.

Rowley, I. and Russell, E. (1991).- Demography of passerines in the temperate southern hemisphere. pp 22–44 *in* Bird Population Studies: Relevance to Conservation and Management, Perrins, C.M., Lebreton, J.-D. and Hirons, G.J.M. (Eds). Oxford University Press, Oxford.

Silverman, B.W. (1986).- Density Estimation for Statistics and Data Analysis. Chapman and Hall, London.

Underhill, L.G., Oatley, T.B. and Harrison, J.A. (1991).- The role of large-scale data collection projects in the study of southern African birds. Ostrich 62: 124–148.

Vernon, C.J. (1986).- Yellow-bellied Bulbul *Chlorocichla flaviventris*. Safring News 15: 56.

Worton, B.J. (1989).- Kernel methods for estimating the utilization distribution in home-range studies. Ecology 70: 164–168.

Marked Individuals in the
Study of Bird Population
J.-D. Lebreton and P. M. North (eds.)

Evaluation of Radio Tagging Marking and Sighting Estimators of Population Size Using Monte Carlo Simulations

Gary C. White

Dept. of Fishery and Wildlife Biology, Colorado State University Fort Collins, CO 80523 USA

Summary. Marking and sighting estimators (Arnason et al. 1991) of population size are effective when used with radio-tagged animals because the number of marked animals in the population is known at the time of sighting surveys. I have used Monte Carlo simulations to evaluate the performance of 3 of these estimators: joint hypergeometric maximum likelihood estimator (JHE) (Bartmann et al. 1987, White and Garrott 1990), an extension of JHE to incorporate immigration and emigration (IEJHE) (Neal et al. Submitted), and the bootstrap estimator (MM) proposed by Minta and Mangel (1989).

Two sets of simulations were used to compare JHE and MM: the basic simulations assumed constant probability of capture and sighting across animals, and thus met the assumptions of the JHE. Heterogeneity simulations assumed variable sighting probabilities across animals, with the sighting probability for each animal drawn from a beta distribution. JHE performed satisfactorily for the basic simulations, but MM had poor confidence interval coverage. For the heterogeneity simulations, JHE 79% confidence interval coverage compared to 70% coverage for MM. Thus, JHE performed better than MM for the basic simulations, where all the assumptions of JHE were met. For the heterogeneity simulations, JHE still performed better than MM because confidence interval coverage was better, although still far from 95%. Coverage of MM is below the 95% level because this estimator is conditioned on the observed number of unmarked animals (K. P. Burnham, Pers. Comm.). Unless extreme heterogeneity is expected, JHE is preferred over MM.

IEJHE estimates both the total number of animals existing on the area of interest during the period surveyed, and the average number of animals on the area during the surveys. IEJHE demonstrated satisfactory performance for the scenarios simulated.

1. Introduction

The estimation of population size using radio tags as marks offers considerable advantages over standard mark-recapture techniques. Arnason et al. (1991) have suggested the phrase "marking and sighting experiments" to identify population estimation techniques where a sample of marked animals are placed in the population, and sightings of marked and unmarked animals are used to estimate population size. When the marked animals carry radio tags, the number of marked animals in the population at any time is known, assuming no censoring from unknown radio failures.

Rice and Harder (1977) first reported using simple Lincoln-Petersen estimates corrected for bias (Chapman 1951) for each of 5 sighting surveys of white-tailed deer (*Odocoileus virginianus*) from a helicopter. The mean of the 5 estimates was used as the overall population estimate, although they also proposed the median of the 5 estimates as an alternative. Marking and sighting also has been used to estimate population size of other ungulates, including mountain sheep (*Ovis canadensis canadensis*) (Furlow et al. 1981; Leslie and Douglas 1979, 1986; Neal et al. Submitted) and mule deer (*O. hemionus*) (Bartmann et al. 1987).

Eberhardt (1990) further investigated the Petersen estimator with the Chapman correction (Chapman 1951) for small population sizes where animals could immigrate/emigrate from the study area. His work was motivated by the attempts of Miller et al. (1987) to estimate black (*Ursus americanus*) and brown (*U. arctos*) bear populations in central Alaska. He concluded that both the mean of the individual estimates and the Petersen estimate constructed from the means of the number of marked animals, unmarked animals, and marked animals sighted across sighting occasions provided satisfactory estimates of the average population size on the area of interest.

White and Garrott (1990) compared 5 methods of estimating the population size from repeated surveys on a closed population. They compared the mean of the estimates, the mean weighted by the reciprocal of variance of each of the estimates, the geometric mean of the estimates, the geometric mean weighted by the reciprocal of variance of each of the estimates, the median of the estimates, and the joint hypergeometric maximum likelihood estimator (JHE), first proposed by Bartmann et al. (1987). White and Garrott (1990) concluded that the JHE provided the least biased estimates with approximately 95% confidence interval coverage, although they did not simulate the very small population sizes of Eberhardt (1990).

Neal et al. (Submitted) and Neal (1990) performed extensive simulations of the JHE, evaluating the robustness of the estimator to heterogeneity of sighting probabilities, and to lack of independence of sightings (aggregation of animals in groups). Aggregation did not bias the estimate but did decrease precision and confidence interval (CI) coverage. Violation of the assumption of equal sighting probabilities among individuals biased the estimates upwards, especially with smaller populations (≤50), but the bias was not large (≈ 8% maximum). The JHE demonstrated increased bias for the smaller population size simulated, typical of the small sample bias of maximum likelihood estimators.

The objective of this paper is to make further comparisons by Monte Carlo simulations of population estimators for mark and sighting surveys. Results from Neal et al. (Submitted) and Neal (1990) will be compared to 2 new estimators. I emphasize the bias of the estimator and its confidence interval coverage as the most important criteria for comparing the estimators simulated.

2. Notation

T_i Number of marked (telemetered) animals in the population at the time of the ith survey, $i = 1,...,k$.

M_i Number of marked animals in the population that are on the area surveyed at the time of the ith sighting survey. For all M_i constant, define $M \equiv M_i$.

n_i Number of animals seen during the ith sighting survey, consisting of m_i marked animals and u_i unmarked animals, so that $n_i = m_i + u_i$.

f_i Number of marked animals seen i times during the k surveys (sighting frequencies). Note that for radio tags, the number of marked animals never sighted, f_0, is known.

$m.$ Total number of sightings of marked animals seen, so that $m. = \Sigma m_i = \Sigma i f_i$.

$u.$ Total number of sightings of unmarked animals seen, so that $u. = \Sigma u_i$.

3. Estimators

Three estimators of population size for marking and sighting experiments will be compared. First is the joint hypergeometric maximum likelihood estimator (JHE) (Bartmann et al. 1987, White and Garrott 1990, Neal 1990, Neal et al. Submitted). JHE is the value of N which maximizes the following likelihood ($\hat{N}$):

$$\mathcal{L}(N \mid M, n_i, m_i) = \prod_{i=1}^{k} \frac{\binom{M}{m_i}\binom{N-M}{n_i-m_i}}{\binom{N}{n_i}} \tag{1}$$

and the terms are defined for all $i = 1$ to k sighting occasions. The estimate $\hat{N}$ can be found by iterative numerical methods. Confidence intervals are determined with the profile likelihood method (Hudson 1971, Venzon and Moolgavkar 1988). This estimator assumes that all the marked animals are on the area surveyed for each survey, i.e., that the population is geographically closed. Hence, the number of marked animals (M) is constant for each survey, although the sighting probability is not assumed to be constant for each survey.

Second, Minta and Mangel (1989) suggested a bootstrap estimator (MM) of population size based on the sighting frequencies of the marked animals, f_i. For unmarked animals, sighting frequencies are drawn at random from the observed sighting frequencies of the marked animals until the total number of captures equals $u.$. The number of animals sampled is then an estimate of the number of unmarked animals in the population, so that M plus the number sampled is an estimator for N. Only samplings where the number of sightings was equal to $u.$ were used, i.e., cases where the cumulative sightings exceeded $u.$ were rejected. Minta and Mangel (1989) accepted the first value where the cumulative sightings equalled or exceeded $u.$. The stopping rule I used results in less bias than the rule used by Minta and Mangel (1989). One-thousand replications were performed to generate the distribution of $\hat{N}$. Minta and Mangel (1989) suggested the mode of the 1000 replicates as the population estimate. Confidence intervals were computed as probability intervals with the 25th and 975th order statistics from this sample of 1000 estimates.

Third, the JHE estimator has been extended to accommodate immigration and emigration (Neal et al. Submitted) through a binomial process. This estimator is labeled IEJHE, and does not assume that the population is geographically closed. Assume that the total population with any chance of being observed on the study area is N^*, and that at the time of the ith sighting survey, N_i animals occur on the study area. I am interested in estimating the mean number of animals on the study area, $\bar{N}$, and possibly N^*. At the time of the ith sighting occasion, a known number of the marked animals (M_i) are on the study area of the possible T_i animals with transmitters. The probability that an individual is on the study area on the ith occasion can be estimated as M_i/T_i, or in terms of the parameters of interest as N_i/N^*. Then the likelihood for the model that includes immigration and emigration is a product of the binomial distribution for emigration/immigration times the joint hypergeometric likelihood of Eq. (1):

$$\mathcal{L}(N^*, N_i \mid T_i, M_i, m_i, n_i) = \prod_{i=1}^{k} \binom{T_i}{M_i} \left(\frac{N_i}{N^*}\right)^{M_i} \left(1 - \frac{N_i}{N^*}\right)^{T_i - M_i} \frac{\binom{M_i}{m_i}\binom{N_i - M_i}{n_i - m_i}}{\binom{N_i}{n_i}} \qquad (2)$$

The parameters N^* and N_i for $i=1$ to k can be estimated by numerical iteration to maximize this likelihood, with the constraints that $N_i > (M_i + u_i)$ and $N^* > N_i$ for $i=1$ to k. Profile confidence intervals can be obtained for the $k+1$ parameters. I was not interested in the k population estimates for each sighting occasion, but rather desired the mean of the N_i estimates. Therefore, I re-parameterized the likelihood to estimate the total population and mean population size on the study area directly, and their profile likelihood confidence intervals. In the re-parameterized likelihood, I used $N_i = \bar{N} + \alpha_i$, where $\Sigma\ \alpha_i = 0$.

4. Simulations

Two sets of simulations were used to compare JHE and MM. The basic simulations assumed constant probability of capture and sighting across animals, and thus met the assumptions of the JHE. Four population sizes (50, 100, 200, 500), 4 sets of sighting occasions (5, 10, 15, 20), 4 sighting probabilities (0.1, 0.3, 0.5, 0.7), and 3 capture probabilities (0.1, 0.3, 0.5) were simulated, yielding 192 scenarios. Heterogeneity simulations assumed variable sighting probabilities across animals, with the sighting probability for each animal drawn from a beta distribution with $\alpha = 3$ and $\beta = 3$. Four population sizes (50, 100, 200, 500), 4 sets of sighting occasions (5, 10, 15, 20), and 3 capture probabilities (0.1, 0.3, 0.5) were simulated, yielding 48 scenarios.

Each scenario was simulated 1000 times. For each scenario, number of marked animals was determined as a random binomial variable for population size and capture probability prior to sighting. Number of marked animals seen on the ith sighting survey was determined as a Bernoulli trial for each animal given its sighing probability, and likewise for unmarked animals.

IEJHE was evaluated with a subset of the basic simulations: 4 population sizes (67, 133, 266, 667), 2 sets of sighting occasions (5, 10), 1 capture probability (0.3) and 2 sighting probabilities (0.5, 0.7). Probability of being on the area of interest at the time of a survey was 0.75, resulting in average densities of 50, 100, 200, and 500.

One restriction for all simulations was that the number of marked animals in the population had to be at least 10, to represent a reasonable radio-tagging effort. Cases with $M < 10$ were rejected. This restriction is most important in interpreting scenarios with $N = 50$ and capture probabilities of 0.1, as at least 10 marked animals were always in the population. M was simulated given an expected capture probability so that results would be useful for designing studies prior to radio tagging, i.e., a trade-off between radio-tagging (capture) effort and sighting effort could be evaluated.

To evaluate the simulations, percent relative bias (PRB), CI length, percent CI coverage, and percent of confidence intervals lying above the true parameter were used. PRB was estimated as

$$\mathrm{PRB} = \frac{(\mathrm{E}(\hat{N}) - N)}{N} \times 100 \ . \tag{3}$$

where $\mathrm{E}(\hat{N})$ is estimated as the mean of the simulated estimates. Overestimation ($\mathrm{E}(\hat{N}) > N$) results when PRB is positive, and underestimation ($\mathrm{E}(\hat{N}) < N$) results when PRB is negative. Percent CI length (PCIL) is the length of the CI divided by the population size x 100 and allows comparisons across population sizes, although this statistic is generally only useful when 95% coverage has been achieved. Confidence intervals were computed as integers for JHE, MM, and for total population size of IEJHE. Confidence interval endpoints were defined to be within the 95% confidence interval, i.e., for true N equal to the lower bound, coverage was assumed. Hypothetically, confidence interval length can be zero. Such a situation might occur when $N = \hat{N}$ and both the upper and lower bound equal $\hat{N}$, for instance, when the population estimate is equal to the number of marked animals plus the maximum number of unmarked animals sighted.

5. Results

Results for the basic simulations for JHE (Table 1) and MM (Table 2) where all assumptions of both estimators are met suggest that neither estimator shows serious bias, although both show significant positive bias ($P < 0.001$). JHE shows greater positive bias than does MM ($P < 0.001$). Bias for JHE ranged from -0.2 to 19.9% for the 192 scenarios, and from -0.1 to 15.6% for MM. Smaller population sizes ($P < 0.001$), lower sighting probabilities ($P < 0.001$), lower capture probabilities ($P < 0.001$), and fewer sighting occasions ($P < 0.001$) lead to increased positive bias for both estimators.

For the basic simulations, coverage for JHE is close to the expected 95% for all cases (mean 95.3%, range 93.6 to 96.7% for 192 scenarios), whereas coverage for MM is far below 95% (mean 70.3%, range 44.0 to 90.3% for 192 scenarios). Coverage was not significantly different among the 192 scenarios ($P = 0.319$) for JHE. For MM, larger population sizes ($P < 0.001$), lower sighting probabilities ($P < 0.001$), lower capture probabilities ($P < 0.001$), and fewer sighting occasions ($P < 0.001$) lead to decreased confidence interval coverage. Lower capture probabilities had the most striking effect on coverage, with the coverage approaching 50% for capture probabilities of 0.1.

Table 1. Percent relative bias (PRB), mean percent confidence interval length (PCIL), and coverage results for JHE with basic simulations, where all assumptions of the mark-resight estimator are satisfied. Above gives the percent of confidence intervals where the lower bound exceeded N.

Subdivision	PRB	SE PRB	Mean PCIL	SE PCIL	Coverage (%)	Above (%)
All variables	1.044	0.029	34.600	0.322	95.34	2.29
Population size						
50	1.766	0.081	53.008	1.015	95.32	2.27
100	1.373	0.068	41.386	0.754	95.35	2.27
200	0.796	0.046	27.764	0.204	95.28	2.36
500	0.239	0.025	16.242	0.074	95.43	2.26
Number of sighting occasions						
5	2.181	0.092	60.468	1.254	95.35	2.31
10	0.917	0.053	32.126	0.213	95.18	2.35
15	0.574	0.040	24.823	0.123	95.43	2.19
20	0.502	0.033	20.982	0.097	95.42	2.32
Sighting probability						
0.1	3.065	0.106	78.964	1.260	95.37	2.29
0.3	0.745	0.041	29.651	0.104	95.32	2.43
0.5	0.264	0.026	18.365	0.058	95.42	2.17
0.7	0.101	0.017	11.418	0.036	95.28	2.27
Capture probability						
0.1	2.025	0.073	57.091	0.886	95.34	2.31
0.3	0.754	0.042	29.264	0.360	95.29	2.29
0.5	0.351	0.025	17.443	0.088	95.40	2.27

Results for the heterogeneity simulations for JHE (Table 3) and MM (Table 4) again suggest that both estimators show slight positive bias ($P < 0.001$), but the bias is not biologically serious (mean 0.6%, range for 48 scenarios of -0.2 to 2.7% for JHE; mean 0.6%, range -0.2 to 2.3% for MM). Smaller population sizes ($P < 0.001$), fewer sighting occasions ($P < 0.046$), and lower capture probabilities ($P < 0.001$) lead to increased bias for both estimators.

For the heterogeneity simulations, confidence interval coverage for both estimators was considerably less than the expected 95% ($P < 0.001$). JHE had better average coverage with 78.8% (range 68.4 to 90.4% for 48 scenarios) compared to 70.1% for MM (range 45.2 to 86.6% for 48 scenarios). For JHE, population size ($P = 0.505$) and capture probabilities ($P = 0.818$) showed little effect, whereas increasing the number of sighting occasions decreased coverage ($P = 0.001$). For MM, larger population sizes ($P < 0.001$), and lower capture probabilities ($P < 0.001$) most affected coverage, with no effect from number of sighting occasions ($P < 0.867$).

Table 2. Percent relative bias (PRB), mean percent confidence interval length (PCIL), and coverage results for Minta and Mangel estimator (1989) with basic simulations, where all assumptions of the mark-resight estimator are satisfied. Above gives the percent of confidence intervals where the lower bound exceeded N.

Subdivision	PRB	SE PRB	Mean PCIL	SE PCIL	Coverage (%)	Above (%)
All variables	0.785	0.028	13.382	0.026	70.26	15.18
Population size						
50	1.379	0.077	20.604	0.070	74.07	13.48
100	0.952	0.063	15.075	0.049	69.81	15.36
200	0.629	0.046	10.874	0.034	68.82	15.85
500	0.180	0.025	6.974	0.022	68.33	16.04
Number of sighting occasions						
5	1.528	0.086	19.503	0.070	69.36	15.65
10	0.761	0.051	13.571	0.048	69.93	15.29
15	0.414	0.038	10.980	0.039	70.65	14.82
20	0.437	0.033	9.472	0.033	71.08	14.97
Sighting probability						
0.1	2.218	0.100	26.331	0.066	68.74	16.40
0.3	0.621	0.041	13.139	0.032	69.19	16.08
0.5	0.234	0.026	8.551	0.021	70.68	14.68
0.7	0.067	0.016	5.505	0.013	72.41	13.57
Capture probability						
0.1	1.743	0.071	14.774	0.050	51.89	24.44
0.3	0.514	0.039	13.692	0.046	73.31	13.73
0.5	0.098	0.024	11.679	0.039	85.57	7.38

Table 3. Percent relative bias (PRB), mean percent confidence interval length (PCIL), and coverage results for JHE estimator from simulations with heterogeneity of individual sighting probabilities ($\bar{x}$ sighting probability = 0.5). Above gives the percent of confidence intervals where the lower bound exceeded N.

Subdivision	PRB	SE PRB	Mean PCIL	SE PCIL	Coverage (%)	Above (%)
All variables	0.595	0.039	18.701	0.064	78.85	10.27
Population Size						
50	0.890	0.102	25.997	0.142	79.27	9.84
100	0.817	0.090	22.113	0.139	78.89	10.12
200	0.434	0.067	16.429	0.102	78.58	10.42
500	0.239	0.041	10.265	0.057	78.67	10.68
Number of sighting occasions						
5	0.907	0.094	27.998	0.174	88.99	5.58
10	0.595	0.076	18.844	0.107	81.49	9.09
15	0.521	0.073	15.094	0.085	74.79	12.31
20	0.357	0.069	12.868	0.071	70.13	14.09
Capture probability						
0.1	1.246	0.095	28.920	0.128	78.59	10.51
0.3	0.431	0.059	16.738	0.083	79.01	10.14
0.5	0.108	0.037	10.444	0.048	78.96	10.16

Table 4. Percent relative bias (PRB), mean percent confidence interval length (PCIL), and coverage results for the Minta and Mangel (1989) estimator from simulations with heterogeneity of individual sighting probabilities ($\bar{x}$ sighting probability = 0.5). Above gives the percent of confidence intervals where the lower bound exceeded N.

Subdivision	PRB	SE PRB	Mean PCIL	SE PCIL	Coverage (%)	Above (%)
All variables	0.562	0.039	13.377	0.027	70.08	15.23
Population Size						
50	0.950	0.102	20.658	0.047	73.87	13.45
100	0.824	0.088	15.079	0.030	69.80	15.52
200	0.309	0.066	10.842	0.019	68.88	15.59
500	0.164	0.041	6.929	0.011	67.79	16.37
Number of sighting occasions						
5	0.727	0.091	15.740	0.063	69.96	15.48
10	0.546	0.076	13.320	0.053	70.23	15.17
15	0.556	0.073	12.469	0.049	70.33	15.22
20	0.418	0.070	11.979	0.047	69.81	15.07
Capture probability						
0.1	1.180	0.093	14.767	0.051	52.07	24.29
0.3	0.384	0.058	13.660	0.047	73.24	13.66
0.5	0.122	0.038	11.705	0.041	84.93	7.75

PCIL for MM was considerably less than for JHE in all cases. This effect is because this estimator conditions on the number of unmarked animals sighted, and the variance of this term (var $u.$) is not accounted for in the estimation of confidence intervals (K. P. Burnham Pers. Comm.). As a result, even with no serious bias, confidence interval coverage is below the expected 95%.

6. Discussion

One interesting result for MM is the closeness of PCIL for Table 1 (no individual heterogeneity) and Table 3 (heterogeneity). I would expect wider confidence intervals for heterogeneity. However, the mean sighting probability is 0.5 in Table 3, and so the heterogeneity simulations do not provide as great a range in sighting probabilities across occasions as do the simulated data in Table 1. Hence, the closeness of the observed PCIL values may be partially an artifact of the parameters simulated.

Coverage of MM can be improved by accounting for the variation of the number of unknown animals sighted. I also incorporated the Robbins-Monro procedure (Buckland and Garthwaite 1990) to provide a more sophisticated approach to computing the MM confidence interval. The population estimate $\underline{\hat{N}}$ was computed as recommended by Minta and Mangel (1989). Then, I used algorithm AS 259 (Buckland and Garthwaite 1990) to compute the upper and lower confidence bounds, where the number of unmarked animals sighted ($u.$) was simulated based on the observed frequencies (in contrast to Minta and Mangel where the observed value was used).

Table 5. Percent relative bias (PRB), mean percent confidence interval length (PCIL), and coverage results for total population (N^*) estimator of IEJHE, where all assumptions of the estimator are satisfied. Above gives the percent of confidence intervals where the lower bound exceeded N^*.

Subdivision	PRB	SE PRB	Mean PCIL	SE PCIL	Coverage (%)	Above (%)
All variables	-0.033	0.047	22.297	0.097	97.09	1.26
Total Population size						
67	0.104	0.138	36.424	0.199	97.80	0.80
133	-0.034	0.097	24.855	0.111	96.97	1.52
267	-0.073	0.067	17.219	0.068	96.85	1.32
667	-0.008	0.041	10.690	0.039	96.72	1.40
Number of sighting occasions						
5	-0.092	0.077	26.200	0.153	96.87	1.40
10	0.087	0.053	18.394	0.101	97.30	1.12
Sighting probability						
0.5	-0.151	0.075	25.440	0.152	96.80	1.49
0.7	0.146	0.055	19.154	0.108	97.37	1.04

Table 6. Percent relative bias (PRB), mean percent confidence interval length (PCIL), and coverage results for average population ($\bar{N}$) estimator of IEJHE, where all assumptions of the estimator are satisfied. Above gives the percent of confidence intervals where the lower bound exceeded $\bar{N}$.

Subdivision	PRB	SE PRB	Mean PCIL	SE PCIL	Coverage (%)	Above (%)
All variables	0.187	0.043	18.364	0.085	94.88	2.07
Average Population size						
50	0.456	0.128	30.150	0.183	95.37	1.47
100	0.251	0.089	20.497	0.106	94.78	2.17
200	0.032	0.060	14.101	0.066	95.12	2.00
500	0.007	0.038	8.708	0.038	94.25	2.62
Number of sighting occasions						
5	0.134	0.071	21.500	0.134	94.30	2.42
10	0.239	0.049	15.229	0.090	95.46	1.71
Sighting probability						
0.5	0.053	0.071	22.176	0.134	95.09	2.17
0.7	0.320	0.049	14.553	0.084	94.67	1.96

Coverage greatly improved for MM with the above procedure (Tables 7 and 8), with the average coverage for the basic simulations 82.8%. However, coverage was still below the expected 95% because the confidence interval is conditional on the observed sighting frequencies of the marked animals. Coverage was 95.0% for the cases where capture probability was 0.5, but was considerably less for capture probabilities of 0.3 (86.8%) and 0.1 (66.6%). I attribute this lack of coverage to the small sample properties of the bootstrap procedure. Specifically, the variation inherent in the sampling process is not show exhibited in the sample when less than 50% of the population is marked. The poor coverage characteristics of MM were not discussed in the original paper (Minta and Mangel 1989), as they did not present coverage results for Monte Carlo simulations. Examples presented by Minta and Mangel (1989) had small numbers (≤16) of marked animals, and hence I would expect poor coverage for these cases if the true population size were known.

Table 7. Percent relative bias (PRB), mean percent confidence interval length (PCIL), and coverage results for Minta and Mangel estimator (1989) using Robbins-Monro confidence intervals (Buckland and Garthwaite 1990) for the basic simulations. Above gives the percent of confidence intervals where the lower bound exceeded *N*.

Subdivision	PRB	SE PRB	Mean PCIL	SE PCIL	Coverage (%)	Above (%)
All variables	0.815	0.028	20.018	0.038	82.81	7.49
Population size						
50	1.358	0.075	31.399	0.098	86.04	5.32
100	1.082	0.066	22.537	0.069	82.40	7.55
200	0.622	0.045	16.012	0.049	81.11	8.61
500	0.199	0.026	10.123	0.031	81.68	8.48
Number of sighting occasions						
5	1.554	0.085	28.558	0.100	82.26	7.86
10	0.716	0.053	20.256	0.070	82.82	7.55
15	0.561	0.040	16.692	0.057	82.87	7.44
20	0.430	0.033	14.564	0.049	83.28	7 12
Sighting probability						
0.1	2.340	0.100	38.154	0.094	81.60	8.94
0.3	0.594	0.040	19.720	0.048	82.57	8.01
0.5	0.231	0.026	13.231	0.032	83.04	7.13
0.7	0.095	0.017	8.967	0.022	84.03	5.88
Capture probability						
0.1	1.799	0.070	22.012	0.071	66.55	15.42
0.3	0.554	0.041	20.467	0.067	86.83	5.34
0.5	0.091	0.024	17.574	0.057	95.04	1.71

As shown in Table 8, 50% of the population must have marks for 95% coverage to be achieved with MM. Because most radio-tagging studies do not mark this large of proportion of the population being studied, I would expect that MM would typically produce a relatively unbiased estimate, but the confidence interval would be too narrow. Likewise, for typical radio-tagging studies, JHE would also produce too narrow of confidence interval because typically animals do not all exhibit the same sighting probability on a particular occasion.

Table 8. Percent relative bias (PRB), mean percent confidence interval length (PCIL), and coverage results for Minta and Mangel (1989) estimator using Robbins-Monro confidence intervals (Buckland and Garthwaite 1990) for simulations with heterogeneity of individual sighting probabilities ($\bar{x}$ sighting probability = 0.5). Above gives the percent of confidence intervals where the lower bound exceeded *N*.

Subdivision	PRB	SE PRB	Mean PCIL	SE PCIL	Coverage (%)	Above (%)
All variables	0.559	0.040	19.966	0.042	82.53	7.46
Population Size						
50	0.968	0.105	31.317	0.066	85.25	5.53
100	0.745	0.089	22.492	0.044	82.12	7.60
200	0.343	0.068	15.972	0.028	80.75	8.43
500	0.178	0.040	10.083	0.016	82.00	8.28
Number of sighting occasions						
5	0.695	0.095	23.310	0.095	82.33	7.58
10	0.599	0.079	19.932	0.081	82.72	7.41
15	0.421	0.072	18.634	0.075	82.79	7.18
20	0.518	0.070	17.990	0.073	82.28	7.67
Capture probability						
0.1	1.126	0.095	21.974	0.077	66.32	15.17
0.3	0.443	0.060	20.391	0.073	86.66	5.36
0.5	0.106	0.039	17.534	0.062	94.61	1.86

The 3 estimators simulated are all relatively complex to compute, compared to, say, the arithmetic mean of the Lincoln-Petersen estimates. However, White and Garrott (1990) showed the additional effort to compute the JHE was worthwhile. The arithmetic mean showed significantly lower coverage than the expected 95% for Monte Carlo simulations, while JHE was very close to 95%. For situations where the arithmetic mean did exhibit 95% coverage, the confidence interval was wider than that of JHE. Program NOREMARK for PC computers is available from the author to compute the 3 estimators discussed in this manuscript.

Acknowledgments. A. K. Neal worked diligently to develop the initial evaluations of the JHE estimator. K. P. Burnham, D. R. Anderson, R. E. Kenward, and R. M. Cormack provided helpful comments on the manuscript. Funding for this work was provided by the Colorado Division of Wildlife through the efforts of R. B. Gill.

References

Arnason, A. N., Schwarz, C. J., and Gerrard, J. M. (1991).-Estimating closed population size and number of marked animals from sighting data. *Journal of Wildlife Management* **55**, 716-730.

Bartmann, R. M., White, G. C., Carpenter, L. H., and Garrott, R. A. (1987).-Aerial mark-recapture estimates of confined mule deer in pinyon-juniper woodland. *Journal of Wildlife Management* **51**, 41-46.

Buckland, S. T., and Garthwaite, P. H. (1990).-Algorithm AS 259 -- Estimation confidence intervals by the Robbins-Monro search process. *Applied Statistics* **39**, 413-424.

Chapman, D. G. (1951).-*Some properties of the hypergeometric distribution with applications to zoological sample censuses.* University of California Publication in Statistics **1**, 131-160.

Eberhardt, L. L. (1990).-Using radio-telemetry for mark-recapture studies with edge effects. *Journal of Applied Ecology* **27**, 259-271.

Furlow, R. C., Haderlie, M., and Van den Berge, R. (1981).-Estimating a bighorn sheep population by mark-recapture. *Desert Bighorn Council Transactions* **1981**, 31-33.

Hudson, D. J. (1971).-Interval estimation from the likelihood function. *Journal of Royal Statistical Society Series B* **33**, 256-262.

Leslie, D. M., Jr. and Douglas, C. L. (1979).-Desert bighorn sheep of the River Mountains, Nevada. *Wildlife Monograph* **66**, 1-56.

Leslie, D.M., Jr. and Douglas, C. L. (1986).-Modeling demographics of bighorn sheep: current abilities and missing links. *North American Wildlife and Natural Resources Conference Transactions* **51**, 62-73.

Miller, S. D., Becker, E. F., and Ballard, W. H. (1987).-Black and brown bear density estimates using modified capture-recapture techniques in Alaska. *International Conference on Bear Research and Management* **7**, 23-35.

Minta, S. and Mangel, M. (1989).-A simple population estimate based on simulation for capture-recapture and capture-resight data. *Ecology* **70**, 1738-1751.

Neal, A. K. (1990).-*Evaluation of mark-resight population estimates using simulations and field data from mountain sheep.* Masters Thesis, Colorado State University, Fort Collins.

Neal, A. K., White, G. C., Gill, R. B., Reed, D. F., and Olterman, J. H. (1992).-Evaluation of mark-resight methods to estimate mountain sheep numbers. *Journal of Wildlife Management*, Submitted.

Rice, W. R. and Harder, J. D. (1977).-Application of multiple aerial sampling to a mark-recapture census of white-tailed deer. *Journal of Wildlife Management* **41**, 197-206.

Venzon, D. J. and Moolgavkar, S. H. (1988).-A method for computing profile-likelihood based confidence intervals. *Applied Statistics* **37**, 87-94.

White, G. C. and Garrott, R. A. (1990).-*Analysis of wildlife radio-tracking data.* New York:Academic Press.

Survival analysis in practice

Marked Individuals in the
Study of Bird Population
J.-D. Lebreton and P. M. North (eds.)

Combining Mark-Recapture Data Sets for Small Passerines

Will J. Peach

British Trust for Ornithology, The Nunnery, Thetford, Norfolk IP24 2PU, United Kingdom

Summary. Thirty-three sets of mark-recapture data collected at fourteen Constant Effort ringing Sites between 1983-91 are analysed using single site and multiple site models provided by program SURGE4. In both types of analyses there was little evidence of time-dependence in either apparent survival or recapture probability. In multiple site analyses there was evidence of site-dependent recapture probability but little evidence of site-dependent apparent survival. Strong evidence of temporal variation in apparent survival was found for only one species (*Sylvia communis*). Despite the relatively small amounts of recapture data available, the precision of survival estimates was similar to or better than that attainable from national ring recovery data for 4 of the 5 species for which there were comparable estimates. The effects of transients in cohort samples and of short-distance emigration on parameter estimates are also considered.

1. Introduction

The Constant Effort Sites Scheme (CES) of the British Trust for Ornithology (BTO) is an extensive programme of standardised mist-netting which aims to monitor changes in population size (adult abundance), productivity (proportion of young birds in the total catch) and adult survival rates (between-year recaptures) for a range of common passerines in Britain and Ireland. The CES Scheme forms part of the BTO's Integrated Population Monitoring (IPM) Programme, which incorporates the Common Birds Census and the Nest Records Scheme, and aims to identify the causes (natural and anthropogenic) of population changes affecting breeding birds in Britain and Ireland (Baillie,1990).

Analyses of long runs of mark-recapture data from intensive study sites have suggested pronounced variation in the annual adult survival rates (strictly return rates) of some small passerines (Peach *et al.*,1990) which in some cases may have a strong influence on population levels (Peach *et al.*,1991; Baillie & Peach,1992). In this paper I estimate annual adult survival rates for a range of common passerines using data collected by amateur ringers on constant effort sites. With the exception of a single site analysis for Willow Warbler *Phylloscopus trochilus* (Pratt & Peach,1991) this is the first attempt to estimate adult survival rates using data collected as part of the CES Scheme.

Since mark-recapture data collected at individual CES sites are often sparse, possibilities for the estimation of survival rates from multiple data sets are considered. With the initiation of multiple site constant effort ringing schemes in Finland, France and North America analyses of the type presented here are likely to be increasingly important in future.

2. Methods

2.1. The Constant Effort Sites Scheme

Most constant effort sites are located in either wet or dry scrub, reedbeds or deciduous woodland. Sites in coniferous woodland are not accepted because rapid tree growth is likely to cause relatively short-term changes in catching efficiency as well as changes to the avian community structure. Potential constant effort sites are proposed by volunteer ringers, and are generally accepted into the national scheme as long as they are located in the target habitats listed above and are not considered to be undergoing major successional changes.

Each year CES ringers are asked to make twelve visits to their site spread evenly between early May and late August. Most adult birds are captured during May and June, whilst most juveniles are trapped in July and August (Baillie *et al.*,1986). A set of standard mist nets is erected in the same positions on each of the twelve main visits. The total length and density of standard netting is decided by the ringer and therefore differs between sites. Additional nets are permitted during main visits, as are extra visits. During 1989 a mean of 3.0 extra visits were made per CES site, with no extra visits being made at 53% of sites. The mean length of additional netting per main visit, considered as a percentage of the standard net length, was 8.6% during 1989. The mean additional netting per main visit was less than 25% of the standard net length at 90% of sites.

Between-year changes in the size of the adult and juvenile catches are calculated for paired main visits at each site, and aggregated across sites to give an overall measure of change (Baillie *et al.*,1986; Peach,1991). Changes in the percentage of young birds in the total catch provides an annual measure of breeding productivity (Du Feu & McMeeking,1991).

The CES Scheme has expanded from an initial 13 sites in 1981 to 110 sites in 1991. However, ringing data for individual birds (i.e. ring numbers) have only been computerised since 1983, thus limiting the scope for mark-recapture analyses to the period 1983-91.

2.2. Mark-Recapture Data

In this paper mark-recapture data collected at 14 CES sites during the period 1983-91 are used to estimate annual adult survival rates for twelve common passerine species. Details of habitat, standard net length and the availability of capture data in addition to that collected during main CES visits are listed in Table 1. In the case of Theddlethorpe Dunes data from two adjacent constant effort sites (within 80m of each other) are aggregated and at Treswell Wood the CES methodology has not been strictly applied, although the effort has remained constant across years.

Table 1. Details of the 14 CES ringing sites from which annual data are available for the years 1983-91. Habitat codes are: DS = dry scrub; WS = wet scrub; DW = deciduous woodland; RB = reedbed. Additional visits or nets indicates that capture data were available in addition to that from the standard set of CES nets during the 12 main visits. Missing visits indicates that fewer than 12 (but rarely fewer than 10) main CES visits were completed in some years.

Site Name	Site Code	Habitat	Standard CES net length (metres)	Notes
Threestone Burn	TS	DS	48.8	additional visits
Great Corby Wood	CO	DW	64.0	
Woodvale	WV	DS	100.6	missing visits
Theddlethorpe Dunes	TH	DS	362.7	additional visits
Treswell Wood	TW	DW	182.9	additional visits and nets
Ballagh Wood	BA	DW	128.0	occasional missing visits
Llangorse Lake	LL	DS	111.6	additional visits
Tewinbury	TE	WS	57.9	
Marsworth	MW	WS	158.5	additional visits
Denford Mill	DF	WS	182.9	
Stodmarsh	ST	RB	85.3	1984-91 only
Westbere	WB	RB	155.4	additional nets; few missed visits
Clanger Wood	CW	DW	274.3	
Ensbury	EN	DS	118.9	

Of the 46 CES sites operated in 1983, 12 (27%) have continued to be operated in every year until 1991. One of these sites failed to produce more than 5 between-year recaptures of adult birds for any species during this 8-year period, and does not therefore contribute to the analyses described here. Three of the 14 sites considered here were only registered as constant effort sites in 1984. For two of these sites data for 1983 were available through the ringers.

Annual sets of mark-recapture data were produced for adults of each species at each site by aggregating data from all main and additional CES visits. This represents an annual sampling period of 4 months although in practice most adults are captured during May and June (Baillie *et al.*,1986; Pratt & Peach,1991). Although methods for estimating survival rates from mark-recapture data assume that the sampling period is short relative to the periods between samples, there is little evidence to suggest that small passerines experience significant mortality during the sampling (breeding) period (e.g. Tiainen (1983)). Smith & Anderson (1987) concluded that survival rates estimated from ring recovery data were unlikely to be biased by a lengthy ringing period as long as survival during the ringing period was high and relatively constant from year to year, and that the distribution of ringing effort within the sampling period remained relatively constant from year to year. The latter is certainly true of data collected at CES sites.

2.3. Analyses using SURGE

2.3.1. Notation

Before considering multiple site analyses, each of the 33 data sets was analysed separately using program SURGE4 (Clobert *et al.*,1987; Pradel *et al*, 1990). The notation used here is similar to that of Pradel *et al.* (1990) in which apparent survival is represented by Φ and recapture probability by P. Time and site dependence in either of these model components is denoted by the subscripts t and s respectively. For each data set the Cormack-Jolly-Seber (CJS) model with time-dependent (annual) survival (Φ_t) and recapture probability (P_t) was taken as a starting model and attempts made to simplify P and then Φ using likelihood ratio tests (LRTs) to compare successive models.

2.3.2. The effect of transient birds on model selection and parameter estimates

Since the inclusion of transient birds (individuals with a lower probability of recapture in subsequent years) in ringed samples is likely to depress estimates of survival in the first year after ringing, an attempt was made to separate newly-ringed cohorts into residents and transients by creating a second 'sample' of birds retrapped at least 10 days after marking (after Buckland & Baillie,1987; Peach *et al.*,1990). An alternative approach to the problem of transients in cohort samples is to exclude recaptures in the first year after ringing when estimating survival rates. However, in short-lived species like warblers this involves the loss of a high proportion of the recapture data with a consequent loss of precision (Peach *et al.*,1990) and I prefer to utilize the within-year recapture information as described above.

Thus, in the second stage of the analysis a modification of the CJS model was fitted to each of the 33 data sets in which two additional parameters were used to represent the probability of 'surviving' (i.e. remaining within the sampling area) and of being recaptured more than 10 days after first being trapped. Both of these probabilities were modelled independently of time, and are effectively dummy age parameters covering the 10 day period following ringing. The notation used to refer to this model is (Φ_{10+t},P_{10+t}). Comparison of parameter estimates from models with and without the pair of dummy age parameters was used to assess the likely influence of transient birds in analyses of this kind.

2.3.3. The effect of short-distance emigration on parameter estimates

In all analyses of mark-recapture data any permanent emigration of marked individuals away from the study area will result in the underestimation of true survival rates. All estimates of 'survival' derived from mark-recapture data should therefore be considered as estimates of apparent survival, reflecting return rates rather than true annual survival rates. A rigorous investigation of the effect of short-distance emigration on estimates of survival derived from mark-recapture data would require intensive data collection at a number of adjacent study areas over a number of years. Models could then be applied which incorporate movement rates between sampling areas, such as those described by Hestbeck *et al.* (1991).

In an analysis of fourteen years of mark-recapture data collected at Wicken Fen in southern England, 7% of retrapped Reed Warblers *Acrocephalus scirpaceus* and 10% of retrapped Sedge Warblers *A. schoenobaenus* that were originally trapped in the main reedbed sampling area, were retrapped for the last time in a small area of habitat outside the main reedbed area. These birds would have been assumed to be dead in any analysis restricted to the reedbed data. Estimates of average annual adult survival derived from the reedbed data only were 13.7% lower for Reed Warblers and 16.9% lower for Sedge Warblers, compared to those derived from the entire data set (Peach *et al.*,1990). Although analyses of this type do not constitute a rigorous appraisal of the effects of short-distance emigration on estimates of survival (differences in Φ and P between sampling areas are confounded, as are the effects of permanent and temporary emigration on Φ and P), they can illustrate the potential effect of short-distance emigration on estimates of survival and thereby highlight useful areas for further research.

During the present study an opportunity arose to consider the effect of increasing the sampling area on estimates of Φ and P. Mark-recapture data were available from two adjacent CES sites at Theddlethorpe Dunes in Lincolnshire. The two sites are situated in dry scrub habitat and have been operated on the same days since 1983. The sites are approximately 80m apart at the closest point, one comprising 216m of netting (site A) and the other 146m (site B). Estimates of Φ and P were compared for five species using mark-recapture matrices derived from combined data from sites A and B, and from site A data only.

2.3.4. Multiple site models

The final stage of the analysis was to consider multiple mark-recapture data sets (i.e. from different sites) within a single model framework, as is now possible using SURGE4 (Pradel *et al.*,1990). Such analyses were carried out for each of seven species for which more than one mark-recapture data set was available. The largest number of data sets were available for Willow Warbler and Blackbird *Turdus merula*

(8 and 5 data sets respectively), though for Willow Warbler multi-site analyses were limited to 7 data set analyses due to a computer-specific compilation constraint. It is not anticipated that this will limit future multi-site analyses to 7 data sets.

In each multiple data set analysis the starting model was one in which both survival and recapture probability were dependent upon time and site with a time-site interaction. With the addition of a pair of parameters for each site representing the probability of 'survival' and recapture more than 10 days after first being trapped in the first year of capture, the fully parameterised starting model is denoted as $(\Phi_{10+t*s}, P_{10+t*s})$. With 8 years of recapture data, each site contributes 18 parameters to the fully parameterised model, and a 7 site analysis therefore has a starting model with 126 parameters.

Attempts were then made to simplify first the recapture probability and then the survival component of the starting model, using likelihood ratio tests between models (after Pradel *et al.*,1990). For each component of the model the first step was to drop the time-site interaction. If this was possible then the time and site factors were dropped (time followed by site in the case of P, and site followed by time in the case of Φ, both on *a priori* reasoning). In cases where likelihood ratio tests permitted the dropping of the site factor from either component of the model, an attempt was also made to replace the site-specific dummy age parameters representing the 10 day period after ringing with a single parameter for all sites. Final models were subjected to goodness-of-fit tests.

3. Results

3.1. The effect of transient individuals on model selection and parameter estimates

Models selected with and without the extra sampling period in the first year of capture, did not differ for 30 of the 33 data sets analysed. In the remaining three cases, time dependence in apparent survival was detected with but not without the extra sampling period.

The effect of creating an extra sampling period of 10 days during the first year of capture was to increase Φ by an average of 0.071 (range 0.003-0.141) and P by an average of 0.082 (range 0.005-0.216) (Table 2). For data sets in which a high proportion of adults were recaptured more than 10 days after first being trapped in the first year of capture (such as Wren *Troglodytes troglodytes*), the effect of creating an extra sampling period was relatively small. For data sets in which relatively few adults were retrapped more than 10 days after first being trapped in the first year of capture (such as Lesser Whitethroat *Sylvia curruca*), the effect of creating an extra sampling period was relatively large (Table 2).

Table 2. The effect of incorporating a pair of dummy age parameters on estimates of survival (Φ) and recapture probability (P). These parameters represent the probability of surviving and being recaptured more than 10 days after first being trapped in the first year of capture as an adult. Numbers of data sets are shown in brackets.

Species	Mean % increase in Φ		Mean % increase in P	
Willow Warbler[1,2,3]	9.6	(8)	10.8	(7)
Blackbird[2]	6.6	(5)	6.8	(5)
Reed Warbler[1]	7.3	(4)	8.6	(4)
Sedge Warbler[3]	7.7	(4)	9.1	(2)
Blackcap[2]	4.7	(3)	10.0	(3)
Whitethroat[2]	7.2	(2)	7.6	(2)
Wren	1.9	(2)	3.6	(2)
Chiffchaff	3.7	(1)	6.3	(1)
Garden Warbler	3.5	(1)	2.8	(1)
Lesser Whitethroat	14.1	(1)	11.6	(1)
Chaffinch[2]	5.7	(1)	7.5	(1)
Dunnock	5.4	(1)	2.8	(1)
Overall Mean	7.1	(33)	8.2	(30)

Notes:
1. Two Willow Warbler and one Reed Warbler data sets were modelled as (Φ,P) without the dummy age parameters and (Φ_t,P) with the dummy age parameters. In these cases the differences between Φ were derived from (Φ,P) models.
2. In five cases the (Φ_t,P) model was selected in both analyses and the differences between the arithmetic means of the annual survival estimates were used in the table.
3. Three cases of P = 1.000 (with one Willow Warbler and two Sedge Warbler data sets) are excluded from the table.

Estimates of both Φ and P were slightly less precise using the model with the extra sampling period. The mean difference between the standard errors from models (Φ_{10},P_{10}) and (Φ,P) was 0.0083 for Φ (n=25) and 0.0070 for P (n=29).

3.2. The effect of increasing the sampling area on estimates of Φ and P

Estimates of Φ and P derived from the combined data from both Theddlethorpe Dunes CES sites (A and B), and from the larger site only (site A) are presented in Table 3. Excluding data from the smaller site B reduced the percentages of adults recaptured in years after ringing for 4 of the 5 species considered, but had a variable effect on estimates of Φ and P (Table 3). In the case of Willow Warbler the between-year recapture rate and the estimate of Φ were relatively unaffected by the exclusion of only a small amount of recapture data from site B. In the cases of Whitethroat *Sylvia communis* and Lesser Whitethroat a reduction in the numbers and percentages of between-year recaptures was associated with reduced estimates of P, and for the latter species a reduced estimate for Φ (Table 3). In the case of Whitethroat the increase in the time-specific estimates of Φ was unexpected and suggests that between-year return rates for birds trapped only at site B have been lower than those for birds trapped at site A. In the case of Blackbird the exclusion of data from site B reduced further the already small numbers of between-year

Table 3. The effect of including captures from beyond the main sampling area on recapture rates and parameter estimates. With the exception of Whitethroat, all parameter estimates are taken from model (Φ_{10}, P_{10}). In the case of Whitethroat an arithmetic mean Φ is calculated from model (Φ_{10+t}, P_{10}). All data were collected at the Theddlethorpe Dunes CES sites in Lincolnshire, England.

Species	ALL DATA (sites A + B)				PARTIAL DATA (site A only)				Difference between parameter estimates (x 100)	
	Nos. of adults trapped (83-90)	Nos. (%) of individuals retrapped in subsequent years (84-91)	Parameter estimates Φ	P	Nos. of adults trapped (83-90)	Nos. (%) of individuals retrapped in subsequent years (84-91)	Parameter estimates Φ	P	Φ	P
Whitethroat	470	81 (17.2)	0.449	0.608	389	58 (14.9)	0.504	0.518	+5.5	-9.0
Lesser Whitethroat	197	40 (20.3)	0.641	0.492	162	24 (14.8)	0.432	0.567	-6.0	-7.4
Willow Warbler	330	29 (8.8)	0.351	0.527	284	25 (8.8)	0.348	0.473	-0.3	-5.4
Dunnock	213	62 (29.1)	0.422	0.790	138	26 (18.8)	0.392	0.884	-3.0	+9.4
Blackbird	236	30 (12.7)	0.392	0.368	172	17 (9.9)	0.172	0.624	-22.0	+25.6

recaptures, increasing markedly the estimate of P (which is based on very few data) and reducing the estimate of Φ (Table 3).

Although these analyses do not allow the possible effects of short-distance emigration on estimates of Φ and P to be quantified, they do indicate that true survival is likely to be underestimated by analyses of the kind presented here because for most species a higher proportion of adults are retrapped in years following ringing when the sampling area is increased.

3.3. Single site analyses

Of the 33 single site analyses in which an extra sampling period was created in the first year of capture, there was no evidence of time dependence in either Φ or P in 25 cases (Table 4). In 8 of the 33 analyses there was evidence of time dependence in Φ, but not in P. Estimates of Φ and associated standard errors are presented in Table 4.

Estimates of P (excluding three instances where P=1.000) varied between 0.152 and 0.818 (mean=0.504, s.d.=0.179), with estimates of less than 0.4 from 10 data sets. Recapture probability varied significantly between sites (one-way ANOVA involving 9 sites each having estimates of P for at least two species, $F_{8,16}=2.9$, $P<0.05$) but not between species (one-way ANOVA involving 4 species each having estimates of P from at least 3 sites, $F_{3,15}=2.25$, $P>0.10$). The arcsine transformation of P was used in these ANOVAs. Estimates of P were not correlated either with the number of adult birds captured during the study period (i.e. the sum of the annual cohorts 1983-90), or with the number of individual adults retrapped in years after marking (r = -0.12 and 0.09 respectively, n=30).

3.4. Multiple site analyses

3.4.1. Model Selection

Models selected in multiple site analyses are summarised in Table 5. With the exception of the five-sites Blackbird analysis, P simplified either to a site dependent model or to a site and time independent model. Site-specific recapture probability is to be expected because the intensity of the capture effort varies between CES sites. Even amongst the data sets with relatively large numbers of between-year recaptures, there was no evidence that P was time dependent. This probably reflects the constant annual capture effort maintained on CES sites.

Table 4. Between-year recapture rates of adult passerines at selected CES sites (1983-91), with annual survival estimates derived from single site SURGE analyses.

Species	Site Code	No. of adults trapped (1983-90)	No. (and %) of individuals retrapped in subsequent years (1984-91)		Results of SURGE analyses Models are either (Φ_{10},P_{10}) or (Φ_{10+t},P_{10})			
					MODEL		Φ	SE
Willow Warbler	TS	269	25	(9.3)	Φ_t	P	0.344	0.199[1]
	CO	106	17	(16.0)	Φ	P	0.450	0.097
	WV	154	34	(22.1)	Φ	P	0.421	0.057
	TH	330	29	(8.8)	Φ	P	0.351	0.070
	BA	113	16	(14.2)	Φ_t	P	0.429	0.221[1]
	LL	137	30	(21.9)	Φ_t	P	0.410	0.160[1]
	DF	195	18	(9.2)	Φ	P	0.263	0.057
	CW	222	30	(13.5)	Φ	P	0.431	0.061
Blackbird	TH	236	30	(12.7)	Φ	P	0.392	0.072
	TW	301	76	(25.2)	Φ	P	0.587	0.043
	BA	122	35	(28.7)	Φ	P	0.632	0.057
	LL	99	24	(24.2)	Φ_t	P	0.458	0.199[1]
	EN	144	24	(16.7)	Φ	P	0.627	0.082
Reed Warbler	TE	101	30	(29.7)	Φ	P	0.482	0.058
	MW	433	71	(16.4)	Φ	P	0.465	0.041
	ST	452	58	(12.8)	Φ	P	0.530	0.046
	WB	418	67	(16.0)	Φ_t	P	0.517	0.124[1]
Sedge Warbler	MW	153	13	(8.5)	Φ	P	0.296	0.081
	DF	226	25	(11.1)	Φ	P	0.190	0.037
	ST	116	7	(6.0)	Φ	P	0.182	0.067
	WB	219	40	(18.3)	Φ	P	0.440	0.055
Blackcap	CO	151	16	(6.5)	Φ	P	0.373	0.097
	TW	297	26	(8.8)	Φ	P	0.529	0.084
	CW	139	9	(10.6)	Φ_t	P	0.425	0.273[1]
Whitethroat	WV	168	39	(23.2)	Φ	P	0.437	0.054
	TH	470	81	(17.2)	Φ_t	P	0.449	0.099[1]
Wren	TH	132	16	(12.1)	Φ	P	0.273	0.082
	MW	102	13	(12.7)	Φ	P	0.360	0.105
Lesser Whitethroat	TH	197	40	(20.3)	Φ	P	0.492	0.064
Garden Warbler	CW	106	30	(28.3)	Φ	P	0.715	0.063
Chiffchaff	CW	157	21	(13.4)	Φ	P	0.380	0.086
Chaffinch	TS	255	51	(20.0)	Φ_t	P	0.489	0.131[1]
Dunnock	TH	213	51	(20.0)	Φ	P	0.422	0.040

1. Estimates of Φ and SEs for year-specific models are arithmetic means. In a few cases year-specific estimates of Φ reached boundary limits (0 or 1) and these are excluded from calculations of average Φ and SE.

For Wren, Reed Warbler, Blackcap *Sylvia atricapilla*, Blackbird and Willow Warbler there was no evidence of either time or site dependence in Φ (Table 5). Only in the case of Sedge Warbler was there evidence of site dependence in Φ. Apart from minor differences in the site specificity of the dummy age parameters, the (Φ_{10},P_{10+s}) model provided an adequate description of the Reed and Willow Warbler data, and of two four-site combinations of the Blackbird data (Table 5). The existence of a significant time-site interaction in P in the five-sites Blackbird analysis (Table 5) suggested heterogeneity between data sets. All four-site combinations were modelled, but only those listed in Table 5 allowed the time-site interaction in P to be dropped.

In the case of Whitethroat there was strong evidence of time-dependence in Φ, the (Φ_{10},P_{10}) model being conclusively rejected in favour of the (Φ_{10+t},P_{10}) model (LRT, χ^2=36.0, 7df, P<0.0001). In an analysis of two relatively sparse Sedge Warbler data sets, there was weaker evidence of time dependence in Φ, the (Φ_{10},P_{10+s}) model being rejected in favour of the (Φ_{10+t},P_{10+s}) model (LRT, χ^2=14.1, 7df, P=0.05).

Table 5. Results of multiple site SURGE analyses. The starting model for all analyses was one in which both survival (Φ) and recapture probability (P) were modelled with site and year effects and a site*year interaction (after Pradel *et al.*, 1990). All models include dummy age parameters (independent of year) representing "survival rates" and recapture probabilities for birds retrapped at least 10 days after being caught for the first time in their first year of capture as adults.

Species	No. Sites	Site codes (from Table 1)	Model selected by likelihood ratio tests		No. of parameters	Site-specific dummy age parameters? (Y=yes, N=no)		Goodness -of-fit test (P)
						Φ	P	
Whitethroat	2	WV,TH	Φ_t	P	12	Y	N	>0.75
Wren	2	TH,MW	Φ	P	4	N	N	>0.95
Reed Warbler	3	TE,MW,WB	Φ	P_s	8	N	Y	>0.10
Sedge Warbler	3	MW,DF,WB	Φ_s	P_s	12	Y	Y	>0.50
	2	MW,DF	Φ_t	P_s	14	Y	Y	>0.75
Blackcap	3	CO,TW,CW	Φ	P	4	N	N	>0.95
Blackbird	5	all	Φ	P_{s*t}	47	N	Y	<0.001
	4	TH,BA,LL,EN	Φ	P_s	10	N	Y	>0.50
	4	TH,TW,BA,EN	Φ	P_s	10	N	Y	>0.05
Willow Warbler	7	all except BA	Φ	P_s	22	Y	Y	>0.50

3.4.2. Parameter Estimates and Precision

Estimates of Φ derived from multiple site analyses in which no significant time or site effects were detected are presented in Table 6. The precision of these estimates improved in analyses in which common model parameters were applied across data sets. For example, in single site analyses of Willow Warbler data standard errors of Φ were 5.7-9.7%, compared to 2.5% in a multiple site analysis (Tables 4 and 6). The precision attainable in analyses of this sort is determined largely by the number of parameters in the model and the numbers of between-year recaptures.

Table 6. Estimates of annual adult survival (and associated precision) derived from mark-recapture data combined from multiple ringing sites, and from recoveries of birds ringed in Britain and Ireland and subsequently found dead (from Baillie & McCulloch, this volume). All survival estimates are derived from time-independent models which fit the data.

Species	Mark-Recapture Analyses				Recovery Analyses		
	No. of data sets combined	No. of individuals retrapped in subsequent years (1984-91)	Average annual survival rate (Φ) (1983-91) and SE[1] from recaptures Φ	SE	No. of recoveries in two age-class recovery matrix	Average annual adult survival rate (1985-90) from recoveries S	SE
Willow Warbler	7	183	0.371	0.025	385[2]	0.554	0.056
Blackbird	4 4	113 165	0.566 0.581	0.036 0.029	1307	0.668	0.020
Blackcap	3	51	0.443	0.057	197	0.534	0.128
Reed Warbler	3	168	0.496	0.026	957[3]	0.558	0.023
Wren	2	29	0.318	0.068	-	-	-
Dunnock	1	62	0.422	0.040	265	0.447	0.050

1. Approximate standard errors of survival estimates were calculated from the SE of logit(Φ) as: $SE_{(\Phi)} = \Phi.(1-\Phi).SE_{logit(\Phi)}$

2. Estimated from recaptures >5km from the place of ringing.

3. Estimates based on recoveries of dead birds were supplemented by recaptures >5km from the place of ringing.

For comparative purposes annual adult survival rates estimated from British ring recovery data (from Baillie & McCulloch (this volume)) are also presented in Table 6. Estimates of average annual survival from multiple site recaptures are between 2.5% and 18.3% (mean 9.0%) lower than those from recoveries (Table 6), although the differences could be a consequence of the differing time-periods covered by the two sets of analyses rather than any underestimation bias affecting estimates from recaptures (i.e. as a consequence of permanent emigration). The precision of the survival estimates from the relatively small amounts of mark-recapture data analysed here is better than that attainable from recoveries for three species (Willow Warbler, Blackcap and Dunnock *Prunella modularis*), similar for one species (Reed Warbler) and poorer for one species (Blackbird). There was no comparable survival estimate for Wren from recovery data.

The mean standard error of the time-specific estimates of Φ was 8.2% (range 5.5-12.2%) in the case of Whitethroat (of which a total of 120 individual adults were retrapped in years after first being trapped at two CES sites) and 7.3% (range 3.1-10.5%) in the case of Sedge Warbler (of which only 38 individual adults were retrapped in years after first being trapped at two CES sites).

4. Discussion

4.1. Model selection considerations

The analyses presented above have involved a large number of likelihood ratio tests between models based upon relatively sparse data sets. Large numbers of tests will increase the chance of committing type I errors (rejecting a true null hypothesis) whilst sparse data will increase the chance of type II errors (accepting a false null hypothesis). In the Whitethroat analysis described above, time dependence was detected in Φ with a high degree of confidence (LRT, $P<0.001$) suggesting that a type I error has not been committed. In the case of Sedge Warbler there is a greater chance that a type I error has been committed because the null hypothesis (that Φ was independent of time) was only just rejected at the 5% level of significance. In several of the multiple sites modelling procedures summarised in Table 5, likelihood ratio tests approached but did not reach statistical significance (i.e. $P<0.1$) suggesting that type II errors might have been committed because of rigid adherence to the 5% significance level.

Possible measures to reduce the likelihood of committing type I and type II errors during model selection are recommended by Lebreton *et al.* (1992), including the use of 15% significance levels to increase the power of likelihood ratio tests to reject models, and the use of Akaike's Information Criterion (Akaike,1973) to reduce the number of tests in complex modelling situations. Relevant biological knowledge of the species in question and practical information relating to study design (such as annual variation in recapture or resighting effort) should also be incorporated into models whenever possible.

4.2. The usefulness of CES mark-recapture data for understanding population processes

Estimates of survival derived from mark-recapture data are likely to underestimate true survival rates because of permanent emigration of animals away from the study area. This was confirmed by the presence of ringed birds outside the main ringing area at Theddlethorpe Dunes (Section 3.2 above) and at Wicken Fen reedbed (Peach *et al.*,1990). The differences between survival estimates derived from ring recovery data and mark-recapture data (Table 6) are also probably partly due to an underestimation bias affecting the latter. In the case of Willow Warbler (for which this difference was largest) information on breeding productivity was used to assess which, if either, of the two adult survival estimates is more realistic. Analyses of BTO nest record cards indicate that the average productivity of Willow Warblers in Britain during the period 1983-89 was 1.74 fledged young per nesting attempt (Baillie & Peach,1992). After allowing for replacement clutches, estimated annual productivity is still less than 2 fledged young per pair. Assuming negligible post-fledging mortality (unlikely), Baillie and McCulloch's (this volume) estimates of first year and adult survival rates for Willow Warbler (0.50 and 0.55 respectively), would require annual productivity rates of approximately 1.8 fledged young per pair to maintain a stable breeding population. With an adult survival rate similar to that estimated from mark-recapture data (0.37) and an assumed and slightly lower first year survival rate (0.30), average annual productivity rates of more than 4 fledged young per pair would be required to maintain a stable breeding population. Although there is evidence from extensive territory mapping that the breeding population of Willow Warblers has declined in Britain since 1989 (Marchant & Musty,1992), numbers were otherwise stable throughout the 1980s. The inescapable conclusion is that the apparent adult survival rate estimated from combined mark-recapture data is seriously underestimating true Willow Warbler survival, whilst the recovery data are providing a much more realistic estimate. It should be noted, however, that the annual survival rate of 0.47 estimated for adult Willow Warblers using mark-recapture data from a single CES site (Pratt & Peach, 1991), is clearly a more reasonable estimate of true survival.

The problem of emigration from mark-recapture study areas merits further investigation, through the operation of several adjacent study sites within a larger habitat. The emigration of marked individuals away from mark-recapture study sites will only prevent the detection of temporal changes in true survival if rates of emigration change over time or in relation to demographic variables such as local population density. As long as parameter estimates are biased in a systematic way, then significant biological processes should still be identifiable.

A general feature of mark-recapture data collected at Constant Effort Sites is the sparsity of between-year recaptures at individual sites, and for most species site-specific estimates of survival lack precision. Program SURGE4 offers the useful facility of fitting common model parameters across data sets (Pradel *et al.*,1990) thus increasing precision and the power of such analyses to detect temporal variation in survival. Although in the multiple site analyses presented here there was no evidence of temporal variation in survival rates for 5 of the 7 species considered, in some cases this may be a consequence of a lack of variation in true survival rates. For example, since 1983 population indices, derived from between-year changes in adult captures on CES sites, have changed much less for species like Reed Warbler, Blackcap and Blackbird (the coefficients of variation (CV) of these indices for these species were 0.13, 0.14 and 0.18 respectively) than for Whitethroat (CV=0.37) for which temporal variation in survival was detected (Table 5).

In cases where temporal variation in apparent survival is detected the next stage of the SURGE analysis is to attempt to explain this variation using environmental or demographic covariates (see Clobert *et al.* (1987), Kanyamibwa *et al.* (1990) and Peach *et al.* (1991) for examples of this). In cases where temporal variation in survival rates cannot be detected, tests for temporal changes in survival might involve comparisons of survival estimates derived from time-independent models for adjacent time periods (Peach *et al.*, 1990). With the degree of precision achieved in the analyses presented here (i.e. about 2.5% for an average survival estimate covering a nine-year period) it should be possible to detect changes in survival of approximately 10% between two nine-year periods. However, as a result of the continued expansion of the Constant Effort Sites Scheme much more mark-recapture data should be available in future for analyses of the sort presented here. Improvements in precision and the power to detect temporal variation in apparent survival rates can therefore be expected.

Acknowledgements. I thank the following CES ringers for collecting the data on which the analyses presented in this paper are based : Mike Boddy (Theddlethorpe Dunes), Sandy Bankier, Mike Carrier, Robin Cole, Dave Fletcher, Jerry Lewis and Llangorse Ringing Group, John McMeeking and Chris Du Feu, Joyce Martin and Jan Pritchard, Chris Reynolds, Sandwich Bay Bird Observatory, Bob and Alison Spencer, Rob Turner, Chris Wilson and Rupert Wilson. Roger Pradel gave invaluable advice on the use of program SURGE4 and Stephen Baillie, Charles Francis and John Tautin helped improve an earlier version of the manuscript. Susan Waghorn and Julie Sheldrake helped with wordprocessing. The author's post at the BTO is funded by the Joint Nature Conservation Committee on behalf of English Nature, the Countryside Council for Wales, and Scottish Natural Heritage, and under a contract from the Department of the Environment for Northern Ireland. The author's attendance at the EURING '92 conference was partly financed by a travel grant from The Royal Society.

References

Akaike, H. (1973).- Information theory and an extension of the maximum likelihood principle *in* International Symposium on Information Theory, B.N. Petran & F. Csaki (Eds) Akadémiai Kiadi, Budapest.

Baillie, S.R. (1990).- Integrated population monitoring of breeding birds in Britain and Ireland. Ibis, 132: 151-166.

Baillie, S.R., Green, R.E., Boddy, M. & Buckland, S.T. (1986).- An evaluation of the Constant Effort Sites Scheme. BTO Research Report No.21, British Trust for Ornithology, Thetford.

Baillie, S.R. & McCulloch, N.M. (this volume).- Modelling the survival rates of passerines ringed during the breeding season from national ringing and recovery data.

Baillie, S.R. & Peach, W.J. (1992).- Population limitation in migrants. Ibis, 134 suppl. 1: 120-132.

Buckland, S.T. & Baillie, S.R. (1987).- Estimating survival rates from organized mist-netting programmes. Acta Orn., 23: 89-100.

Clobert, J., Lebreton, J.-D. & Allaine, D. (1987).- A general approach to survival rate estimation by recaptures or resightings of marked animals. Ardea, 75:133-142.

Du Feu, C.R. & McMeeking, J.M. (1991).- Does constant effort netting estimate juvenile abundance? Ringing & Migration, 12:118-123.

Hestbeck, J.B., Nichols, J.D. & Malecki, R.A. (1991).- Estimates of movement and site fidelity using mark-resight data of wintering Canada Geese. Ecology, 72: 523-533.

Kanyamibwa, S., Schierer, A., Pradel, R. & Lebreton, J.-D. (1990).- Changes in adult annual survival rates in a western European population of the White Stork *Ciconia ciconia.* Ibis, 132: 27-35.

Lebreton, J.-D., Burnham, K.P., Clobert, J. & Anderson, D.R. (1992).- Modeling survival and testing biological hypotheses using marked animals: A unified approach with case studies. Ecological Monographs, 62: 67-118.

Marchant, J. & Musty, L. (1992).- Common Birds Census Report 1990-91. BTO News, 182: 9-12.

Peach, W.J. (1991).- The Constant Effort Sites Scheme *in* Britain's Birds in 1989/90 : the conservation and monitoring review, D.A. Stroud & D. Glue (Eds) BTO/NCC, Thetford.

Peach, W.J., Buckland, S.T. & Baillie, S.R. (1990).- Estimating survival rates using mark-recapture data from multiple ringing sites. The Ring, 13:87-102.

Peach, W.J., Baillie, S.R. & Underhill, L. (1991).- Survival of British Sedge Warblers *Acrocephalus schoenobaenus* in relation to west African rainfall. Ibis, 133: 300-305.

Peach, W.J. & Baillie, S.R. (1992).- Population changes on Constant Effort Sites 1990-91. BTO News, 179: 12-13.

Pradel, R., Clobert, J. & Lebreton, J.-D. (1990).- Recent developments for the analysis of capture-recapture multiple data sets. The Ring, 13:193-204.

Pratt, A.M. & Peach, W.J. (1991).- Site tenacity and annual survival of a Willow Warbler *Phylloscopus trochilus* population in southern England. Ringing & Migration, 12: 128-134.

Smith, D.R. & Anderson, D.R. (1987).- Effects of lengthy ringing periods on estimators of annual survival. Acta Orn., 23: 69-76.

Tiainen, J. (1983).- Dynamics of a local population of the Willow Warbler in southern Finland. Ornis Scandinavica, 14: 1-15.

Marked Individuals in the
Study of Bird Population
J.-D. Lebreton and P. M. North (eds.)

Modelling the Survival Rates of Passerines Ringed during the Breeding Season from National Ringing and Recovery Data

Stephen R. Baillie and Neil McCulloch

British Trust for Ornithology, The Nunnery, Thetford, Norfolk IP24 2PU, United Kingdom

Summary. Ringing and recovery data for 16 species of passerines ringed in Britain and Ireland between April and September from 1985 to 1990 were analysed using the models developed by Brownie *et al.* (1985). Reduced parameter models were also used where appropriate. Reporting rates varied with time for only three species. There was no consistent difference in reporting rates of first year and of adult birds across species. Data for 12 species could be described by models which lacked age-specific reporting rate parameters, including some of those with large datasets. Survival rates of first year birds were lower than those of adults in 12 species and similar in three species. The survival estimate for first year Swallows was higher than that for adults but this may have been an artefact. The number of recoveries of migrant passerines that are available for analysis can be increased greatly if controls (recaptures more than 5km from place of ringing) as well as dead recoveries are included in the analysis. We summarised these data as separate matrices of dead and controlled recoveries and then attempted to fit models with common survival rates for both recovery categories. We were able to fit a model that gave common survival estimates for Reed Warbler but not for Sand Martin or Willow Warbler. Willow Warbler survival estimates based on controls appeared to be more biologically reasonable than those based on dead recoveries.

We used the parameter estimates from our data to investigate relationships between sample size and precision for these types of passerine ringing programmes. We conclude that useful estimates of average survival rates can be obtained for many species but that there are only a few species for which such data can be used to study annual variation in survival rates.

1. Introduction

The British Trust for Ornithology (BTO) is developing an Integrated Population Monitoring Programme which aims to monitor the numbers, breeding performance and survival rates of a wide range of terrestrial breeding birds in order to detect population changes and their likely causes (Baillie, 1990, 1991). Where possible, annual estimates of survival are needed to detect changes over time and to carry out analyses of the population dynamics of individual species (Baillie, 1990, Peach *et al.*, 1991, Baillie and Peach, 1992). Where it is not possible to estimate survival rates for individual years we need at least to be able to test for changes in survival between blocks of years. We are currently developing approaches to this problem based both on ringing recoveries (this paper) and on mark-recapture data (Peach, this volume).

A major problem with carrying out analyses of survival rates using British ringing data, or that from most other European ringing schemes, is that the cohort sizes contributing to the ringing and recovery matrix are unknown. This is because ringing data have not been computerised routinely, although recovery data are generally computerised. We believe that this is the main reason why the Brownie models (Brownie *et al.*, 1985), that were first published over 15 years ago, have not been used more widely in Europe. Several ringing schemes, including that operated by the BTO, are now developing systems for the computerisation of ringing data but it will be several years before sufficient data are available for survival analyses.

The BTO Ringing Scheme introduced Age-specific Totals Lists in 1985 to collect cohort sizes from the summer ringing of 22 passerines in advance of full computerisation (Baillie and Green, 1987). We report here on an analysis of the first six years of data from this programme. Our aims were to assess which models were appropriate for the analysis of these data and the precision of the survival estimates that could be obtained. This information provides a starting point for assessing the numbers of birds that need to be ringed if our monitoring objectives are to be achieved. Our analyses also allowed us to test for age-specific and temporal variation in the recovery reporting rates, thus providing further information on the reliability of analyses that have assumed that reporting rates are constant.

2. Methods

Numbers of nestlings, juveniles and adults ringed throughout Britain and Ireland for the months of April to September inclusive were reported by ringers for each year from 1985 to 1990. The six month ringing period is a compromise between the need to have an adequate sample of ringed birds and the desirability of keeping the ringing period as short as possible. This relatively long ringing period should not have biased the results provided that the seasonal pattern of ringing was similar each year (Smith and Anderson, 1987), which is generally the case with volunteer ringers, especially in a large scheme where individual variations are averaged out. This analysis is restricted to birds ringed as adults or juveniles. The forms include data on the region of ringing and on whether the birds were ringed at a coastal site (which would be likely to attract more migrants). These categories are not considered here but will be examined in future analyses once larger samples of recoveries have accumulated.

Only recoveries of birds that were dead or dying were included in the main analyses. Recoveries were rejected for birds that were sick, transported or held captive when ringed (such birds would not have been included in the totals), where the ring only was reported or where information supplied by the finder suggested that the finding date was not known to within 50 days. We included records where the date was determined from the date of the letter or postmark. Controls are defined as live recaptures more than 5 km from the place of ringing. Many of these records result from ringing programmes which are not sufficiently large or intensive to generate data for more formal mark-recapture analysis. Recaptures under 5 km from the place of ringing are not gathered centrally unless the data form part of the Constant Effort mist-netting sites Scheme (Peach, this volume).

Recoveries reported from April 1985 to June 1991 were used to construct the recovery matrices. For the main analyses based on recoveries of dead or dying birds recovery years ran from 1 July to 31 June. In the year of ringing all recoveries from 1 April onwards were included. Thus, in their first year of ringing, birds were at risk of recovery for between 15 months and 9 months depending on the exact time of ringing. Birds at risk for longer than 12 months will to some extent have compensated for those at risk for less than 12 months. Given that the available ringing totals were amalgamated over a six month period, this system of assigning recovery years allowed us to use as many of the recoveries as possible and to avoid the need for separate survival or reporting rate parameters for the year after ringing. We used models in which reporting rates differed between the year after ringing and subsequent years (Brownie *et al.*, 1985, Baillie and Green, 1987) to check that this system of classification was not likely to bias our estimates severely.

Data were available for 22 species, for each of which totals of dead recoveries and controls were calculated for adults and juveniles separately (Table 1). Survival analyses using dead recoveries were carried out for the 17 species which had more than 100 such recoveries. For one of these species, Wren, the simplest model which provided a statistically adequate description of the data had a large number of parameters and did not provide estimates that were biologically sensible. Wren populations are known to fluctuate markedly from year to year and this problem may result from the combination of strong temporal fluctuations in survival and reporting rates combined with sparse data. The Wren data are not considered further in this paper. Thus 16 species are included in the main analyses of survival and reporting rates presented below.

There were few controls for most resident species but for most migrants there were generally substantially more controls than dead recoveries (Table 1). We therefore investigated the possibility of using data from controls as well as from dead birds to estimate survival rates for three of these species: Sand Martin, Reed Warbler and Willow Warbler. Recovery matrices for controls were constructed using the last recapture of each individual (Brownie *et al.*, 1978, page 175,

Mardekian and McDonald, 1981). Very few birds were controlled more than once so this procedure resulted in little loss of information compared with the use of Jolly-Seber models. We restricted the data to controls made between April and September as nearly all controls of migrants were made during this period. Calendar years were used to define the recovery years for these data.

Table 1 Ringing and recovery data available for this study. Birds were ringed in Britain and Ireland between April and September in each year from 1985-1990. Recoveries of these birds reported up to 30 June 1991 were included in the main analyses.

Species[1]	Ringed as adults			Ringed as juveniles		
	Ringed	Recovered dead	Controlled	Ringed	Recovered dead	Controlled
Sand Martin	48,011	23	407	83,009	33	1,402
Swallow	20,240	48	17	121,101	221	300
Pied Wagtail	2,983	31	5	11,089	92	28
Wren	15,835	34	7	41,463	101	15
Dunnock	18,081	73	3	39,503	192	19
Robin	16,580	105	8	54,072	431	35
Blackbird	30,966	547	15	36,010	760	10
Song Thrush	11,681	95	1	13,815	144	10
Sedge Warbler	24,124	22	106	74,230	84	537
Reed Warbler	28,146	41	262	62,287	93	561
Blackcap	22,089	58	41	54,635	139	171
Willow Warbler	77,942	69	117	130,851	133	268
Pied Flycatcher	12,358	20	140	2,305	5	8
Blue Tit	33,291	143	32	93,714	491	166
Great Tit	18,640	83	11	34,979	192	57
Starling	16,123	326	2	24,813	442	22
House Sparrow	4,281	42	0	4,319	31	3
Chaffinch	25,639	119	16	17,081	71	16
Greenfinch	56,348	602	242	18,110	112	83
Redpoll	5,466	22	21	5,001	12	32
Bullfinch	14,856	73	11	14,021	86	22
Reed Bunting	10,450	19	31	14,521	39	66

1 Scientific names of species referred to in this paper are listed in Appendix 1.

The data were analysed initially using program BROWNIE (Brownie *et al.*, 1985). Reduced parameter models and models involving separate matrices for dead recoveries and for controls were fitted using program SURVIV (White, 1983). For dead recoveries we parameterised our SURVIV models in terms of the reporting rate (λ) rather than the recovery rate (F), where $F = \lambda\ (1\text{-}S)$. We prefer to parameterise our models in terms of λ because it does not include survival rate (S). F is useful mainly for studies of hunted species because it provides an index of harvest rate. Models involving controls were parameterised in terms of α_i, which we define as the probability that a bird was controlled for the last time in year i. Mardekian and McDonald (1981) use p rather than α but we have not done so here to emphasise that α is not the same as the recapture probability of the Jolly-Seber model which is generally designated as p. The models we have used to analyse data for controls are the same as the Brownie models because $F_{i+1} = S_i\ \alpha_{i+1}$ (Mardekian and McDonald 1981). These alternative parameterisations in terms of λ or α have no effect on the estimates of S or the goodness-of-fit statistics.

The terminology used here to describe models and parameters follows that currently adopted for mark-recapture studies (Lebreton *et al.*, 1992). Additive models (for example incorporating time and age effects without interaction) were not used because they have not been implemented in the software that is currently available for recovery analysis.

The main types of parameters are survival rates (S), reporting rates (λ) and final recapture probabilities (α). Upper case letters as subscripts (or as superscripts within model descriptions) are use to qualify them as follows:

AD - Adult; 1Y - First year; D - Dead; C - Control.

Subscripts are used to indicate the parameters included in a model as follows:

a - age specific, first year and adult; t - time specific, different values for each year; c - circumstance specific, different values for found dead and controls; f - different adult reporting rates for year after ringing and subsequent years.

Thus, for example, the simplest two age class Brownie model for the analysis of dead recoveries (model H_{01} of Brownie *et al.*, 1985) is described as model $S_a\lambda_a$ with expected numbers of recoveries from the first year of ringing written as:

	Recovery periods following ringing (years)		
Number ringed	1	2	3
N_1	$N_1(1\text{-}S_{AD})\lambda_{AD}$	$N_1S_{AD}(1\text{-}S_{AD})\lambda_{AD}$	$N_1S_{AD}S_{AD}(1\text{-}S_{AD})\lambda_{AD}$
M_1	$M_1(1\text{-}S_{1Y})\lambda_{1Y}$	$M_1S_{1Y}(1\text{-}S_{AD})\lambda_{AD}$	$M_1S_{1Y}S_{AD}(1\text{-}S_{AD})\lambda_{AD}$

where N_1 is the number of adults ringed at the beginning of recovery period 1 and M_1 is the number of juveniles ringed at the beginning of recovery period 1.

The equivalent model for controls is described as model $S_a\alpha_a$ with expected numbers of recoveries from the first year of ringing written as:

Number ringed	Ringing period 2	3	4
N_1	$N_1S_{AD}\alpha_{AD}$	$N_1S_{AD}S_{AD}\alpha_{AD}$	$N_1S_{AD}S_{AD}S_{AD}\alpha_{AD}$
M_1	$M_1S_{1Y}\alpha_{1Y}$	$M_1S_{1Y}S_{AD}\alpha_{AD}$	$M_1S_{1Y}S_{AD}S_{AD}\alpha_{AD}$

where N_1 is the number of adults ringed between April and September of year 1 and M_1 is the number of juveniles ringed between April and September of year 1.

For a mixed model combining two datasets using the models just described but with a common adult survival rate we have model $S^{AD}S_c^{1Y}\lambda_a\alpha_a$ with expected numbers of recoveries from the first year of ringing written as:

Number ringed		Recovery period 1	2	3
N_1	Dead recoveries	$N_1(1\text{-}S_{AD})\lambda_{AD}$	$N_1S_{AD}(1\text{-}S_{AD})\lambda_{AD}$	$N_1S_{AD}S_{AD}(1\text{-}S_{AD})\lambda_{AD}$
	Controls	$N_1S_{AD}\alpha_{AD}$	$N_1S_{AD}S_{AD}\alpha_{AD}$	$N_1S_{AD}S_{AD}S_{AD}\alpha_{AD}$
M_1	Dead recoveries	$M_1(1\text{-}S_{1YD})\lambda_{1Y}$	$M_1S_{1YD}(1\text{-}S_{AD})\lambda_{AD}$	$M_1S_{1YD}S_{AD}(1\text{-}S_{AD})\lambda_{AD}$
	Controls	$M_1S_{1YC}\alpha_{1Y}$	$M_1S_{1YC}S_{AD}\alpha_{AD}$	$M_1S_{1YC}S_{AD}S_{AD}\alpha_{AD}$

where N_1 is the number of adults ringed and M_1 is the number of juveniles ringed as above.

Except where otherwise indicated we present arithmetic means of year-specific parameters with standard errors adjusted to take account of sampling correlations (Brownie *et al.* 1985, page 19). Weighted means and their standard errors were calculated using the formula given by Peach *et al.* (1990).

3. Results

3.1 Modelling reporting rates of dead recoveries

We first carried out a series of analyses which allowed us to compare first year and adult reporting rates (Table 2). Thus, we only considered models with separate reporting rate parameters for first year birds and adults. We started by testing the hierarchy of models as proposed by Brownie *et al.*, (1985):

$S_{at}\lambda_{at}$	Survival and reporting rates age and year specific
$S_a\lambda_{at}$	Survival age specific, reporting rate age and year specific
$S_a\lambda_a$	Survival and reporting rates age specific

Table 2 Estimates of first year and adult reporting rates (λ).

Species	Model	Likelihood ratio test P[1.]	Goodness -of-fit test P	First year reporting rate (λ_{1Y})		Adult reporting rate (λ_{AD})	
				Estimate	Standard Error	Estimate	Standard Error
Swallow	$S_a\lambda_a$	0.581	0.734	0.00181	0.000259	0.00269	0.000394
Pied Wagtail	$S_a\lambda_{at}$	0.063	0.101	0.00810	0.00128	0.0120	0.00225
Dunnock	$S_a\lambda_a$	0.574	0.615	0.0054	0.000682	0.00468	0.000568
Robin	$S_a\lambda_{at}$[2]	0.445	0.327	0.0108	0.00101	0.00770	0.00778
Blackbird	$S_a\lambda_a$	0.596	0.589	0.0310	0.00350	0.0250	0.00132
Song Thrush	$S_a\lambda_a$	0.096	0.128	0.0132	0.00233	0.0115	0.00146
Sedge Warbler	$S_a\lambda_a$	0.773	0.672	0.00133	0.000244	0.00103	0.000225
Reed Warbler	$S_a\lambda_a$	0.254	0.705	0.00166	0.000264	0.00197	0.000357
Blackcap	$S_{at}\lambda_a$	0.076	0.116	0.00280	0.000383	0.00273	[3]
Willow Warbler	$S_a\lambda_a$	0.379	0.784	0.00113	0.000155	0.000954	0.000116
Blue Tit	$S_a\lambda_{at}$	0.180	0.233	0.00644	0.000488	0.00527	0.000584
Great Tit	$S_a\lambda_a$	0.625	0.580	0.00640	0.000791	0.00492	0.000549
Starling	$S_a\lambda_a$	0.326	0.214	0.0184	0.00136	0.0268	0.00173
Chaffinch	$S_a\lambda_a$	0.653	0.427	0.00475	0.00120	0.00583	0.000595
Greenfinch	$S_a\lambda_{af}$	0.548	0.261	0.00650	0.000985	0.00947	0.000844 [4]
Bullfinch	$S_a\lambda_a$	0.262	0.740	0.00831	0.00219	0.00546	0.000650

1. Test between this and the next most complex model.
2. A simpler model was marginally acceptable, see text.
3. Standard error computed by SURVIV was unrealistically small.
4. λ_{AD} is that for all years except that immediately after ringing.
   λ_{AD} for the year following ringing estimated separately as 0.0146 $\pm$0.00136.

Model $S_{at}\lambda_{at}$ was not chosen as the final model for any of our datasets although it was selected initially for Blackcap (below). The data for Pied Wagtail, Robin and Blue Tit were best described by $S_a\lambda_{at}$. For Robin the LRT between $S_a\lambda_{at}$ and $S_a\lambda_a$ was not quite significant at the 5% level (P=0.051) but we adopted a cautious approach by using the more complex model. The data for 11 species were well described by model $S_a\lambda_a$, all goodness-of-fit tests having P>0.1.

The model required for the Blackcap data could not be simplified below $S_{at}\lambda_{at}$ using the standard hierarchy (LRT $S_{at}\lambda_{at}$ vs. $S_a\lambda_{at}$, P=0.0058). However it did not seem sensible to use a model with 22 parameters for a dataset containing only 197 recoveries. We therefore attempted to simplify the model by removing annual variation in reporting rates while allowing survival rates to vary between years. Model $S_{at}\lambda_a$ gave an adequate fit to the Blackcap data (Table 2).

Program BROWNIE includes a test of whether adult reporting rates in the year following ringing differ from those in subsequent years. This test was significant for Greenfinch (χ^2_4=10.096, P=0.039), and we therefore examined the hierarchy of models $S_{at}\lambda_{atf}$, $S_a\lambda_{atf}$ and $S_a\lambda_{af}$. The simplest of these models with only 5 parameters provided a good fit to the Greenfinch recoveries (Table 2).

Thus there was only evidence of year-specific variation in reporting rates for three of the 16 species examined. Average reporting rates were higher for first year birds in 10 species and for adults in 6 species. A paired t-test provided no evidence of a consistent age-related difference in reporting rates across species (t=0.0128, n.s.). We concluded that there was no *a priori* reason for insisting that age-specific reporting rates should be included in survival models for our study species. Further modelling suggested that age-specific reporting rates could be removed from the models for all but four species (below).

3.2 Modelling survival rates from dead recoveries

Estimates of first year and adult survival rates are presented in Table 3. The values for Blackcap are means of the year-specific survival estimates weighted by the reciprocal of the variance. For all other species estimates of first-year and adult survival were obtained directly from the models.

We used likelihood-ratio tests to determine whether the models presented in Table 2 could be simplified further by the removal of age-specific reporting rate parameters. Model $S_a\lambda_{at}$ could be simplified to model $S_a\lambda_t$ for Pied Wagtail, Robin and Blue Tit, although for Pied Wagtail we retained model $S_a\lambda_{at}$ for the survival analysis due to the relatively poor fit of $S_a\lambda_t$ (P=0.065). Recoveries of ten of the eleven species for which model $S_a\lambda_a$ was used for the reporting rate analyses could be modelled by $S_a\lambda$ (Table 3). The only exception was Starling (LRT, χ^2_1=10.080, P=0.0015). The model used for the Blackcap data ($S_{at}\lambda_a$) could not be simplified further. There was no biologically sensible strategy for further simplification of the Greenfinch model ($S_a\lambda_{af}$).

Table 3 Estimates of first year and adult survival rates.

Species	Model	Likelihood ratio test P[1]	Goodness -of-fit test P	First year survival rate (S_{1Y})		Adult survival rate S_{AD}		Body mass[2]
				Esti-mate	Standard Error	Esti-mate	Standard Error	
Swallow	$S_a\lambda$	0.172	0.592	0.489	0.0345	0.351	0.0483	18.7
Pied Wagtail	$S_a\lambda_{at}$[3]	0.063	0.101	0.349	0.0187	0.497	0.0692	22.0
Dunnock	$S_a\lambda$	0.398	0.652	0.282	0.0376	0.447	0.0503	21.3
Robin	$S_a\lambda_t$	0.472	0.580	0.368	0.0282	0.482	0.0379	19.3
Blackbird	$S_a\lambda$	0.106	0.534	0.537	0.0207	0.668	0.0197	102.0
Song Thrush	$S_a\lambda$	0.575	0.185	0.408	0.0495	0.657	0.0479	76.0
Sedge Warbler	$S_a\lambda$	0.436	0.698	0.238	0.0556	0.362	0.0872	11.2
Reed Warbler	$S_a\lambda$	0.541	0.694	0.321	0.0592	0.586	0.0747	14.0
Blackcap	$S_{at}\lambda_a$	0.076	0.116	0.404	0.0468	0.413	0.0477	17.5
Willow Warbler	$S_a\lambda$	0.405	0.829	0.261	0.0423	0.309	0.0483	8.6
Blue Tit	$S_a\lambda_t$	0.549	0.323	0.328	0.0256	0.486	0.0354	11.0
Great Tit	$S_a\lambda$	0.161	0.569	0.272	0.0360	0.381	0.0442	19.0
Starling	$S_a\lambda_a$	0.326	0.214	0.317	0.0328	0.649	0.0261	82.0
Chaffinch	$S_a\lambda$	0.518	0.469	0.552	0.0583	0.548	0.0472	20.0
Greenfinch	$S_a\lambda_{af}$	0.548	0.261	0.329	0.0685	0.488	0.0408	27.8
Bullfinch	$S_a\lambda$	0.154	0.618	0.400	0.0582	0.398	0.0491	21.8

1. Test between this and the next most complex model.
2. From Hickling (1983) except for Blackbird and Blue Tit which are from unpublished data.
3. Model $S_a\lambda_t$ just fits the data for Pied Wagtail (P = 0.065).

Comparison of first year and adult survival rates across species showed that adults generally had higher survival rates than first year birds (paired t-test, t=4.014, P<0.005). Survival rates of adults were markedly higher than those of first year birds in 12 of the 16 species. Survival rates of first year and adult Blackcaps, Chaffinches and Bullfinches were similar. The estimated survival rate of first year Swallows was markedly higher than that of adults. This may be an artefact caused by a large difference in the timing of ringing of the two age-classes, and requires more detailed investigation.

There was a significant correlation between average survival rate and log10(mean body mass) for adults (r=0.786, P<0.001) but for first year birds it did not reach statistical significance (r=0.433, P<0.1). However, these correlation coefficients are not significantly different (t_s = 1.523, n.s.) and the combined estimate of the correlation coefficient confirms the presence of a significant correlation ($r = 0.642$, $P<0.001$).

3.3 Modelling controls and dead recoveries

Sand Martin, Reed Warbler and Willow Warbler were selected for an examination of the possible use of controls for survival estimation. Sample sizes of controls used for these analyses were smaller than those given in Table 1 because only summer controls from the year after ringing onwards were used (see methods). Sand Martin had too few dead recoveries to qualify for inclusion in the analyses presented above but 921 controls could be included in the survival analysis, potentially making the third largest dataset in this study. Analysis of the data for controls of birds ringed as adults using program ESTIMATE indicated that model $S_t\alpha_t$ adequately described the data ($\chi^2_3=1.91$, $P=0.592$), giving a mean adult survival rate of 0.443 ± 0.0598. Analysis of the control data for birds ringed as adults and juveniles with program BROWNIE indicated that model $S_{at}\alpha_{at}$ provided a good description of the data ($\chi^2_{10}=8.05$, $P=0.624$). However, although the average adult survival rate estimated under this model (0.468 ± 0.0561) appeared to be realistic, the estimate of average first year survival was not biologically reasonable (0.943 ± 0.106). This dataset is clearly not amenable to relatively straightforward analysis and was not investigated further.

There were 540 Reed Warbler controls that could be used for estimating survival rates. Model selection for the combined analysis of dead recoveries and controls is shown in Fig. 1. When dead recoveries and controls were analysed separately the model for dead recoveries could be reduced to $S_a\lambda$, while the model for controls could be reduced to $S_a\alpha$. The adult survival estimates from the two analyses were very similar (dead recoveries 0.586 ± 0.0747; controls 0.565 ± 0.0245) but the first year survival calculated from controls (0.488 ± 0.0490) was considerably higher than that obtained from dead birds (0.321 ± 0.0592). However, a single model with common adult and first year survival rates provided a good fit to the data ($\chi^2_{60}=54.646$, $P=0.671$; parameter estimates in Table 4).

Table 4 Survival and reporting rates for Reed Warblers derived from a mixed model involving dead recoveries and controls.

Parameter	Estimate	Standard Error
Adult survival rate (S_{AD})	0.558	0.0228
First year survival rate (S_{1Y})	0.426	0.0342
Reporting rate (λ)	0.00184	0.000164
Final recapture rate (α)	0.00735	0.000659

Figure 1 Model selection for recoveries of dead and controlled Reed Warblers. Probability values for likelihood ratio tests between models are given against arrows. Probability values for goodness-of-fit tests to models are given in parenthesis under each model.

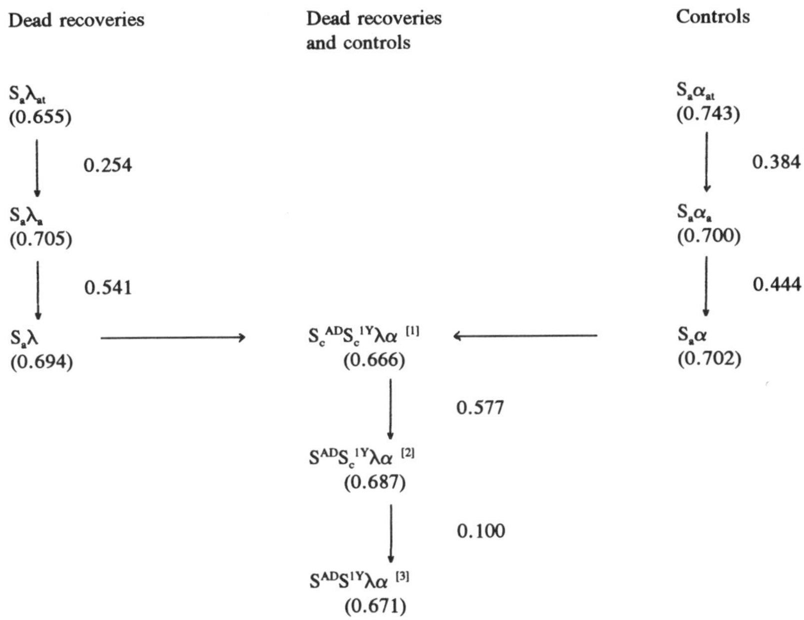

1. Adult and juvenile survival are circumstance specific.
2. Only juvenile survival is circumstance specific.
3. Survival is age-specific only.

A similar analysis was carried out for Willow Warbler using 204 controls and 202 dead recoveries. The recovery data for dead birds could be modelled by $S_a\lambda$ (Table 3) while the simplest model providing an adequate description of the control data was $S_a\alpha_a$ (goodness-of-fit; $\chi^2_{26} = 22.444$, $P = 0.664$; LRT, $S_a\alpha_a$ vs $S_a\alpha$; $\chi^2_1 = 7.193$, $P = 0.0073$). Average survival estimates for adults were 0.309 ± 0.0482 from dead recoveries and 0.554 ± 0.0563 from controls while those for first year birds were 0.261 ± 0.0423 for dead birds and 0.498 ± 0.100 for controls. Estimates from the two methods were so different that we did not attempt to combine the two datasets. The estimates of adult survival based on dead recoveries appear to be unrealistically low in relation to the reproductive rate of the species and to mark-recapture estimates of Willow Warbler survival (Pratt and Peach, 1991, Peach, this volume). The reporting rate for dead recoveries of this species is very low (Table 2) and it seems likely that this sample of dead birds is biased with respect to age.

4. Sample size calculations for ringing of adults

The parameter estimates obtained from this study (Tables 2, 3 and 4) allow a provisional assessment of relationships between numbers of birds ringed and the precision of the resulting survival estimates. Such information is needed for planning the development of ringing in Britain and other parts of Europe. We used the analytical formulae for adult sample size calculations derived by Brownie *et al.* (1985, page 186) and implemented by PROC SAMPLE SIZE of program SURVIV (White, 1983). These calculations are based on model $S_{at}\lambda_{at}$ and more precise estimates might be obtained from a given number of birds ringed if simpler models proved appropriate, as is suggested by the results presented above. The procedures given for calculating sample-sizes for first-year survival rates did not work properly and we have therefore restricted the results presented here to adult survival.

Typical curves relating numbers of birds ringed to the standard errors of the average survival estimates are presented for Blackbird (Fig. 2a) and Reed Warbler (Fig. 2b). Brownie *et al.* (1985) recommend that for most studies a coefficient of variation of $<= 0.05$ is needed, which for the two species considered here is equivalent to a standard error of about 0.03. With current ringing effort such precision should be obtained after about six years for the Blackbird but would require 24 years of Reed Warbler data if the analysis was restricted to recoveries of birds found dead. Increasing the length of the study period from 6 to 12 years results in a reduction of about 50% in the standard error. Further increases in the length of the study period give a proportionately smaller increase in precision (Fig. 2a). Longer study periods have the disadvantage of increasing the probability that large changes in survival will be encountered, in which case it might not be appropriate to average survival over the whole period.

The Reed Warbler data show that a substantial improvement in precision can be achieved by increasing the reporting rate (Fig. 2b), in this case through the inclusion of controls in the analysis.

5. Discussion

A conspicuous feature of the present analysis was that most of the datasets could be described using very simple models. This may be partly a consequence of small sample sizes. However, this does not appear to be the whole explanation because for five species over 500 dead recoveries were available from the six year study period. In the context of ring recovery analyses these are large datasets. Only two of these species required models with year-specific reporting rates and none required models

Figure 2 Precision of average adult survival estimates against number of adults ringed per year.

(a) Blackbirds showing study periods of 6, 12, 18 and 24 years.
(b) Reed Warblers for dead recoveries only (λ=0.002) and for dead recoveries plus controls (λ=0.015), for study periods of 12 and 24 years.

Dotted lines on both graphs indicate the current level of ringing in Britain and Ireland.

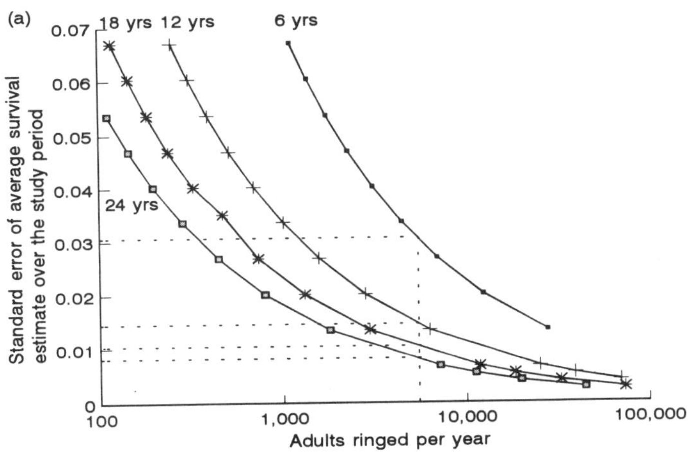

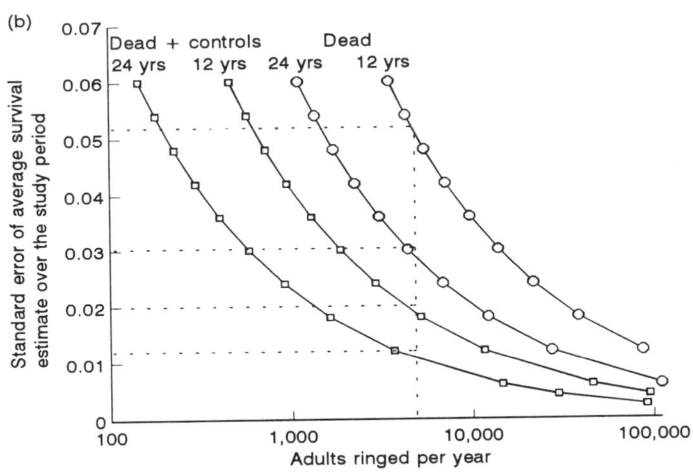

with year-specific survival rates. Temporal variation in reporting rates appears to be less marked that was suggested by earlier analyses (Baillie and Green, 1987) at least over short periods of years. A lack of severe winters during the study period may have contributed to the lack of variation in the survival rates of resident species.

Data for 10 of the 16 species could be modelled by $S_a\lambda$ and for these species at least it might be argued that the same survival rates could be obtained without inclusion of the ringing data. However, there was no way in which those species which were modelled with a constant λ could be identified in advance of the analysis. As longer datasets are gathered it is likely that models involving time or age-specific variation in λ will be needed for more species.

The failure to detect systematic differences in the reporting rates of first year birds and adults for 12 of the 16 species contrasts with earlier studies (Anderson *et al.*, 1981, 1985). However a high proportion of studies, particularly those in North America, have concerned quarry species where age-specific differences in vulnerability to hunting are likely to occur. Furthermore, because the Brownie models are parameterised in terms of F rather than λ, the assumption that λ differs between age-classes is rarely tested. If survival rates differ between first year birds and adults then F ($=\lambda(1\text{-}S)$) will inevitably be age-specific.

It may be that for passerines, which are either relatively sedentary or where both young birds and adults undertake long-distance migrations to the same areas, age-specific differences in reporting rates are smaller than for other groups. Patterns of age-specific variation in λ within the passerines studied here appear to be species specific; for example λ is age-specific for Starlings but not for Blackbirds or Song Thrushes. The reporting rates of young Starlings are lower than those of adults, probably because the young birds join feeding flocks in rural areas where they are less likely to be recovered (Feare, 1984). The models identified by this study should indicate those species for which the results from earlier studies, which assumed λ does not vary with age, might still be used, albeit with caution (e.g. Dobson, 1985). However, although the assumption that λ does not vary with respect to age is perhaps the most important restriction of models based on birds marked as young only, other properties mean that their use should be avoided wherever possible (Burnham and Anderson, 1979, Lakhani and Newton, 1983, Anderson *et al.*, 1985). The assumption that reporting rates are not age-specific can be tested only if both adults and young are ringed and if the numbers ringed are known.

Simple survival estimates based on ring recovery data have provided useful information on the pattern of interspecific variation in the survival rates of birds (Dobson, 1990). Future studies need to concentrate on how variation in survival rates affects the population dynamics of individual species, which requires measures of temporal and age-specific variation in survival rather than simple point estimates for individual species. Survival studies should focus increasingly on identifying environmental correlates of survival such as weather or food availability (North and Morgan, 1979, Peach *et al.*, 1991), on determining whether survival is density-dependent and on determining the extent to which changes in the survival rates of young birds or adults contribute to changes in population size (Baillie and Peach, 1992).

The precision of the survival estimates obtained from this analysis is relatively poor (Table 3). However, Fig. 2a shows that the precision for most species is at a point on the curve of precision against numbers ringed where achievable increases in ringing effort could provide a valuable increase in precision. Only six years of age-specific ringing totals were available for this analysis. Once a further six years of data have been gathered the precision of average survival estimates over the whole study period will be improved substantially. There are no species for which it can be argued that too many birds are being ringed and many for which more ringing effort is needed if long-term shifts in survival rates are to be measured. This paper only considers the summer ringing of adults and juveniles. When data from the ringing of nestlings and from birds ringed at other times of year are brought into the analyses further information will be obtained and the precision of some survival estimates should be increased.

Precision can also be increased by making use of recaptures by ringers. Mark-recapture analysis offers one approach (Peach, this volume) but can only be applied to intensive long-term trapping programmes. While the promotion of such programmes should be a high priority for ringing schemes, it is also important to make the best use of data from less rigorously planned ringing. Our analyses of controls suggest an approach that is worth developing further. In principle such analyses might also make use of recaptures from general ringing (within 5km of the ringing site). The central collation of such data may be a more cost-effective means of improving our ability to detect changes in survival than increasing the numbers of birds ringed. It is likely that in practice both approaches will be needed.

Acknowledgements. This work was carried out under a contract from the Joint Nature Conservation Committee on behalf of English Nature, the Countryside Council for Wales and Scottish Natural Heritage and under a separate contract from the Department of the Environment for Northern Ireland. Dr. Jeremy Greenwood, Dr. Will Peach and two anonymous referees provided constructive reviews of the manuscript. Susan Waghorn provided invaluable assistance by word-processing the manuscript. We particularly thank the hundreds of volunteer ringers who gathered the data on which this paper is based and the staff of the BTO Ringing Unit who collated the data.

References

Anderson, D.R., Burnham, K.P. and White G.C. (1985).- Problems in estimating age-specific survival rates from recovery data of birds ringed as young. Journal of Animal Ecology 54: 89-98.

Anderson, D.R., Wywialowski, A.P. and Burnham, K.P. (1981).- Tests of the assumptions underlying life table methods for estimating parameters from cohort data. Ecology 62: 1121-1124.

Baillie, S.R. (1990).- Integrated population monitoring of breeding birds in Britain and Ireland. Ibis 132: 151-166.

Baillie, S.R. (1991).- Monitoring terrestrial breeding bird populations. In Goldsmith, F.B. (ed.) Monitoring for Conservation and Ecology, pp. 112-132. Chapman and Hall, London.

Baillie, S.R. and Green, R.E. (1987).- The importance of variation in recovery rates when estimating survival rates from ringing recoveries. Acta Ornithologica 23: 41-60.

Baillie, S.R. and Peach, W.J. (1992).- Population limitation in Palaearctic-African migrant passerines. Ibis 134 (Suppl. 1): 120-132.

Brownie, C., Anderson, D.R., Burnham, K.P. and Robson, D.S. (1978).- Statistical inference from band recovery data - a handbook. Resource Publication No. 131 (First edition). United States Department of the Interior Fish and Wildlife Service, Washington.

Brownie, C., Anderson, D.R., Burnham, K.P. and Robson, D.S. (1985).- Statistical inference from band recovery data - a handbook. Resource Publication No. 156 (Second edition). United States Department of the Interior Fish and Wildlife Service, Washington.

Burnham, K.P. and Anderson, D.R. (1979).- The composite dynamic method as evidence for age-specific waterfowl mortality. Journal of Wildlife Management 43: 356-366.

Dobson, A.P. (1985).- In Morgan, B.J.T. and North, P.M. (eds.). Age-dependent mortality rates of some common British birds. Statistics in Ornithology pp. 275-288. Springer-Verlag, Berlin.

Dobson, A.P. (1990).- Survival rates and their relationship to life-history traits in some common British birds. Current Ornithology 7: 115-146.

Feare, C. (1984).- The Starling. Oxford University Press, Oxford.

Hickling, R. (Ed.) (1983).- Enjoying Ornithology. T. & A.D. Poyser, Calton.

Lakhani, K.H. and Newton, I. (1983).- Estimating age-specific bird survival rates from ring recoveries - can it be done? Journal of Animal Ecology 52: 83-91.

Lebreton, J.-D., Burnham, K.P., Clobert, J. and Anderson, D.R. (1992). Modelling survival and testing biological hypotheses using marked animals: a unified approach with case studies. Ecological Monographs 62: 67-118.

Mardekian, S.Z. and McDonald, L. (1981).- Simultaneous analysis of band-recovery and live-recapture data. Journal of Wildlife Management 45: 258-262.

North, P.M. and Morgan, B.J.T. (1979).- Modelling heron survival using weather data. Biometrics 35: 667-681.

Pratt, A. and Peach, W.J. (1991).- Site tenacity and annual survival of a Willow Warbler *Phylloscopus trochilus* population in southern England. Ringing & Migration 12: 128-134.

Peach, W.J. (This volume).- Combining mark-recapture data sets from small passerines.

Peach, W.J., Baillie, S.R. and Underhill, L.G. (1991).- The survival of British Sedge Warblers *Acrocephalus schoenobaenus* in relation to west African rainfall. Ibis 133: 300-305.

Peach, W.J., Buckland, S.T. and Baillie, S.R. (1990).- Estimating survival rates using mark-recapture data from multiple ringing sites. The Ring 13: 87-102.

Smith, D.R. and Anderson, D.R. (1987).- Effects of lengthy ringing periods on estimators of annual survival. Acta Ornithologica 23: 69-76.

White, G.C. (1983).- Numerical estimation of survival rates from band-recovery and biotelemetry data. Journal of Wildlife Management 47: 293-305.

Appendix 1 Scientific names of species referred to in the text

Sand Martin	Riparia riparia
Swallow	Hirundo rustica
Pied Wagtail	Motacilla alba
Wren	Troglodytes troglodytes
Dunnock	Prunella modularis
Robin	Erithacus rubecula
Blackbird	Turdus merula
Song Thrush	Turdus philomelos
Sedge Warbler	Acrocephalus schoenobaenus
Reed Warbler	Acrocephalus scirpaceus
Blackcap	Sylvia atricapilla
Willow Warbler	Phyloscopus trochilus
Pied Flycatcher	Ficedula hypoleuca
Blue Tit	Parus caeruleus
Great Tit	Parus major
Starling	Sturnus vulgaris
House Sparrow	Passer domesticus
Chaffinch	Fringilla coelebs
Greenfinch	Carduelis chloris
Redpoll	Carduelis flammea
Bullfinch	Pyrrhula pyrrhula
Reed Bunting	Emberiza schoeniclus

Marked Individuals in the
Study of Bird Population
J.-D. Lebreton and P.M. North (eds.)

Approaches to Estimation of Survival in Two Arctic-Nesting Goose Species.

M.C. Bell[1], A.D. Fox[1], M. Owen[1], J.M. Black[1] and A.J. Walsh[2]

[1]*The Wildfowl & Wetlands Trust, Slimbridge, Gloucester, GL2 7BT, UK and* [2]*National Parks and Wildlife Service, Wexford Wildfowl Refuge, North Slob, Wexford, Ireland*

Summary. Survival rates are estimated in two populations of arctic-nesting geese: Svalbard-breeding Barnacle Geese *Branta leucopsis*, wintering on the Solway Firth, southwest Scotland; Greenland White-fronted Geese *Anser albifrons flavirostris*, wintering on the Wexford Slobs, southeast Ireland. The Barnacle Goose population is discrete, with virtually no exchange with Greenland- and Russian-breeding populations. Annual survival rates are estimated from resightings of marked birds; the effects of sample size and resighting strategy are considered, and conclusions are drawn about the design of mark-resighting schemes. The Greenland White-fronted Goose population is neither discrete nor closed: there is significant exchange between population components. Annual survival rates are estimated from resightings and recoveries of marked birds, and also from census statistics.

1. Introduction

Eight species or sub-species of arctic-nesting geese have regular wintering grounds in Britain and Ireland. Most of these are legal quarry for hunters over at least part of their ranges. There is good information on the size of wintering goose populations, yet little is known of the processes by which numbers are regulated. This paper considers what is perhaps the most important population process, that of mortality. We examine methods for the estimation of survival in two arctic-nesting goose species, comparing estimates obtained from recoveries or resightings of marked birds with those obtained from population size and productivity data.

The population of Barnacle Geese *Branta leucopsis* (Bechstein) which winters on the Solway Firth in south-western Scotland and breeds in the Svalbard archipelago is probably the best studied goose population in the Western Palaearctic. The population has been studied on the wintering grounds and periodically on the breeding and staging grounds since 1973. During this time numbers of Barnacle Geese have risen from about 4,000 to 10-12,000. The population is virtually closed, with little or no interchange with Russian- and Greenland-breeding Barnacle Geese (Owen & Black, 1989a).

Over 6,000 Svalbard-Solway Barnacle Geese have been marked with plastic leg rings since 1973, and the resighting rate of birds alive and ringed has been estimated to be in excess of 95% (Owen & Black, 1989a). Owen (1982) calculated that only 0.14% of living birds were missed for more than one year. Owen (1982) and Owen & Black (1989a) estimated rates of mortality in cohorts of ringed Barnacle Geese under the simple assumption that a goose not seen in two successive seasons died soon after the date of its last resighting (see also Owen & Black, 1989b). In this paper we compare survival rates estimated using this *ad hoc* approach with maximum-likelihood estimates obtained using a full Jolly-Seber approach in which both survival and resighting rates are estimated. We also address the question of how much marking and resighting data are required for the data-set to retain its essential information content with regard to annual variations in survival.

The Greenland White-fronted Goose *Anser albifrons flavirostris* Dalgety & Scott breeds in western Greenland, and winters in Britain and Ireland (Salomensen, 1950). This paper examines survival amongst geese wintering at the single most important site, the Wexford Slobs in southeast Ireland, where up to 37% of the world population is concentrated. Population size at the Slobs has risen from 6,000 geese in the early 1970's to over 8,000 after the introduction of protective legislation in Britain and Ireland. The population is neither discrete nor closed: about 15% of Wexford geese change wintering sites between years, most commonly being seen at the other major resort, Islay in western Scotland (Wilson *et al.*, 1991). We compare survival rates estimated from population census statistics with those estimated from a marking programme: both recoveries and resightings of marked birds are considered.

2. Methods

2.1. Barnacle Geese

2.1.1. Marking of birds

Captures are by rounding up of flightless birds during the post-breeding moult (July and August) on the breeding grounds on Svalbard and by cannon-netting of birds on the wintering grounds at the Caerlaverock wildfowl refuge on the Solway Firth. Worn or lost rings are replaced on recaptured birds. Geese are resighted by regular observers throughout the wintering period, October to March. Leg rings are individually identifiable by three-character codes, which can be read at distances of up to 250m with the aid of a telescope. See Owen (1982) and Owen & Black (1989a,b) for an overview of the Barnacle Goose population study.

2.1.2. Estimation of survival

Annual survival rates were estimated firstly by the *ad hoc* method, assuming that a bird not seen in two successive seasons did not live to return to its wintering grounds after the date of its last resighting, and secondly from Jolly-Seber models of survival and resighting probabilities. In both cases the survival year was between winters, nominally October to October, but the effective resighting period differed between methods. In the *ad hoc* approach, resightings at any time of the year were considered as evidence of a bird being alive. In the Jolly-Seber approach the resighting period was limited to October to March, in order to reduce sampling time as against time between samples; entry of summer-ringed birds into the resighting matrix was conditional on survival to October - surviving birds were included with the following winter's cohort.

We considered 15 years of ringing from winter 1973/74, involving 4640 ringed birds. Ringing totals varied from 0 to 1634 birds per year. Maximum-likelihood estimates were obtained using program SURGE4 (Clobert *et al.*, 1987; Pradel *et al*, 1990). A complete hierarchy of time-dependent models from S_t, P_t to S, P was estimated (model notation after Kanyamibwa *et al.* (1990) and Peach *et al.* (1991) - S denotes survival probabilities, P denotes resighting probabilities, *t* indicates time-dependence in each case). For the sake of simplicity, survival and resighting probabilities were assumed not to be age-dependent (although it might be biologically more realistic to consider survival of first-years and adults separately, as with the Greenland White-fronted Geese below). Logistic constraints were applied to the estimated parameters (logit(θ), where θ is a parameter value), so that estimated survival and resighting probabilities were constrained in the range 0 to 1. Standard errors and confidence intervals around estimates were based on $logit^{-1}$ transformation of the transformed values (Pradel *et al.*, 1990), thus confidence intervals are asymmetric. Models were compared through likelihood-ratio tests, by which were selected the most parsimonious and biologically meaningful models adequately to describe the data. Annual survival probabilities estimated under these models were compared with *ad hoc* survivorship estimates averaged over cohorts for any one year.

The effects of reducing ringing effort were explored through random selection (without replacement) of birds from the ringing totals for each year, i.e. random selection of capture-resighting histories. Ringing totals were reduced either to a fixed percentage of their original values, or to a fixed number of new rings in each year. We also considered the effects of restricting the resighting period. Models were estimated for resighting matrices constructed using resightings restricted to single months, October to March. In each case estimates were compared with those obtained for the full data set.

2.2. Greenland White-fronted Geese

2.2.1. Census statistics

Monthly counts were made of the Wexford Slobs each winter, 1970/71 to 1989/90. We used the peak mid-winter count to define the overall wintering population size after the end of the shooting season. Adult survival rates were calculated from estimates of population size and productivity by the method of difference (Owen, 1982; Ebbinge *et al.*, 1991): percentage survival (*S*) between winters *i*-1 and *i* is given by

$$S_{i\text{-}1 \text{ to } i} = 100 \times (N_i\text{-}N_iP_i)/N_{i\text{-}1}$$

where N_i and P_i are total population size and proportion of first year birds in winter *i* respectively.

2.2.2. Marking of birds

Between October 1983 and March 1990 706 geese have been captured at Wexford using cannon-nets. Birds were marked with a white plastic leg ring bearing three characters and an orange plastic neck collar bearing the same code in black. Collar codes may be read with the aid of a telescope at distances of up to 800m (Warren *et al.*, 1991). There is a regular resighting programme at the Wexford Slobs, and systematic checking of geese for rings and collars at the other major wintering site, Islay in western Scotland.

2.2.3. Recoveries of marked birds

Maximum-likelihood estimates of survival rates were obtained from recoveries of collared geese using the model framework of Brownie *et al.* (1985). It is known that survival is lower among juveniles than adults (Kampp *et al.*, 1988) and so a hierarchy of two-age-class models in the BROWNIE-set was used: H1 - time-dependence in both survival and recovery rates; H02 - time-dependent recovery rates, constant survival in each age-class; H01 - constant annual rates of survival and recovery in each age-class. Model comparisons were through likelihood-ratio tests by which was selected the most parsimonious model. Goodness-of-fit χ^2 tests were applied to models, with expected values from the recovery matrix, and pooling within cohorts (rows) to give expected values >2. All calculations were performed using program MULT (Conroy *et al.*, 1988).

In two years of the study (1985/86 and 1989/90) regulated shooting was permitted on the Wexford Slobs, coinciding with the period of marking and resighting. However, because marked geese were resighted frequently throughout the winter it was possible to construct a post-hunting-season recovery matrix, based on a January to January survival year. We removed from the banding vector any bird never seen again after December 31st in its winter of banding; similarly, birds recovered before December 31st in their winter of banding were removed from the banding vector and recovery matrix. Adjusted banding totals ranged from 28 to 100 adults and from 33 to 75 first-years. Geese collared in 1983/84 were excluded from the analysis because their low recovery rate precluded estimation of the full hierarchy of models.

2.2.3. Resightings of marked birds

Survival rates of adult and first-years were also estimated from resightings of collared birds. Data were restricted to birds marked and resighted in County Wexford. Maximum-likelihood estimates of survival and resighting probabilities were obtained using program SURGE4. Resighting matrices for birds collared as first-years and birds collared as adults were entered separately into the program, with equality constraints between data sets on the parameters fitted, as appropriate for each model. A starting model was adopted in which both survival and resighting rates were time- and age-dependent, with age-dependence limited to the distinction between first-year and adult birds. Estimates were obtained for a full hierarchy of models in the set S_{a2*t},P_{a2*t} to S,P (*a2* indicates age-dependence of parameters limited to two age-classes - first-years and adults). Goodness-of-fit χ^2 tests were applied to final models, with expected values pooled according to time of last resighting in each cohort (Clobert & Lebreton, 1985; Lebreton & Clobert, 1986). In age-dependent models expected values were pooled separately for adults and first-year birds for the first year of each cohort, otherwise pooling was across data sets.

Resighting matrices were constructed in a similar fashion to the recovery matrix, in order to avoid coincidence of shooting with sampling: geese never resighted after December 31st in their winter of ringing were removed from the banding vector; only resightings occurring after December 31st in each winter were considered. The survival year is thus January to January, in line with both recovery and census analyses. Adjusted banding totals ranged from 15 to 99 adults and from 29 to 78 first-years.

3. Results

3.1. Barnacle Geese

3.1.1. Comparison of methods

Maximum-likelihood estimates of survival rates correspond quite closely to those estimated by the *ad hoc* method (Figure 1). However, estimates from the latter method tend to be slightly lower, indicating that the assumption of 100% resighting probability over two years is not entirely justified. These estimates differ from maximum-likelihood estimates by an average of 1.5%, which is not large in relation to annual survival in the order of 90%, but may cause significant bias in the corresponding mortality estimates. Mean resighting rate estimated by maximum-likelihood is 70.7%, which is much lower than expected, even considering the more restricted resighting period (October to March) compared with the *ad hoc* method (all year).

3.1.2. Ringing effort

The yearly pattern of survival rates estimated by the maximum-likelihood method is largely retained, even at greatly reduced ringing effort (Figure 1), but increased parameter variance means that estimates from different years cannot be statistically distinguished. As would be expected, the mean estimates for each year from repeated (five) random selections of capture histories did not differ markedly from those using the full resighting matrix, but the coefficient of variation between estimates increased from about 2% at 75% of the full ringing effort, to about 20% at 5% effort.

Reduction of ringing effort also has implications for model selection. Considering fixed ringing totals for a five year subset of the data, hypotheses of constancy of survival and/or resighting rates are firmly rejected at and above the threshold of 30 new rings per year (likelihood-ratio tests, $P<0.001$). Presumably, rather larger ringing totals would be required to detect age-dependence in addition to time-dependence.

Figure 1. Annual survival rates of Barnacle Geese estimated from resightings. *Ad hoc* estimates: ———. Maximum-likelihood estimates from model S_t,P_t: □ all rings; + 20% ringing effort; ◇ 5% ringing effort. Vertical lines represent 95% confidence limits. See text for explanation.

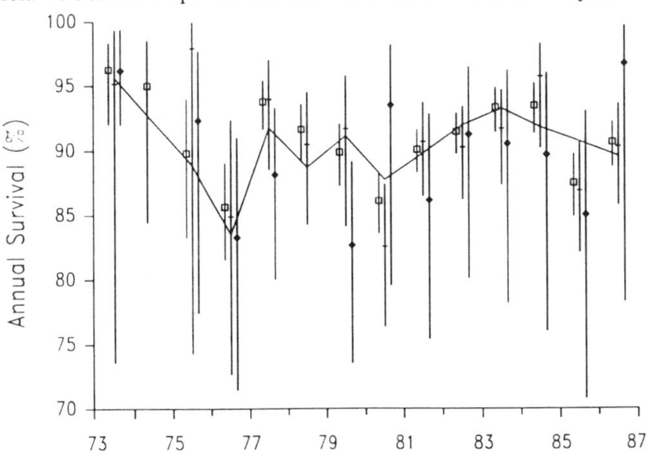

Figure 2. Distribution through the winter season of resightings of Barnacle Geese on the wintering grounds: mean and standard error of percentage of birds first resighted in each month.

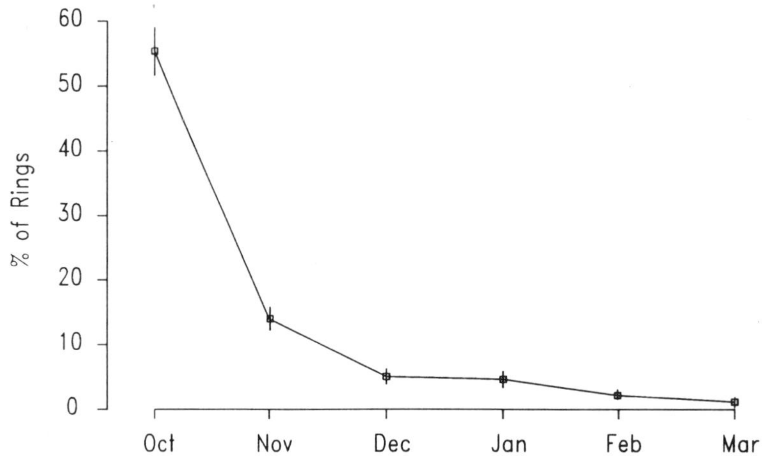

Figure 3. Maximum-likelihood estimates of annual survival rates of Barnacle Geese from model S_t, P_t: comparison between estimates based on an October to March resighting period (———) and estimates based on single month resighting periods. □ October; + November; ◇ December; △ January; × February; ▽ March. See text for explanation.

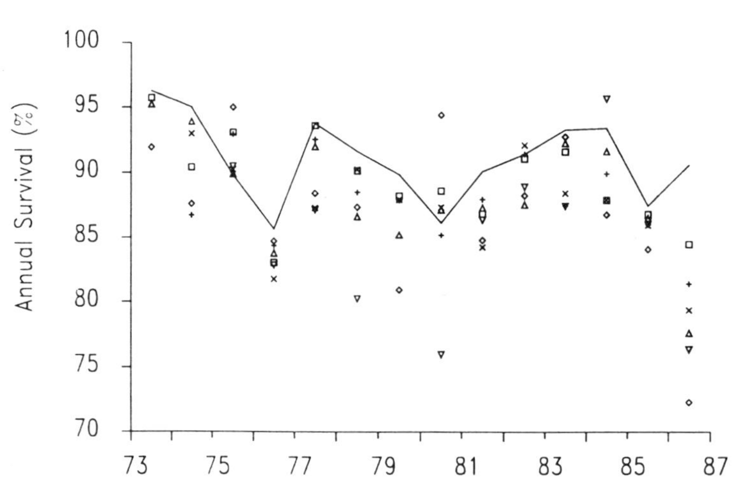

Table 1. Mean between-winter survival and resighting rates for Barnacle Geese, estimated from model S_t,P_t: comparisons between different sampling periods. Comparisons limited to 11 years for which survival and resighting rates were separately identifiable for all sampling periods.

Sampling Period	Survival Rate Mean	Survival Rate S.E.	Resighting Rate Mean	Resighting Rate S.E.
October-March	0.9029	0.0086	0.7342	0.0263
October	0.8882	0.0103	0.6292	0.0473
November	0.8796	0.0108	0.5209	0.0556
December	0.8716	0.0188	0.3611	0.0550
January	0.8748	0.0127	0.4311	0.0671
February	0.8681	0.0114	0.3987	0.0654
March	0.8525	0.0180	0.2687	0.0559

3.1.3. Timing of resighting effort

The between-winter survival year dealt with so far is based on an October to March resighting period, which clearly violates the model assumption of negligible duration of sampling as against time between samples. On average, over half of all rings to be resighted in any one winter are seen by the end of October (Figure 2); this offers the possibility of a much shorter resighting period, such as the month of October or a similarly short period.

The yearly pattern of estimated survival differs considerably depending on which month is chosen for marking/resighting (Figure 3), which may be due in part to variations in within-winter survival. However, mean survival rates estimated using each single month, October to March, as the mark-resighting period are also different (Table 1), although not significantly so ($F_{5,50}$=1.80, P=0.130). Mean estimates of survival and resighting rates appear to be correlated between sampling periods (Pearson r=0.950, d.f.=4, P=0.004), the reasons for which are unclear, but it may result from heterogeneity of resighting and/or survival probabilities. The October results most closely resemble those for the full resighting period. Clearly, however, the time chosen for sampling strongly influences the modelling of survival, so that it is vital to define the survival year very clearly.

3.2. Greenland White-fronted Geese

3.2.1. Census statistics

Figure 4 shows annual survival rates of adult Greenland White-fronted Geese estimated by the method of difference. These ranged from 68.4% to 97.7% for the period 1970/71 to 1989/90, mean 84.4%.

3.2.2. Recoveries of marked birds

A BROWNIE format recovery matrix for birds collared as adults and first-years is shown in Table 2. Recovery rates were low so that the information content of the recovery matrix is correspondingly low. The null hypothesis of constant annual

Figure 4. Annual survival rates of Greenland White-fronted Geese wintering at Wexford, estimated from census statistics.

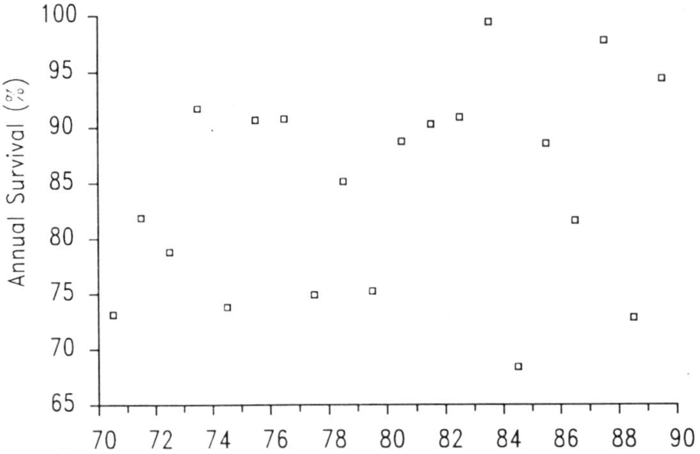

Table 2. Recovery matrix for Greenland White-fronted Geese marked with neck collars as adults and first-years at Wexford 1984 to 1989, adjusted for a January to January survival year.

Year Marked	Number Marked	Year of Recovery 1985	1986	1987	1988	1989	1990
		Geese marked and released as adults					
1984	100	9	1	2	5	2	
1985	53		1	1	1	1	3
1986	30			3			
1987	28				2	1	
1988	69					6	1
1989	33						1
		Geese marked and released as first-years					
1984	61	7	2		2	2	
1985	75		4	4	3		
1986	60			4	2	3	1
1987	62				2	1	1
1988	38					4	
1989	33						1

Table 3. Annual survival and recovery probabilities for Greenland White-fronted Geese estimated under BROWNIE-set model H01 from recoveries of geese neck banded at Wexford, 1984 to 1989. See text for explanation.

Parameter	Estimate	S.E.	95% C.L.	
First-Year Survival	0.5964	0.1521	0.2983	0.8945
First-Year Recovery	0.0699	0.0141	0.0424	0.0975
Adult Survival	0.7244	0.0723	0.5827	0.8661
Adult Recovery	0.0542	0.0723	0.5827	0.0742

survival rates within age-classes was not rejected (H1 vs H02, X^2=3.51, d.f.=8, P=0.898), nor was that of constant annual recovery rates (H01 vs H02, X^2=14.8, d.f.=10, P=0.138). The final model, H01 constant annual survival and recovery rates within each age-class, was a good fit to the data (χ^2=20.2, d.f.=20, P=0.447). Parameter estimates from this model are shown in Table 3.

3.2.3. Resightings of marked birds

Comparisons of models in the hierarchy S_{a2*t},P_{a2*t} to S,P are shown in Table 4. Null hypotheses of no age- or time-dependence of resighting probability were accepted, but it was not possible to simplify survival probabilities beyond age- and time-dependence. Estimates of survival and resighting probabilities under the final model S_{a2*t},P are given in Table 5. This model fits the data very well (χ^2=11.1, d.f.=27, P=0.997). Very low survival was estimated among the last cohort of first-years (20%). Inspection of the raw resighting data reveals that this was due to two groups of closely related collared juveniles which failed to return to Wexford after their first winter. The most likely fate of these birds is that they were shot as family groups on migration, probably in Iceland. Clearly, the fundamental assumption of independence in the behaviour of marked individuals is violated.

Table 4. Comparisons between models for survival rates of Greenland White-fronted Geese estimated from resightings of collared birds. n.p.: number of identifiable parameters; d.f.: degrees of freedom. See text for explanation.

Model	n.p.	Deviance	Likelihood-ratio test	d.f.	Outcome
Modelling resighting rate					
(1) S_{a2*t},P_{a2*t}	26	2792.89	—	—	—
(2) S_{a2*t},P_t	19	2795.42	(2)-(1)=2.53	7	(2) accepted, P=0.925
(3) S_{a2*t},P	15	2800.45	(3)-(2)=5.03	4	(3) accepted, P=0.284
Modelling survival rate					
(4) S_t,P	8	2845.07	(4)-(3)=44.6	7	(4) rejected, P<0.001
(5) S_{a2},P	3	2850.83	(5)-(3)=50.4	12	(5) rejected, P<0.001

Table 5. Annual survival and resighting probabilities for Greenland White-fronted Geese estimated under model S_{a2*t},P from resightings of collared birds. See text for explanation.

Parameter	Estimate	S.E.	95% C.L.	
Survival of first-years:				
1983-1984	0.7769	0.0888	0.5607	0.9047
1984-1985	0.6803	0.0651	0.5421	0.7927
1985-1986	0.7653	0.0569	0.6368	0.8585
1986-1987	0.6947	0.0633	0.5590	0.8033
1987-1988	0.8309	0.0573	0.6885	0.9161
1988-1989	0.5175	0.0880	0.3497	0.6815
1989-1990	0.1990	0.0817	0.0834	0.4042
Survival of adults:				
1983-1984	0.5722	0.1379	0.3072	0.8013
1984-1985	0.7159	0.0439	0.6227	0.7937
1985-1986	0.8604	0.0352	0.7762	0.9163
1986-1987	0.6659	0.0345	0.5953	0.7297
1987-1988	0.8332	0.0299	0.7663	0.8839
1988-1989	0.7456	0.0310	0.6802	0.8016
1989-1990	0.7716	0.0405	0.6828	0.8413
Resighting:				
	0.7852	0.0142	0.7561	0.8117

3.2.4. Comparison between estimates

Survival rates of adult Greenland White-fronted Geese were estimated to be higher from census statistics than from resightings or recoveries of marked birds. Mean annual survival of adults was estimated at 84% by the method of difference, whilst resighting and recovery analyses gave mean estimates in the range 72-74% (for slightly differing periods). Presumably the lower estimates from at least the resighting scheme result from the compounding of mortality with emigration of marked birds, estimated by Wilson *et al.* (1991) to be about 15% per year. However, in a geographically open population, conclusions about annual survival rates drawn even from census statistics depend strongly on the assumption of no net emigration; in practice it is difficult to distinguish between patterns of survival and site-fidelity.

Both resighting and recovery analyses estimated lower mean annual survival amongst first-years than adults, although the difference was not consistent across all years from the resighting analyses (see Table 5). Recovery analyses failed to pick up time-dependence of survival, presumably owing to the very low information content of the recovery matrix (Table 2). Estimates from BROWNIE-set model H1 are nonsensical, with confidence limits encompassing survival probabilities of 0 and 1. Estimates from resightings bear little relation to equivalent estimates from the method of difference (Pearson correlation coefficient, r=0.013), which may result from variation in the annual rate of emigration of marked birds, as well as from over- or under-estimates of population size and/or productivity.

4. Discussion

4.1. Methods of survival estimation

In essence, the two approaches used in this paper to the estimation of Barnacle Goose survival rates from resightings of ringed birds are the same. The *ad hoc* method is a Jolly-Seber method, but with resighting rate assumed to be 100% over two years. Confidence intervals around estimates could have been constructed using binomial probabilities. In the light of the relatively low annual resighting probabilities estimated by maximum-likelihood, albeit for a shorter resighting period, the assumption of 100% resighting probability over two years needs in future to be more closely examined.

The more rigorous statistical framework of the maximum-likelihood approach is clearly preferable, particularly considering the availability of software such as SURGE4 providing flexible and efficient estimation of models and testing of hypotheses. Nevertheless, considering the use made of the *ad hoc* method in drawing conclusions about patterns of mortality (Owen, 1982; Owen & Black, 1989a), it is encouraging that the two approaches give comparable results. Owen (1982) provided evidence for the reliability of the *ad hoc* method through comparison of annual mortality estimates derived by this method with those obtained from population size and productivity data: correspondence between the estimates was good, but estimates from census data were considerably more variable. Owen concluded that small errors in counts or age-ratio assessments make a large difference in apparent mortality, but that such errors are largely self-compensating over time so that the mean over a number of years is reliable. Ebbinge *et al.* (1991) came to similar conclusions regarding estimation of survival of Russian Barnacle Geese from census statistics. Greater reliance may thus be placed on individual annual estimates derived from resightings.

Despite problems of non-independence of resighting histories (discussed below), the final model from Jolly-Seber analysis of resightings of Greenland White-fronted Geese is the most biologically sensible, or at least that which would be intuitively expected given our knowledge of the population - resighting probability constant between years, survival probabilities differing between years and between adult and first-year birds. Resighting effort at the Wexford Slobs is so high that resighting probability is effectively constant - most geese that are seen at all are seen repeatedly during the winter season. We already know that survival of first-years tends to be lower than that of adults (Kampp *et al.*, 1988), and it is suspected that winter shooting mortality, which is very variable between years, is a significant addition to natural mortality. In a forthcoming publication (Bell *et al.*, in prep.) the effects on annual survival of winter shooting mortality and hunter-induced emigration are examined through the incorporation of external variables into Jolly-Seber type models. However, despite this plausible modelling of survival, the survival rate estimates themselves are very different from those derived from census statistics. This must partly be because the resighting approach suffers from the compounding of emigration with mortality - the magnitude of the difference in estimates between methods appears to be accounted for by what we know about exchange of birds between sites. Also, as has been noted above, the estimation of survival from census statistics depends on their being no net immigration or emigration, a probably unrealistic assumption for a geographically open population. It would be possible to include emigration probability as a third element in resighting models, but the proliferation of parameters to be estimated is likely to place unreasonable demands on the amount and quality of data collected. A preferable alternative for species such as Greenland White-fronted Geese concentrated at a few major sites, would be to consider data from several sites in the same analysis, examining exchange of birds between sites. This will soon be possible for the Greenland White-fronted Geese since, in addition to the Wexford marking scheme, wintering birds are now being marked with neck collars on Islay, the second most important winter resort.

Dead recoveries of marked birds appears to offer an advantage over resightings in that there is no confusion of emigration with mortality. However, data from recoveries tend to be sparse for two reasons: (1) recovery rates tend to be low, and not all recovered birds are reported; (2) each marked bird contributes information only once over the study period, in being recovered or not recovered, in contrast to a resighting scheme where a bird contributes data on each sampling occasion. In the case of the Greenland White-fronted Geese the information content of the recovery matrix was clearly insufficient to draw conclusions about temporal patterns of survival. It is probable that in future both recoveries and resightings of birds will be incorporated into a unified modelling approach for marked birds, but at present it is clearly preferable to deal with resightings than recoveries. The major practical drawback is the investment of time and resources involved in a resighting programme, but there are great advantages in terms of more controlled collection of information on the fate of marked birds.

4.2. Sample size for mark-resighting

The information content of the resighting matrix appears to be preserved at even quite low ringing totals. Time-dependence of survival and resighting probabilities over a five year period were detected at ringing totals as low as 30 per year, although at this level of effort, confidence intervals around estimates were so wide as to make interpretation hazardous. Clearly, it would be unreasonable to suggest that 30 new rings per year is an adequate ringing effort to allow survival estimation from a resighting programme, but it is still encouraging that pattern could be detected at this level. Ringing totals for Greenland White-fronted Geese at Wexford are in the region of 30-100 new rings per year each of adults and first years; this level of effort is fairly typical of other goose marking schemes in the British Isles, e.g. 50-100 Greylag Geese *Anser anser anser* per year on North Uist, up to 300 Pink-footed Geese *Anser brachyrhynchus* per year at Martin Mere in Lancashire (C. Mitchell, pers. comm.), although resighting effort is usually not so high as for the Greenland White-fronted Geese.

4.3. Sources of bias

In one respect at least, the comparison of randomly selected resighting histories with a complete resighting matrix of a similar size is not valid. Random selection results in a group of birds which are more-or-less independent of one another in their resighting histories. Captures of Greenland White-fronted Geese on their wintering grounds tend to have a higher proportion of first-years than the population as a whole

(Wilson *et al.*, 1991), owing to the dominance of family groups over single birds and pairs without young in the competition for feeding resources (Boyd, 1953). Some implications of this are clearly seen in the 1989/90 cohort of first-year birds, 27 out of 30 of which were not resighted in the following winter. Inspection of the ringing records reveals that these were a few groups of siblings, the fates of which were, presumably, not independent. There is evidence that Greenland White-fronted Geese remain associated with parents and siblings in extended family parties, even after their first year (Warren *et al.*, 1992), so that the implications of non-independence of related individuals are not confined to first-year birds. The survival estimates are not, of course, statistically biased in relation to the population of marked individuals, but parameter variances will be underestimated. A computer-intensive method, such as resampling of capture histories, could be used to indicate the true size of confidence intervals.

In order to draw conclusions about patterns of mortality from resightings of marked birds it is necessary to assume that the marked sample is representative of the population at large, and that the method of marking has no effect on the subsequent fate of marked individuals. As has been noted above, at least among Greenland White-fronted Geese, and possibly among wildfowl in general, the luring of birds to bait results in the capture of a biased sample of the study population, owing to competition for food resources. In an analysis designed to test whether neck collars had an adverse effect on breeding in Greenland White-fronted Geese, Warren *et al.* (1991) found that collared birds were more successful than uncollared birds owing to the biased nature of the captured sample. A further study of the effects of collars on Greenland White-fronted Geese is now underway, in which some captured first-year birds are marked with leg rings only, which are assumed to have no effect on survival or breeding success. In future, it should be possible to test for differences in survival between collared and uncollared birds. At present, one should be cautious about extending conclusions drawn about mortality among collared geese to the population at large.

Other potential sources of bias in the estimation of survival and resighting parameters and their variances from mark-resighting schemes include mark loss and misreading of mark codes. The effects of the latter are difficult to quantify. Ebbinge *et al.* (1991) discussed the problems of observer variability in some detail; these authors estimated that for Russian Barnacle Geese wintering in the Netherlands and northwestern Germany the annual probability of an undetected misreading of coloured plastic leg rings was 0.2%. In the case of the Svalbard-Solway Barnacle Geese and the Wexford Greenland White-fronted Geese misreading of ring and collar codes is thought not to be a great problem. In these two populations most geese which are seen at all during a winter season are seen repeatedly. Many errors are detected and removed from the database: the reporting of non-existent codes, the reporting of codes of birds known to be dead, or single isolated reports of birds which have not been seen for many years and are assumed to be dead. The extent of misreading is unknown, but it is probable that if all birds seen only once during a winter season were excluded from analysis then the problem could be almost entirely eliminated.

Loss of marks is potentially a more serious source of bias, leading to underestimation of survival rates. Owen (1982) estimated a ring loss rate of 0.14% per year in the Svalbard-Solway Barnacle Geese. The loss of collars in the Wexford Greenland White-fronted Geese has been found to be higher in years with winter-shooting, owing to the fracturing of collars by lead shot (Wilson *et al.*, 1991); overall, the annual rate of collar loss has been estimated at up to 9.4% (Wilson *et al.*, 1991).

However, the marking of these geese with both leg rings and neck collars, coupled with the high resighting effort, ensures that collar loss does not contribute much to apparent mortality. Only one instance is known of a collared bird losing a leg ring (Wilson *et al.*, 1991). Collar loss could introduce heterogeneity of resighting probabilities owing to the much lower visibility and legibility of leg rings, but in practice this is probably offset by the high resighting effort at Wexford.

A drawback to the marking of geese on the wintering grounds is that this generally coincides with the winter shooting season. In the drawing up of recommendations for the management of goose populations it is very often the effects of this shooting in which we are most interested. An obvious solution to this problem, that we regularly catch and mark geese on their arctic breeding grounds is too often precluded through expense and practical difficulties. In this paper we have attempted to avoid the problem by adjusting the resighting or recovery matrix for a post-hunting-season sampling period, but this is far from a satisfactory solution, and could potentially introduce biases.

4.4. Approaches to survival estimation

It is clear that none of the approaches considered here yields estimates of survival rates that are uninfluenced by other population processes or biases introduced by the application of the methods themselves. What we have estimated is *apparent* survival, the meaning of which changes according to the approach adopted. In order to draw reliable inferences about patterns of survival in wintering goose populations information from both marking and population census schemes are required. At present, improvement in survival estimation for arctic-nesting geese lies not with the development of new modelling techniques, but with the design and execution of marking and census schemes. In particular, there needs to be reliable estimates of the rates of other population processes and the extent of potential biases, and collection of adequate data on other sources of mortality.

Acknowledgements. We are indebted to the many observers who have assisted with ringing or provided observations of marked geese, and to all who have helped with the Greenland White-fronted Goose and Barnacle Goose projects. In particular we thank Dave Norriss, Paddy O'Sullivan, Paul Shimmings, Stephanie Warren and John Wilson. The paper was improved after the comments of Dr John Tautin and an anonymous referee.

References

Bell, M.C., Fox, A.D. & Walsh, A.J. (in prep.).- Changes in patterns of mortality amongst Greenland White-fronted Geese *Anser albifrons flavirostris* under protective legislation in Ireland.

Boyd, H. (1953).- On encounters between wild White-fronted Geese in flocks. Behaviour, 5: 85-129.

Brownie, C., Anderson, D.R., Burnham, K.P. & Robson, D.S. (1985).- Statistical inference from band recovery data - a handbook. US Fish and Wildlife Service, Resource Publication nº156, 2nd edition.

Clobert, J. & Lebreton, J.D. (1985).- Dépendence de facteurs de milieu dans les estimations de taux de survie par capture-recapture. Biometrics, 41: 1031-1037.

Clobert, J., Lebreton, J.D. & Allaine, D. (1987).- A general approach to survival rate estimation by recaptures or resightings of marked birds. Ardea, 75: 133-142.

Conroy, M.J., Hines, J.E. & Williams, B.K. (1988).- Procedures for the analysis of band recovery data and user instructions for program MULT. Patuxent Wildlife Research Center.

Ebbinge, B.S., van Biezen, J.B. & van der Voet, H. (1991).- Estimation of annual adult survival rates of Barnacle Geese *Branta leucopsis* using multiple resightings of marked individuals. Ardea, 79: 73-112.

Kampp, K., Fox, A.D. & Stroud, D.A. (1988).- Mortality and movements of the Greenland White-fronted Goose *Anser albifrons flavirostris*. Dansk Ornitologisk Forenings Tidsskrift, 82: 43-55.

Kanyamibwa, S., Schierer, A., Pradel, R. & Lebreton, J.D. (1990).- Changes in adult annual survival rates in a western European population of the White Stork *Ciconia ciconia*. Ibis, 132: 27-35.

Lebreton, J.D. & Clobert, J. (1986).- User's manual for program SURGE. Montpellier, France: C.E.P.E./C.N.R.S.

Owen, M. (1982).- Population dynamics of Svalbard barnacle geese, 1970-1980. The rate, pattern and causes of mortality as determined by individual marking. Aquila, 89: 229-247.

Owen, M. & Black, J.M. (1989a).- Factors affecting the survival of barnacle geese on migration from the breeding grounds. Journal of Animal Ecology, 58: 603-617.

Owen, M. & Black, J.M. (1989b).- Barnacle Goose. pp 349-362 *in* Lifetime Reproduction in Birds, I.Newton (Ed.) Blackwell Scientific Publications, Oxford.

Peach, W., Baillie, S. & Underhill, L. (1991).- Survival of British Sedge Warblers *Acrocephalus schoenobaenus* in relation to west African rainfall. Ibis. 133: 300-305.

Pradel, R., Clobert, J. & Lebreton, J.D. (1990).- Recent developments for the analysis of capture-recapture multiple data sets. An example concerning two blue tit populations. The Ring. 13: 193-204.

Salomensen, F. (1950).- Grønlands Fugle. Munksgaard, Copenhagen.

Warren, S., Fox, A.D. & Owen, M. (1991).- A preliminary examination of the effects of marking with neck collars on the survival and breeding success of Greenland White-fronted Geese. Wildfowl & Wetlands Trust, Internal Report.

Warren, S., Fox, A.D., Walsh, A. & O'Sullivan, P. (1992).- Extended parent-offspring relationships in Greenland White-fronted Geese (*Anser albifrons flavirostris*). Auk, in press.

Wilson, H.J., Norriss, D.W., Walsh, A., Fox, A.D. & Stroud, D.A. (1991).- Winter site fidelity in Greenland White-fronted Geese *Anser albifrons flavirostris*, implications for conservation and management. Ardea, 79: 287-294.

Marked Individuals in the
Study of Bird Population
J.-D. Lebreton and P.M. North (eds.)

Modelling Raptor Populations: to Ring or to Radio-tag?

Robert E. Kenward

Institute of Terrestrial Ecology, Furzebrook Road, Wareham, Dorset BH20 5AS, Great Britain.

Summary. Raptors are typically uncommon, wide-ranging and very vulnerable to human activities. This makes it difficult, but important, to assess their populations. Traditional techniques include estimation of breeding density and productivity by a visual search for nests, and estimation of mortality by ringing. Both methods are vulnerable to bias, especially if nests are hard to detect, and if ring recoveries are most likely after particular types of death or at particular ages, or if rings may be lost during the long lives of larger species.

Compared with ringing, radio-tagging can provide comparatively rapid and less biased assessment of mortality for large raptors. Since radio-tagging also helps to find nests in atypical habitat, and to record early breeding failures, it is also a useful adjunct to purely visual techniques for monitoring productivity. It provides more detailed information than ringing on individual performance, especially on ranging behaviour and dispersal of raptors. The parameters of individual performance (PIPs) can be used directly to assess management schemes, such as re-introductions, and also to build dynamic models of animal populations from the bottom up. Radio-tagging can provide survival data more economically than ringing where voluntary effort is scarce.

The radio-tagging of 350 northern goshawks (*Accipiter gentilis*) in Sweden, and of 60 common buzzards (*Buteo buteo*) in southern England, gave first-year survival estimates considerably higher than estimates obtained previously by ringing. This may be a result of bias in ring estimations, because causes and timing of goshawk deaths differed from contemporary ring recoveries. However, ringing data from rare species accumulates for many years before analysis, so survival estimates obtained rapidly by radio-tagging may also reflect reduced destruction of raptors by humans in recent years. If high first-year survival rates are relatively typical of healthy raptor populations, such that many adults do not attempt to breed, then monitoring nests is a comparatively insensitive method of monitoring population change. Radio-tagging, complemented by ringing, will eventually lead to models which predict how behaviour and density of scarce bird species will change in response to altering landscapes and climate.

1. Introduction

Raptorial birds, including the Falconiformes and Strigiformes, occur at the end of many terrestrial and aquatic food chains. Living at the top of trophic pyramids, they are uncommon relative to other birds, although they sometimes gather in tens or even hundreds at breeding colonies, on migration or at rich sources of food. Their predatory habits have often brought raptors into conflict with humans, as competitors for fish or game, or through the killing of domestic poultry and pigeons. They were therefore killed by humans to the extent that some large species vanished from wide areas of suitable habitat in the last century. In Britain, for example, the osprey *Pandion haliaetus*, sea-eagle *Haliaeëtus albicilla* and northern goshawk *Accipiter gentilis* became extinct, while the common buzzard *Buteo buteo* and red kite *Milvus milvus* became restricted to small parts of their previous distribution (Newton, 1979).

The deliberate killing of raptors may be less common nowadays, but they remain vulnerable to other human activities. Toxic chemicals, especially organo-chlorines, accumulate along food chains and can devastate raptor populations. Raptors are highly mobile, which enables a rapid response to changes in prey abundance but also puts them at risk of collision with wires, vehicles and other constructions. Still more insidious are the impacts of changing land-use, which may remove hunting habitats and nest sites, or reduce the numbers of prey, either live or as carrion.

The risk to raptors from human activities makes it particularly important to move from reactive to predictive management of their populations. This requires not only knowledge of their productivity and survival in typical conditions, but also an understanding of how these parameters may respond to perturbations. In other words, we require dynamic models of raptor populations.

Raptor populations have been studied mainly by visiting nests, to record productivity and to ring the young for subsequent estimation of survival and dispersal rates. However, natural raptor nests are often hard to find and difficult to reach, especially for rare species nesting in forests or on cliffs, so relatively few raptors are ringed. It can therefore take many years to accumulate enough ring recoveries for survival analyses, even if all the records for a whole country are added together. Thus, when all the data for raptors in Britain were added together, only kestrel *Falco tinnunculus* and sparrowhawk *Accipiter nisus* exceeded 100 ring recoveries before 1954, when legal protection started, with only kestrel and buzzard exceeded 100 recoveries in the following 15 years (Glue, 1971). Such limited data are not really suitable for analyses of local and temporal variation in survival rates which could be used to produce dynamic population models.

A second problem is the possibility of recovery bias in favour of those ringed raptors which impact human constructions or are killed deliberately, and a reporting bias against recoveries from killed birds which are protected by law. Prior to legal protection, 45-92% of ring recoveries in Finland were from raptors which had been shot or trapped (Nordström, 1963; Haukioja and Haukioja, 1970), and 41-60% of kestrels and sparrowhawks in Britain (Glue, 1971). After protection, the proportion recorded as killed was 10-16%, and the proportion reported to have died in collisions rose from 13-17% to 28-32%. Since the young and inexperienced raptors are not only much more frequently trapped or shot at game farms than adults (Marcström and Kenward, 1981a), but may also be more prone to accidents, their rings may be more likely to be recovered than those of adults which die in the wilder areas. Such bias could lead to over-estimation of first-year mortality, which was at least 60% in 17 of 26 analyses for 13 diurnal raptor species (Newton, 1979). Skuas and large gulls, representing other predatory birds which can equal raptors in longevity, have first-year mortality estimates of only 20-30% (Cramp and Simmons, 1983; Furness, 1978).

During the last 30 years, radio-tags have evolved in size and reliability to become an alternative to rings in raptor demography studies, with the following advantages:

(1) Survival rates can be estimated rapidly in a small area.
(2) Juvenile survival is estimated without awaiting recoveries from older birds.
(3) Bias can be minimised in estimating survival, productivity and causes of death.
(4) Individual movements and other behaviour can be recorded in great detail.

However, compared with using ring recoveries to estimate survival and dispersal parameters, radio-tagging also has disadvantages, which include:

(1) Problems with lost signals, especially with unreliable tags.
(2) Radio tags do not last as long as rings.
(3) Tagging itself may affect survival.
(4) Expense.

This paper considers these advantages and disadvantages in more detail, drawing examples from studies of northern goshawks in Sweden and common buzzards in southern Britain. The concluding section discusses how radio-tagging may best complement ringing in future studies of bird population dynamics.

2. The benefits of radio-tagging

2.1. Rapid results

In the Swedish goshawk study, 352 goshawks were radio-tagged during 7 years in a 650 km^2 area on the Baltic island of Gotland, an average of 50/year of which about 30/year were young hawks at nests. Similarly, a study in 150 km^2 of Dorset in southern Britain has radio-tagged 58 young buzzards at nests in 1990-91. It is quite practical for one person to tag and radio-track at least 30 birds annually for sedentary raptors, except for species with very dispersed breeding sites, such as eagles. This is comfortably above the 20 individuals/cohort considered by Pollock *et al.* (1989) to be the minimum for inter-cohort comparisons.

In contrast, only 5 of 19 diurnal raptor and owl species have provided more than 30 ring recoveries from the whole of Britain and Ireland in each of the last 3 years (Table 1). Since only 50-70% of the recoveries are likely to have come from birds in their first year, there are only enough data from sparrowhawk, kestrel, barn owl and tawny owl to study variation in first-year survival rates.

Table 1. Mean numbers of raptors ringed, and recovered, in Britain and Ireland each year, 1988-90. Data from Mead and Clark (1989, 1990, 1991).

Species	*N* ringed	*N* recovered
Red kite *Milvus milvus*	46	3
White-tailed eagle *Haliaeëtus albicilla*	5	1 (Total)
Marsh harrier *Circus æruginosus*	50	0
Hen harrier *Circus cyaneus*	338	17
Montagu's harrier *Circus pygargus*	21	0
Northern goshawk *Accipiter gentilis*	171	6
Sparrowhawk *Accipiter nisus*	1892	188
Common buzzard *Buteo buteo*	298	17
Golden eagle *Aquila chrysaëtos*	45	1
Osprey *Pandion haliaetus*	59	4
European kestrel *Falco tinnunculus*	1956	134
Merlin *Falco columbarius*	741	42
Hobby *Falco subbuteo*	36	2
Peregrine falcon *Falco peregrinus*	337	26
Barn owl *Tyto alba*	1093	156
Little owl *Athene noctua*	195	9
Tawny owl *Strix aluco*	1147	71
Long-eared owl *Asio otus*	134	8
Short-eared owl *Asio flammeus*	114	7
Totals for residents:migrants	8512:166	686:6 (8%:4%)

Despite the exceptional ringing effort in Britain, where the British Trust for Ornithology now supervises the ringing of more than 800,000 birds each year, the data are inadequate for studying annual variations in the survival rates of large raptors. Moreover, since annual variation is likely to depend on fluctuations in weather and food supplies, which themselves vary regionally, combining data from the whole of Britain is less satisfactory than working in one or more local study areas. Tags to provide data on survival and movements for at least a year can now be built to study even the smallest British raptors.

Even before tags were available with the present performance, their ability to monitor small numbers of animals thoroughly for short periods in specific areas had conservation value in re-stocking programs. Radio-tags were used in the 1970s to monitor movements and survival for the first few weeks after release of peregrine falcons in North America (Sherrod *et al.*, 1981) and goshawks in Britain (Kenward *et al.*, 1981).

2.2. First year data

Estimates of first-year survival rates from ring recoveries depend on recoveries in subsequent years, and are highly sensitive to departures from assumptions which underlie the survival models (Lakhani and Newton, 1983). This is particularly unfortunate for population models, because density dependent factors are most likely to affect survival in the first year (Goss-Custard and Durell, 1987), and errors in estimating first-year survival can severely affect estimates of adult population size. Accurate assessment of first-year survival is also important for conservation studies of declining raptors, because some problems in the environment may be most apparent in this year of natal dispersal, when birds are least experienced.

Radio-tagging also provides survival rate estimates for first-year birds with less marking bias than for adults. Although tagged chicks from the same nest are not a random sample, marking bias can be minimised by radio-tagging young at every nest in a study area or at randomly selected nests, provided that all nests are known and accessible. Marking bias at nests is similar for both radio-tagging and ringing studies, as is the greater bias if raptors are trapped away from the nest: goshawks which were live-trapped in winter had lower weights than shot hawks (Marcström and Kenward, 1981b), and those caught in compartment traps baited with pigeons were less likely to be resident in an area than those caught in spring nets on kills (Kenward *et al.*, 1983).

2.3. Recovery bias

During the radio-tracking of goshawks in Sweden, 67 deaths were recorded. At that time, goshawks could be killed legally at farms to protect poultry, and 24 of the hawks (36%) had been killed (Kenward *et al.*, in press). Over the same period, 50 other ring recoveries included 24 (48%) from hawks reported to have been killed; the difference is in the expected direction but not significant (P=0.19). Recoveries peaked in the same months for ringed and radio-tagged hawks that were killed by man (Fig. 1a), but there was a significant seasonal bias in ring recoveries of hawks "found dead". Whereas the mortality rate of radio-tagged hawks that were not killed by humans gave an expected median recovery date in October, 22 of 30 ringed hawks were found later in the winter (binomial test, two-tailed P=0.018), with a median recovery date in February (Fig. 1b). Ground cover, followed by snow, probably reduced the detection of dead hawks in late summer and autumn, so that many were not found until children and ramblers again took to the woods in spring.

Figure1. The number of rings returned in different seasons from goshawks on Gotland in 1980-88, compared with expected recovery estimates for each period from deaths of radio-tagged hawks. Hawks reported killed by humans (a) are shown separately from those found dead (b).

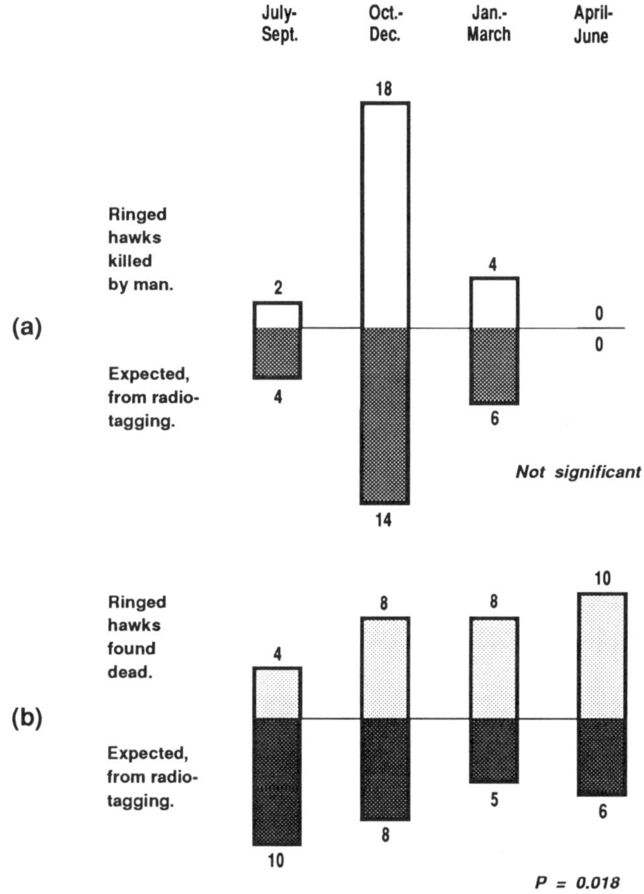

As well as providing relatively unbiased data on mortality, radio-tagging helps to find nests. This not only saves time which would otherwise be needed for visual searches, but also helps to detect nests in unusual sites. On Gotland, at least 2 nests were found in pinewoods which would normally have been considered too young, thus reducing bias both in marking young and in describing nest habitats. Reducing bias in assessment of habitat-use is important for forest conservation programs, for example to remove the misconception that the hawks nest only in old-growth stands. Moreover, in population modelling it is important to record birds whose breeding fails so early that their attempt might not otherwise have been detected.

2.4. Individual performance

Another advantage of radio-tagging over ringing is that radios allow detailed data to be collected on movements and other behaviour. Radio-tracking has shown that even when buzzards have apparently dispersed from their natal areas, they still travel long distances back towards the nest for short periods each spring. Such short-term phenomena can only be detected by analysing very large numbers of ring recoveries, but rapidly become obvious in radio-tracking. If dispersers generally seek to breed in their natal area when they are old enough, it would explain why some large raptors are slow to recolonise areas from which they have become extinct, thus making reintroductions a conservation priority.

Radio-tagging can have value in reintroductions beyond the recording of immediate survival, by virtue of its ability to record many other parameters of individual performance. Parameters of individual performance (PIPs), such as range size and structure, habitat use or sociality, can be estimated by the systematic intermittent sampling of only 30-50 radio-locations (Kenward, in press). In theory at least, PIPs which predict dispersal, death or breeding failure (Predictors of Impending Peril!) could be used for rapid assessment of an area's suitability for re-stocking, by releasing very few radio-tagged animals. For example, if animals in a healthy population prove unsuccessful when they have large range cores, the recording of large range cores for all the individuals released at a new site would cast early doubts on its suitability. PIPs also have a future role in the production of sophisticated bottom-up population models, for instance by linking survival to the use of particular habitats (Aebischer *et al.*, this volume).

3. Problems with radio-tagging

3.1. Censoring

Estimating survival with radio-tags is complicated by unexplained loss of tags, through undetected emigration or component failure. If loss of signal occurs without prejudice to the survival of the animal, it is reasonable in interval-based survival estimates (*eg* Trent and Rongstad, 1974, Bunck and Pollock, this volume) to censor that individual from the interval in which it was lost and from subsequent intervals, but to include it as a survivor in previous intervals. However, if impacts or predators sometimes destroy the tag as well as the tagged animal, then the survival rate is over-estimated by the censoring.

There is a way round the censoring problem if there are (i) resightings and recoveries of animals from which signals were lost, and (ii) if tags that were known to fail without prejudice to survival can be separated from tags lost inexplicably. Among juvenile goshawks, for example, the former category included 70 birds which moulted their tags, or for which signal changes indicated that tags were failing, and 17 (24%) of these were later recaptured or found dead. There were also 68 tags that were lost without explanation, 16 (24%) of these were recorded again well after the loss. There was thus no evidence of a reduced recovery rate for hawks that were lost inexplicably: they must have been lost mainly because of undetected failures, and not because hawks and their tags were destroyed at the same time. The observed survival rates (with censoring of all tag losses) could therefore be considered representative of the tagged cohort. The function $(1-d_i/r_i)$ could be used to estimate survival during each interval i, in which d_i deaths were recorded from r_i birds at risk.

If, on the other hand, some of u_i birds lost without explanation during interval i had been destroyed with their tags, the true survival would have been less than $(1-d_i/r_i)$. Unless all u_i tags had been destroyed with their wearers, the survival would also have been greater than $(1 - (d_i + u_i)/(r_i + u_i))$. These two functions represent maximum and minimum estimates of the true survival. Given the destruction of some tags with their wearers, such that the recovery rate for u animals with unexplained tag losses (f_u) is less than the recovery rate (f_e) for e animals with explained tag losses, then the quantity $\zeta = (1-f_u/f_e)$ can be used to estimate a corrected survival by interpolation as

$$s^*_i = \exp(\ln s_{max} - \zeta.(\ln s_{max} - \ln s_{min})) = (1 - d_i/r_i)(r_i/(r_i + u_i))^{\zeta}$$

Confidence limits for this estimate are derived by N.J. Aebischer in the Appendix. Note that $\zeta = 1$ if no animals are recaptured after unexplained tag loss, which implies that all were destroyed with their tags ($f_u = 0$). Only sampling error should cause f_u to exceed f_e, in which case ζ can be set to 0 and survivorship is estimated by $(1-d_i/r_i)$.

3.2. Tag performance

Tag reliability and tracking techniques improved during the goshawk study, such that signals were lost without explanation from only 6 of 40 tags on juveniles in the last year of the study, compared with 21 of 31 at the start. As a result of further improvements in the life, range and reliability of tags, only 4 of 51 long-life tags have been lost on buzzards, with 3 of the birds known to have emigrated from the study area and 1 lost during a conference at the peak of the dispersal period. Only 2 radios are known to be faulty; nevertheless, their intermittent signals have provided survival data for up to 20 months. High tag reliability will reduce the need to correct for censoring in future studies of raptor demography. However, increased tag reliability will also reduce the accuracy of f_e, by reducing the number of explained signal losses. Censoring will therefore remain a problem if tags face frequent destruction with their wearers, for example on species vulnerable to mammal predators.

The life of tags with primary cells has also improved greatly during the last decade. The goshawk tags weighed 15 g, were detectable at up to 20 km and transmitted for 9-12 months. The latest buzzard tags weigh 25-30 g, are detectable at 40 km and last 4 years. Long tag life not only provides data to an age at which survival rates are probably fairly constant but also records age of first-breeding, and breeders can be trapped at nests to change their tags. Although radio tags do not last as long as rings, large birds can already be radio-tagged long enough to provide age-specific data on survival and productivity well into adulthood, and further improvements are likely.

3.3. Impact of tagging

Unfortunately, long-life radio tags can only be attached to nestling raptors as back-packs, and this technique itself has been shown to bias behaviour and fitness in several studies (review by White and Garrott, 1990). Adverse effects have not been reported for tail-mounted tags, but these tags can only be attached when tail feathers are full-grown; birds must therefore be caught after fledging, which is laborious and risks capture bias. Of 187 surviving nestling goshawks, 18 were not captured to attach tail-mount tags. In contrast, despite greater experience and effort, only 17 of 34 buzzards could be trapped post-fledging (χ^2=32.2, df=1, P<0.001, S.S. Walls and R.E. Kenward, unpublished). Fortunately, the survival of buzzards with back-packs was the same as those with tail-mounts, with no evidence from activity records that back-packs hindered flight. However, variation of only 5% above or below the ideal size of harness loops was found to cause loss of feathers or even chafing of skin on the back. It is crucial that tags are attached correctly if they are not to affect fitness.

3.4. Economics

The cost of 30 tags a year is £2000-£3000, depending on type and source, with a minimal cost of £1,500 for receiving equipment at the start of the project. The cost of monitoring will depend on the quality of data required. However, if cause of death were not required in more detail than from ringing, 2 man-days per month would be adequate for goshawk and buzzard studies, apart from a 1 month dispersal period, when checks would be advisable twice a week. The total is about 35 man-days per year for a carefully trained tracker, assuming access to a light aircraft for up to 10 hours to re-locate any birds that disperse long distances.

A very rough estimate of the cost of comparable ringing data is the effort required for 30 recoveries. With an 8% recovery rate (Table 1), 375 raptors must be ringed for 30 recoveries. At best, a team of 2 can find and ring about 5 young each day, so ringing 375 would require 150 man-days. This compares with about 15 man-days for placing harnesses on nestlings, which need not take much longer than ringing, bringing the total to 50 man-days for the radio-tagging project. Even allowing for the cost of tags, aircraft hire and other travel during radio-tracking, the 100 days of additional labour would make ringing the more expensive approach, if it were not a voluntary effort.

4. A goshawk model from radio-tagging

The best estimate for the first-year mortality rate of radio-tagged male goshawks was 51%, significantly higher than the 34% estimated for females, and 21% for adults of both sexes. Using these values, and the productivity recorded for radio-tagged hawks, a simple model could be constructed for the goshawk population on Gotland (Kenward *et al.*, 1992). Although the model initially predicted a population decline, it stabilised if the adult mortality rate was reduced from 21% to 18%. This seems a reasonable correction, because adults had to be live-trapped for tagging, which biases towards marking the poorer hawks. The breeding population remained stable through the study, and ring recoveries showed that there was little emigration (*ca.* 2%). As shown in Figure 2, hawks never even attempted to breed in their first year, and half the adult females did not breed. This contrasts with depressed populations of goshawks in Germany, where up to 35% of breeding pairs have contained first-year hawks (Link, 1986).

Figure 2. The number of goshawks per 1000 km² on Gotland in spring, showing those in each age and sex class that bred successfully, nested but failed to rear young and made no attempt to breed.

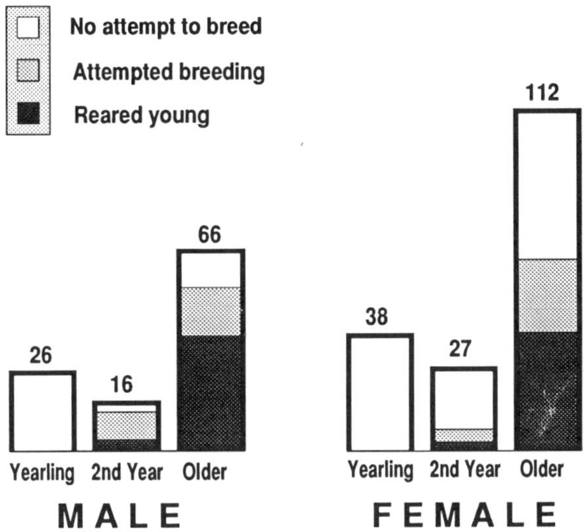

In buzzards too, the first-year mortality estimate in Dorset is about 30%, despite some being killed illegally (R.E. Kenward and S.S. Walls, unpublished). If mortality factors were additive, in both goshawks and buzzards the first-year mortality without human intervention would have been 20-30%, as in skuas and large gulls. Why, then, have estimates of 51-64% (Saurola, 1976; Tubbs, 1974) been produced by ringing these species?

Three possible explanations involve (i) ring loss, which was found both on Gotland and in German goshawks (Ziesemer, 1981), (ii) recovery bias, in which killed hawks provided an excess of juvenile recoveries, and (iii) genuinely higher first-year mortality rates in years when up to 90% of rings were returned from killed hawks, and when populations were affected by pollution. Further study should show which explanations are most likely for goshawks and buzzards, and whether similar problems have influenced estimates for other raptors. If so, then healthy populations of large raptors may have a much larger non-breeding component than was previously suspected, in which case the monitoring of breeding pairs may be a relatively insensitive way to detect some population problems.

5. Conclusion

Radio-tagging may prove a useful complement to ringing, by providing correction factors (Lakhani, 1987) or as an aid during density estimation by mark-resighting (White, this volume). However, radio-tracking can also provide the main method in studies of bird population dynamics, with ringing the complementary technique to provide a correction factor for censoring after radio signals have been lost. Radio-tagging can provide more detailed data more rapidly than ringing on mortality and movements, and may even be the cheaper technique for work on uncommon species, such as large raptors. Although ringing will remain an essential technique in most studies of birds, radio-tags have already been used to estimate parameters for simple population models of large raptors, and their importance will increase as a means of collecting data for bottom-up models to predict future changes in bird populations.

Acknowledgements. Data on goshawks were collected in collaboration with V. Marcström and M. Karlbom, with funding from the Swedish Sportsmens Association and a private foundation. Data on goshawks were gathered in collaboration with S.S. Walls, with support from Biotrack. Both projects were also funded by the Natural Environment Research Council. Early manuscripts were improved by comments from N.J. Aebischer, C.M. Bunck, J.D. Lebreton, M.G. Morris, I. Newton, S.S. Walls and an anonymous referee.

References

Cramp, S. and Simmons, K.E.L. (1983).- The birds of the Western Palaearctic, Volume 3. University Press, Oxford.

Furness, R.W. (1978).- Movements and mortality rates of great skuas ringed in Scotland. Bird Study, 25: 229-238.

Glue, D.E. (1971).- Ringing recovery circumstances of small birds of prey. Bird Study, 18: 137-146.

Goss-Custard, J.O. and Durell, S.E.A. le V. dit (1987).- Age-related effects in oystercatchers, *Haematopus ostralegus*, eating mussels, *Mytilus edulis*. III. The effect of interference on overall intake rate. J. Anim. Ecol., 56: 549-558.

Haukioja, E. and Haukioja, M. (1970).- Mortality rates of Finnish and Swedish goshawks *(Accipiter gentilis)*. Finn. Game Res., 31: 13-20.

Kenward, R.E. (1992).- Quantity verses quality: programming for collection and analysis of radio tag data. *In* Wildlife telemetry: remote monitoring and tracking of animals, Priede, I.G. and Swift, S.M. (eds), pp. 231-246. Ellis Horwood, Chichester.

Kenward, R.E., Marquiss, M. and Newton, I. (1981).- What happens to goshawks trained for falconry. J. Wildl. Manage., 45: 802-806.

Kenward, R.E., Karlbom, M. and Marcström, V. (1983).- The price of success in goshawk trapping. Raptor Res., 17: 84-91.

Kenward, R.E., Marcström, V. and Karlbom, V. (1992).- The goshawk as predator and renewable resource. Gibier Faune Sauvage.

Kenward, R.E., Marcström, V. and Karlbom, V. (in press).- Causes of death in radio-tagged northern goshawks. *In* The status of biomedical research on raptors, Cooper, J.E., Hahn, T.R. and Redig, P.T. (Eds.) Raptor Research Foundation, Minnesota.

Lakhani, K.H. (1987).- Efficient estimation of age-specific survival rates from ring recovery data of birds ringed as young, augmented by field information. J. Anim. Ecol., 56: 969-987.

Lakhani, K.H. and Newton, I. (1983).- Estimating age-specific bird survival rates from ring recoveries - can it be done? J. Anim. Ecol., 52: 83-92.

Link, H. (1986).- Untersuchungen am Habicht (*Accipiter gentilis*). Deutscher Falkenorden Schriftenreihe 2, Blomberg.

Marcström. V. and Kenward, R.E. (1981a).- Movements of wintering goshawks in Sweden. Swedish Wildl. Res., 12: 3-35.

Marcström. V. and Kenward, R.E. (1981b).- Sexual and seasonal variation in condition and survival of Swedish goshawks *Accipiter gentilis*. Ibis, 123: 311-327.

Mead, C.J. and Clark, J.A. (1989).- Report on Bird Ringing for 1988. Ringing Migration, 10: 158-196.

Mead, C.J. and Clark, J.A. (1990).- Report on Bird Ringing for 1989. Ringing Migration, 11: 137-176.

Mead, C.J. and Clark, J.A. (1991).- Report on Bird Ringing for Britain and Ireland for 1990. Ringing Migration, 12: 139-176.

Newton, I. (1979).- Population ecology of raptors. Poyser, Berkhamsted.

Pollock, K.H., Winterstein, S.R., Bunck, C.M. and Curtis, P.D. (1989).- Survival analysis in telemetry studies: the staggered entry design. J. Wildl. Manage., 53: 7-15.

Nordström, G. (1963).- Einige Ergebnisse der Vogelberingung in Finnland in der Jahren 1913-1962. Orn. Fenn., 40: 81-124.

Saurola, P. (1976).- Kanahaukan kuolevuus ja kuolinsyyt. Suom. Luonto, 6: 310-314.

Sherrod, S.K., Heinrich, W.R., Burnham, W.A., Barclay, J.H. and Cade, T.J. (1981).- Hacking: a method for releasing peregrine falcons and other birds of prey. The Peregrine Fund, Ithaca.

Trent, T.T. and Rongstad, O.J. (1974).- Home range and survival of cottontail rabbits in southwestern Wisconsin. J. Wildl. Manage., 38: 459-472.

Tubbs, C.R. (1974).- The buzzard. David and Charles, London.

White, G.C. and Garrott, R.A. (1990).- Analysis of wildlife radio-tracking data. Academic Press, New York.

Ziesemer, F. (1981).- Habichte (*Accipiter gentilis*) verlieren Ringe. Corax, 8: 211-212.

APPENDIX

Using recoveries of censored animals to correct survival rates estimated from radio-tagging.

Nicholas J. Aebischer

The Game Conservancy, Fordingbridge, Hants SP6 1EF, Great Britain

Following the notation in section 3.1, the corrected interval-based survival rate s^*_i must lie, for $0 \le \zeta \le 1$, between the original, minimum survival estimate $s_i=1-d_i/r_i$ and the maximum estimate $1-(d_i+u_i)/(r_i+u_i)$. Hence, by geometric interpolation:

$$s^*_i = s_i \alpha_i^\zeta, \quad \text{where} \quad \alpha_i = r_i/(r_i+u_i)$$

Because tag losses are usually too few in any given interval to give workable estimates of ζ, the recovery rates f_u and f_e are calculated from the total losses $u=\Sigma u_i$ and $e=\Sigma e_i$. This requires either that sufficient time has elapsed after the radio-tracking for all recoveries to have been received, or that recoveries from the losses e_i and u_i in each interval i were recorded during an equal period after each interval.

The asymptotic variance of s^*_i is estimated by Taylor series linearisation, observing that s^*_i is a function of four parameters whose variances and covariances are given by the binomial or multinomial: $\text{Var}(s_i)=s_i(1-s_i)/r_i$, $\text{Var}(\alpha_i)=\alpha_i(1-\alpha_i)/(r_i+u_i)$, $\text{Var}(f_u)=f_u(1-f_u)/u$, $\text{Var}(f_e)=f_e(1-f_e)/e$, $\text{Cov}(s_i,\alpha_i)=\text{Cov}(s_i,f_u)=\text{Cov}(s_i,f_e)=\text{Cov}(\alpha_i,f_e)=\text{Cov}(f_u,f_e)=0$ and $\text{Cov}(\alpha_i,f_u)=\alpha_i(1-\alpha_i)f_u/u$. Then:

$$\text{Var}(s_i^c) = (s_i^c)^2[\text{Var}(s_i)/s_i^2+\zeta^2\text{Var}(\alpha_i)/\alpha_i^2+(1-\zeta)^2(\ln \alpha_i)^2(\text{Var}(f_u)/f_u^2+\text{Var}(f_e)/f_e^2)- 2\zeta(1-\zeta)(\ln \alpha_i)\text{Cov}(\alpha_i,f_u)/(\alpha_i f_u)].$$

The correction extends naturally to the overall survival rate s given by the product of the interval-based survival rates s_i. Let s^* be the corrected overall survival rate:

$$s^* = \Pi s^*_i = s\alpha^\zeta, \quad \text{with} \quad s = \Pi s_i \quad \text{and} \quad \alpha = \Pi\alpha_i$$

From Pollock *et al.* (1989), Var(s) may be estimated as $s^2\Sigma d_i/(r_i(r_i-d_i))$. The asymptotic variance of s^* is then estimated by:

$$\text{Var}(s^*) = (s^*)^2[\text{Var}(s)/s^2+\zeta^2(\Sigma\text{Var}(\alpha_i)/\alpha_i^2)+(1-\zeta)^2(\ln \alpha)^2(\text{Var}(f_u)/f_u^2+\text{Var}(f_e)/f_e^2)- 2\zeta(1-\zeta)(\ln \alpha)(\Sigma\text{Cov}(\alpha_i,f_u)/(\alpha_i f_u))].$$

Approximate 95% confidence limits of the corrected survival rates are given by:

$$s^*_i \pm 1.96[\text{Var}(s^*_i)]^{1/2} \quad \text{and} \quad s^* \pm 1.96[\text{Var}(s^*)]^{1/2}$$

Marked Individuals in the
Study of Bird Population
J.-D. Lebreton and P. M. North (eds.)

A Comparison of Survival Rate Estimates from Live Recaptures and Dead Recoveries of Lesser Snow Geese

Charles M. Francis[1] and Fred Cooke[2]

[1]*Center for Tropical Conservation, Duke University, Durham, NC 27705, USA*
[2]*Biology Department, Queen's University, Kingston, Ontario, Canada K7L 3N6*

Summary. Sound statistical methods are available for analysis of demographic parameters based on data from either live recaptures or dead recoveries. We evaluated these models using 12000 recaptures and 10000 recoveries from about 80000 marked Lesser Snow Geese, *Anser c. caerulescens* to determine their suitability for estimating survival rates of these geese. Many of the resultant estimates of apparent survival rates based on recaptures differed significantly from those based on recoveries. Apparent survival rates estimated from recaptures were generally lower than those from recoveries, especially for males. Analyses of recovery data indicated a long-term increase in adult survival rates, while recapture analyses suggested a decline. Additional data indicate that apparent survival estimates from recaptures underestimated true survival rates due to emigration from the study area. The estimates were also biased by heterogeneity in site fidelity and capture probabilities. In contrast, recovery estimates were relatively unbiased (with some exceptions). These results highlight the need for caution when interpreting recapture data in relation to survival, especially when additional data (such as recoveries) are not available to measure emigration. Particular care must be taken in expanding or contracting populations to avoid confounding changes in site fidelity or sampling probabilities, with true changes in survival rates.

1. Introduction

Estimating survival rates of wild populations of birds (or mammals) generally involves marking a large number of animals, and then subsequently resampling them, either through recovering tags when the animals die or through regularly recapturing or resighting animals that are still alive. Sound statistical methods (with associated computer programs) have been developed for analysis of survival rates from either type of data (e.g. live recaptures: Pollock et al., 1990, Lebreton et al., 1992; dead recoveries: Brownie et al., 1985). However, a critical assumption of all of these methods is that all individuals (at least within identifiable classes) are equally likely to be re-encountered. Because of differences in sampling properties, this assumption may be violated in different ways for live recaptures and dead recoveries.

Live recaptures (or resightings) are usually arranged by the researcher at or near the original marking site. As a result, emigration is largely confounded with mortality, because animals that leave the study area are no longer available for recapture. Capture probabilities may also vary among individuals if, for example, individuals differ in the proportion of their time spent within the study area (thus affecting capture rates) or if the sampling effort is not uniform over the study area.

In contrast, dead recoveries are usually reported by the general public (mostly hunters in the case of game birds), and are not restricted to the original study area. As a result, they are less likely to be affected by emigration, although there may be some bias due to regional variation in the reporting rate (the probability that a dead bird will be found and reported). Also, the reporting rate varies with cause of death, which must be considered in the choice of analysis model (e.g. birds that are shot are much more likely to be retrieved and their ring reported than birds that die of natural causes).

The consequences of these sampling characteristics for estimating survival rates can be evaluated by comparing the results of analyses based upon both recaptures and recoveries for the same population. Data from a long-term study of Lesser Snow Geese, *Anser caerulescens caerulescens*, nesting in northern Manitoba, Canada, provide an ideal opportunity for this comparison. Over 80000 geese have been ringed, with about 12000 recaptures available from the breeding colony, and about 10000 recoveries from hunters along the migration routes and wintering areas. Data are available from a 20-year period, allowing comparisons of estimates of annual fluctuations and long-term trends in addition to mean survival rates. Temporal changes are of particular interest as the estimated population size has increased from just over 2000 pairs in 1970, to about 10000 pairs in 1988 (Cooch et al., 1991).

2. Methods

The population of Lesser Snow Geese nesting around La Pérouse Bay near Churchill, Manitoba has been under study by Dr. Fred Cooke and co-workers since 1968. Details of the field methods were described by Finney and Cooke (1978) and Rockwell et al. (1983), and only methods relevant to survival analysis are described here. In late July or early August, when the adults were moulting their flight feathers, and before the young could fly, geese were rounded up and captured for ringing. Mean annual ringing totals averaged 3200 geese from 1970 to 1978, and increased to 5300 from 1979 to 1989 due to increased effort. Geese were classified as adults (at least 2 years old), yearlings or goslings, based upon plumage, and their sex was determined by cloacal eversion. All new birds were marked with numbered aluminium rings, and most birds also received one or more coloured plastic leg rings. The ring numbers of already marked birds were recorded, and any worn or damaged rings were replaced. Most geese were captured in their foraging areas on the salt flats around La Pérouse Bay itself. Since the early 1980's, increasing numbers of geese have been observed foraging in areas 10-20 km away from the bay (Cooch et al., 1992). Starting in 1985, additional geese were captured in these areas, but they have been excluded from the present recapture analyses.

Recoveries were obtained through hunters who had shot marked geese and reported their ring numbers to the wildlife service. Some geese were recovered that had died of other causes, but these constituted less than 5% of dead recoveries, so we excluded them to allow interpretation of recovery rates in relation to hunting pressure.

Recapture data were analysed based on the Cormack-Jolly-Seber models and derived models (Clobert et al., 1987, Lebreton et al., 1992), using the computer programs SURGE 4.1 (Lebreton et al., 1992), RELEASE 2.4 (Burnham et al., 1987), and JOLLYAGE (Pollock et al., 1990). Recovery data were analysed with the models of Brownie et al. (1985), using their computer programs BROWNIE and ESTIMATE, as well as program MULT (Conroy et al., 1989). Terminology generally follows those papers. Survival rate is the proportion of birds alive at the time of ringing in one year that were still alive at the same time the following year. The term 'apparent survival rate' is used where the model estimate is likely to be biased by emigration or violation of model assumptions. Capture rate is the proportion of birds alive at the time of ringing one year that were captured that year. Recovery rate is the proportion of birds alive at the time of ringing in one year that were shot and reported during the following hunting season.

Model selection was based on the principles of choosing the model with the fewest parameters that adequately fits the data, as judged from goodness of fit tests and likelihood ratio tests. In all cases, models were used that allowed survival rates and recovery or capture rates to vary with time and age.

For recovery analyses, these rates were assumed to differ only between immatures and adults. Yearlings have significantly higher recovery rates and/or lower survival rates than older geese (Francis et al., 1992b), but too few geese were captured as yearlings at La Pérouse Bay to obtain explicit estimates of the differences using three-age-class models. However, the two-age-class models were not rejected by standard goodness-of-fit tests (Brownie et al., 1985), and simulations indicate that the differences between yearlings and older geese were too small to affect any of our conclusions about patterns of variation in survival and recovery rates (Francis et al., 1992a). We also used models assuming that survival and/or recovery rates of newly ringed adults did not differ from those of adults ringed in previous years, despite evidence that there were some differences (Francis et al., 1992b). Again, models allowing such variation can not provide explicit estimates of any relevant parameters except recovery rates of newly ringed birds (Brownie et al., 1985:34). Fortunately, the differences between newly ringed and previously ringed birds were sufficiently small that models ignoring this variation were not rejected by goodness-of-fit tests. Also, survival estimates based on models allowing recovery rates (but not survival rates) to differ for newly ringed birds (models M0 or H2), showed essentially identical trends.

For recapture analyses of geese ringed as adults the standard Cormack-Jolly-Seber model was strongly rejected by the goodness-of-fit tests in Burnham et al. (1987), largely because newly marked birds were much less likely to be recaptured than birds marked in previous years. As a result, we used models providing separate estimates of apparent survival rates for newly ringed and previously ringed birds. For females, though not for males, the best-fitting models also allowed capture probabilities to differ in the first year after capture and in subsequent years. For estimating apparent survival rates of birds ringed as goslings, models were required allowing different recovery rates for yearlings, two-year-olds, three-year-olds and older birds due to differences in age-specific breeding propensities (Cooke et al., 1984). However, survival rates were assumed to differ only between goslings and older birds because too few birds were captured as yearlings to estimate their survival separately. As a result, this model may underestimate first-year survival if yearlings have lower survival than adults.

Recovery models were based on birds ringed from 1970 through 1988, including all recoveries reported before September 1989. Recapture models were based on birds ringed or recaptured from 1970 through 1990. Mean survival rates over the duration of the study period were estimated from models allowing survival rates to vary from year to year. Standard errors of the mean were calculated from the variances and covariances of the individual estimates using the equations of Brownie et al. (1985:19).

3. Results and discussion

3.1 Newly captured versus previously captured geese

For geese ringed as adults, estimated apparent survival rates were significantly lower during the first year after ringing than in subsequent years (females: 57% $\pm$ 2% versus 76% $\pm$ 2%; males: 38% $\pm$ 2% versus 63% $\pm$ 2%; for both sexes, $\underline{P} < 0.0001$ for likelihood ratio tests comparing these models with models assuming survival rates were the same). These differences could be due to heterogeneity in survival rates, site fidelity, or capture probability, or changes in any of these parameters after the first capture.

Survival differences might occur if, for example, first time breeders (which are represented only among newly ringed birds) have lower survival than experienced breeders. This can be tested by comparison with recovery data, which are less affected by site fidelity or capture probabilities. Analyses of recoveries also indicated that newly ringed geese were significantly more likely to be recovered and/or had lower survival rates than geese that had survived at least one year (Francis et al., 1992b). Unfortunately, as already mentioned, explicit estimates of both recovery and survival rates cannot be obtained in this case using only recovery data (Brownie et al., 1985:34). However, the differences could be explained by a mean difference in recovery rates of less than 1%, with no more than 2-3% difference in survival rates (Francis et al., 1992b). This difference is much smaller than that observed for recaptures. Analyses of apparent survival rates based on recaptures of birds ringed as goslings also failed to find any significant difference between first-time breeders and older breeders (Viallefont et al., unpublished manuscript).

This indicates that the differences between newly ringed and older birds were due mainly to other factors such as heterogeneity in site fidelity or capture probabilities, perhaps because some birds are transients, or have different movement patterns than others. Both of these processes bias downwards estimates of apparent survival for newly ringed birds because birds with low fidelity or capture probabilities are rarely recaptured and appear to have died (Buckland, 1982). These possibilities will be discussed further in a separate section below.

3.2 Mean adult survival rates

Apparent survival rates estimated from recaptures (excluding estimates for newly ringed birds) were lower for males than for females (Table 1). However, recovery estimates of male survival rates were actually slightly higher than those for females, although the difference was not significant (Table 1). These sexual differences in apparent survival rates estimated from recaptures were due mainly to emigration of males away from the breeding colony (Cooke and Sulzbach, 1978). Snow geese select their mates on the wintering grounds in the southern United States, and males return to the natal breeding colony of their female (Cooke et al., 1975). Most geese mate for life, so males continue to return to the colony as long as their mate is alive. However, if the female dies, the male selects a new mate, and returns to her natal colony. Because the La Pérouse Bay colony is small relative to other colonies around Hudson Bay, most males from this colony are likely to return elsewhere with their new mates, thus disappearing from the population (Cooke et al., 1975). If deaths of males and females are independent, assuming no emigration, the expected return rate of males, as measured by recaptures, should be roughly equal to the product of the true male and female survival rates. Assuming recovery data provide unbiased estimates of true survival, this should be 0.81 x 0.82 = 0.66, which is close to the observed value of 0.64 (Table 1). In practice, however, this model is rather simplified for several reasons. Male and female deaths may not be independent, as pairs remain together on migration (making them potentially vulnerable to being shot together). Some males may select new mates while their female is still alive (Cooke and Sulzbach, 1978), while others may select new mates that are also from La Pérouse Bay. Finally, there may be some emigration of pairs away from the bay, as suggested by the lower apparent survival estimates for females based on recaptures compared with recoveries. These mean numbers must also be treated rather cautiously in view of the opposite long term changes in apparent survival estimates based on both recaptures and recoveries (see below). Proper statistical testing and measurement of these differences, requires joint analysis of recapture and recovery data using models such as that proposed by Burnham (this volume).

Table 1. Mean apparent survival rates (%) estimated from recaptures and recoveries for Lesser Snow Geese ringed at La Pérouse Bay, Manitoba from 1970-1989.

		Recaptures		Recoveries	
		Mean	SE	Mean	SE
Adults					
	Females	77.1	0.6	80.7	1.2
	Males	66.0	3.0	82.5	0.8
Immatures					
	Females	37.0	1.4	42.3	1.9
	Males	-	-	34.5	1.5

3.3 Mean immature survival rates

Mean apparent survival rates of immature females estimated from recaptures were slightly, but significantly, lower than those estimated from recoveries (Table 1). Too few immature males return to the colony to estimate any sort of apparent survival rate from recaptures. Recovery estimates of first-year survival rates are significantly lower for males than for females (Table 1). However, there is good reason to believe that survival rate estimates of males from recovery models are biased by emigration as well (Francis and Cooke, 1992a). A critical assumption of the recovery models is that recovery rates of birds ringed as young, once they become adults, are the same as those originally ringed as adults (Brownie et al., 1985). Virtually all young males emigrate to new colonies, whereas only those adults that lost their mates switch colonies. Because birds from different colonies follow migration routes with differing hunting pressure and recovery rates (Francis and Cooke, 1992b), the assumption of similar recovery rates is violated. Analyses of the number and distribution of recoveries during the first hunting season, as well as survival analyses based on subsets of the recovery data from common wintering areas suggest that true survival rates of immature males and females are similar (Francis and Cooke, 1992a).

3.4 Temporal changes in adult survival rates

We tested for long-term changes in survival rate by using likelihood ratio tests to compare models allowing survival to change linearly with time with models assuming constant survival over time. For recovery data, such comparisons indicated a highly significant long-term increase in adult survival rates (Figure 1; $X^2 = 13.3$, df = 1, $P = 0.0003$). The increase corresponds to a change in mortality from about 22% in 1970, to only 12% in 1987 (Francis et al., 1992a). In contrast, apparent survival rates estimated from recaptures showed a long-term decline (Figure 2). This decline is highly significant for the first year after ringing ($X^2 = 22.9$, df = 1, $P < 0.00001$), but not significant for apparent survival estimates in subsequent years ($X^2 = 1.03$, df = 1, NS). For recapture data, a model allowing apparent survival to vary from year to

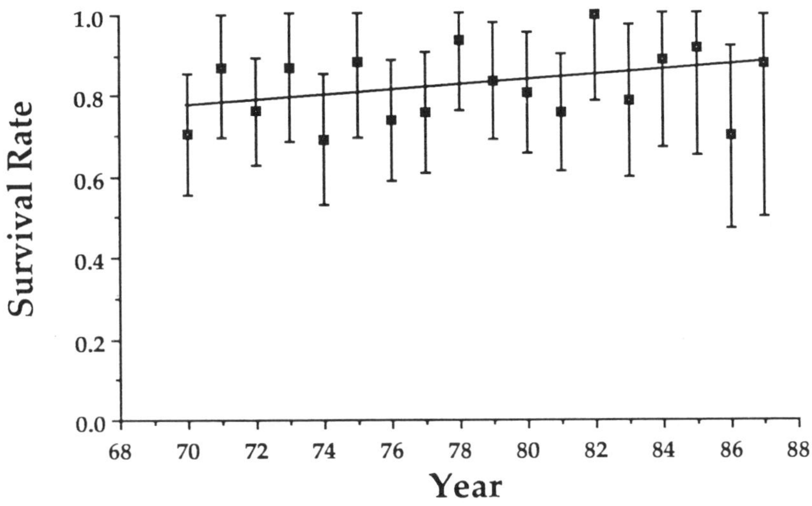

Figure 1. Estimated survival rates from recovery data for adult Lesser Snow Geese ringed at La Pérouse Bay between 1970 and 1987 (after Francis et al. 1992a). The long-term increase in survival is highly significant. Vertical lines indicate 95% confidence limits.

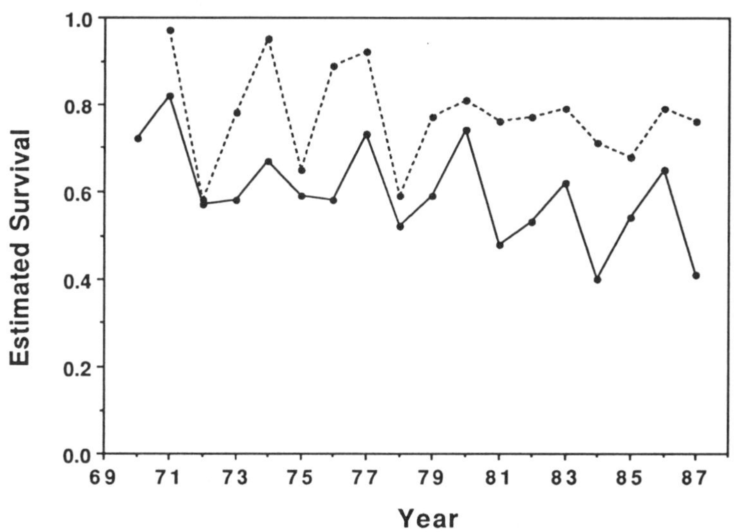

Figure 2. Estimated apparent survival rates from recapture data for adult female Lesser Snow Geese ringed at La Pérouse Bay between 1970 and 1987 during the first year after ringing (solid line) and in subsequent years (dashed line).

year was a significantly better fit than the regression model, indicating substantial additional variation in apparent survival not explained by a simple long term decline. These discrepancies between the results of recapture and recovery analyses could be due to emigration, or to biases in either method. Additional information is required to determine which estimates most likely reflect real changes in survival rates.

Increasing adult survival rates are consistent with evidence of declining hunting pressure on the population (Francis et al., 1992a). Recovery rates have declined substantially over the study period, and independent data from hunter surveys, permit sales, and population counts, indicate this is due mainly to a decline in the proportion of geese being shot, rather than any change in reporting rate (Francis et al., 1992a). Given that hunting is the major cause of mortality for adult geese, an increase in adult survival rates would be expected from a reduction in hunting. If so, this suggests that the decline in apparent survival rates, as measured by recaptures, is due to changes in other factors such as the degree of heterogeneity of site fidelity or capture probabilities, or overall increases in emigration rates.

The possibility that heterogeneous capture rates are affecting estimated recapture survival rates seems plausible based on the biology of the geese. All geese used in this analysis were captured on the salt flats within La Pérouse Bay, where traditionally most geese foraged with their broods while they were moulting. Recently, an increasing number of geese have been foraging away from the bay. If some geese foraging in new areas occasionally choose to forage within the bay, they may be captured but less frequently than more philopatric geese that forage within the bay every year. Similar variation in capture probability might be expected from observations at another colony that some geese spend most of the summer foraging within a limited area, while others travel extensively, foraging in many different areas (J. Hughes, unpublished data). Because La Pérouse Bay has a limited capacity for foraging geese, the long-term increase in population size would probably result in an increasing proportion of geese with low capture probabilities because they spend relatively little time in the bay.

Similarly, heterogeneity in site fidelity might be expected if some individuals, perhaps those that have returned to a site for many years, have very high site fidelity regardless of environmental conditions (Cooch et al., 1992), while others, perhaps those using the area for the first time, will emigrate from the area if conditions are poor. An increase in the latter component would be expected from the long-term decline in foraging conditions within La Pérouse Bay due to over-grazing by the geese (Cooch et al., 1989; Cooch et al., 1992). Trauma from the first capture could also cause some individuals to be less likely to return in subsequent years.

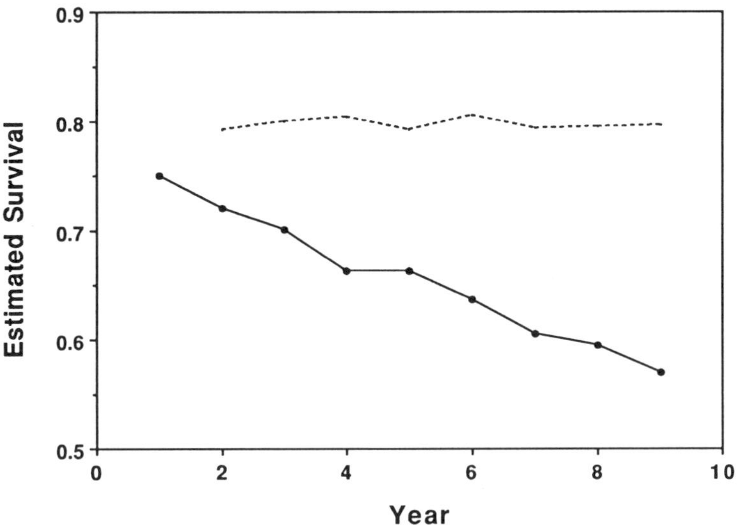

Figure 3. Estimated survival rates from simulated recapture data, showing the influence of increasing heterogeneity of capture probabilities on estimated survival during the first year after ringing (solid line) and in subsequent years (dashed line). Note that true survival in all years was 80%.

3.5 Simulations of heterogeneity in capture probabilities

To estimate the impact heterogeneity could have on survival rate estimates, we developed a simple simulation assuming that capture probabilites are high for some geese (0.3), and low for others (0.1). We further assumed that the proportions of geese in each category were equal initially, but that the number of geese with low capture probabilities increased three-fold over a 10-year period. The initial population size was set at 50000, to minimize the impact of rounding errors, and the expected distribution of capture histories was generated based on a constant survival rate of 0.8.

Apparent survival rates estimated from recapture models for this data set differed between newly ringed birds, and birds surviving from previous years (Figure 3). Estimates for newly ringed birds appeared to decline over time, while estimates for subsequent years were relatively unbiased. Additional simulations (not shown here), using other survival rates or capture probabilities produced similar patterns.

Mathematically, constant heterogeneity in capture probabilities can not be distinguished from heterogeneous changes in capture probabilities after the first capture. Heterogeneity in site fidelity could be modelled as changes in capture probabilities for some birds after their first capture. The case of permanent emigration can be considered as the limiting case where capture probabilities become zero. In this case, apparent survival rates after the first year should be completely unbiased, as all birds with low fidelity have disappeared. However, as we have just seen (Figure 3), apparent survival rates from recapture models after excluding the first capture are essentially unbiased whenever capture probabilities of some birds are much lower than those of others, even if they are greater than zero (see also Pradel, this volume). Thus distinguishing between hypotheses requires additional data.

For the snow geese, some additional data are available from recent ringing efforts in areas away from the bay. We have not yet tried to use models such as those of Hestbeck et al. (1991) to estimate actual exchange rates, but preliminary analyses show that all movements are unidirectional. Of 10784 birds ringed in the bay between 1985 and 1991, 69 were captured in subsequent years outside the bay, while of 4673 birds captured outside the bay during the same period, none was ever recaptured in the bay. These data indicate that heterogeneity of emigration probabilities is a primary factor explaining the patterns of apparent survival estimated from recaptures, but further analyses are still needed to measure the possible role of other factors.

An important result of the simulations is that even in the case of heterogeneous capture probabilities, estimated survival rates based on birds captured at least twice were essentially unbiased. This suggests that the indications of an apparent decline (or at least lack of an increase) in recapture survival rate estimates for previously ringed birds (Figure 2) may be due to changes in site fidelity even for geese that traditionally foraged within the bay.

3.6 Temporal changes in immature survival rates

In contrast to measures of adult survival, annual estimates of immature female survival rates are remarkably consistent from both recaptures and recoveries (Figure 4). Although, as already noted, estimates from recaptures average slightly lower than those from recoveries, the annual fluctuations are nearly identical from the two data sets. Considering the discrepancies in estimates of adult survival, it initially seems surprising that similar bias is not apparent in the estimated immature survival rates. However, if site fidelity is influenced by brood-rearing success, then high return rates of surviving young females might be predicted as, by definition, all were reared in successful areas. Furthermore, slight bias in the adult estimates due, for example, to reduced fidelity in future years, would have little influence on the estimated immature survival rates.

Given that the recapture and recovery estimates are largely independent (as they are based on different samples), the observed large fluctuations in survival rates clearly cannot be a sampling artifact, and must represent real variation in survival rates. Much of this variation is correlated with the timing of the breeding season, and with gosling growth rates (Francis et al., 1992a, Cooke and Francis this volume). In years when breeding is delayed by poor weather, goslings grow more slowly (Cooch et al., 1991), and may not be in suitable condition to fly south at the end of the summer. The long-term decline in survival is probably related to deterioration of the habitat around the colony leading to slow growth rates and goslings in poor condition (Cooch et al., 1991).

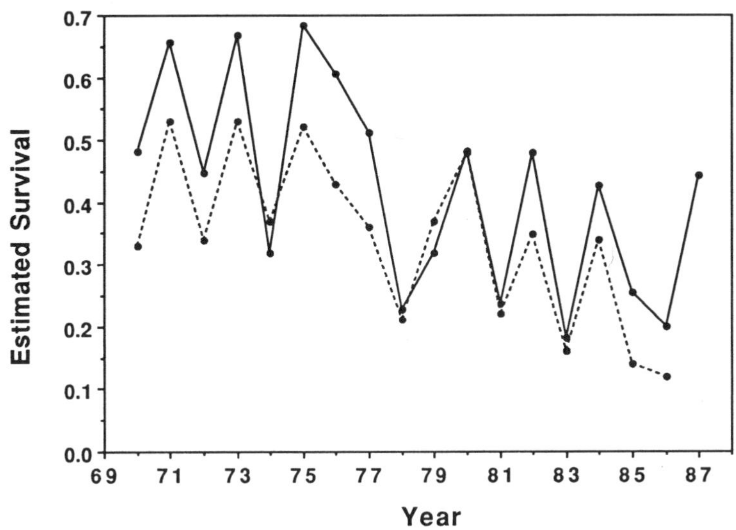

Figure 4. Apparent first-year survival rates of immature female Lesser Snow Geese ringed at La Pérouse Bay between 1970 and 1987 estimated from recaptures (dashed line) and recoveries (solid line).

4. General Discussion and Conclusions

As with all statistical analysis, valid inference from mark-recapture or recovery models for estimating survival rates requires that the model assumptions be met. In this paper, through comparisons between models and goodness of fit tests, we demonstrate that violation of the assumption of homogeneous capture and recovery probabilities can cause significant bias in survival estimates. We also show that differences in emigration rates or site fidelity can affect apparent survival estimates, particularly for comparisons between the sexes, or interpretation of long term trends.

Few previous studies have compared survival rate estimates from separate analyses of recaptures and recoveries. Herrmann and Holz (1988) found lower survival estimates based upon recoveries than recaptures for Ringed Plovers, *Charadrius hiaticula*. However, their recovery estimates were not based upon the models of Brownie et al. (1985), and may have been strongly biased by ring loss (Herrmann and Holz 1988). Also, the recoveries did not come from the same sample of ringed birds as the recaptures. Perrins (1971) obtained lower survival estimates based upon recoveries than recaptures for Swifts, *Apus apus*, attributing the difference to restriction of the latter sample to breeding adults which may have higher survival.

Again, however, the estimates were not based upon modern statistical methods, and various biases may have affected the estimates. In contrast, survival estimates for common British passerines based upon recaptures (Peach, this volume) and recoveries (Baillie, this volume) were quite similar, although the latter were based upon a much larger sample of ringed birds than the recapture analyses.

Problems of heterogeneous capture probabilities are likely to be present in many studies. In studies of territorial birds, individuals whose territories are contained wholly within a particular study area may have higher capture probabilities than birds living at the edge of the study area. Furthermore, even birds with territories outside of the study area may be captured occasionally if, for example, they had left their territory in search of extra pair copulations (e.g. Gibbs et al., 1990). In many studies of hole-nesting species, individuals nesting in boxes are more likely to be captured than individuals using natural cavities (e.g. Blondel et al., 1992). If some individuals consistently prefer one type of nest site, or if birds tend to return to the same site in subsequent years, this will result in heterogeneity of capture probabilities. As discussed above, such heterogeneity cannot easily be distinguished from changes in capture rate after first capture, or the presence of transients that remain only briefly in the study area (Pradel, this volume).

The presence of heterogeneity can usually be detected by goodness-of-fit tests (Burnham et al., 1987), and may be reduced or even eliminated by omitting the first capture event for each bird in the analysis. Unfortunately, this results in substantial loss of data, and may hide real differences in survival or emigration rates between, for example, first-time breeders (which may all be included in the newly-ringed category) and more experienced birds. As suggested by Peach (this volume), the problem can be partially solved with a study design analogous to the robust design of Pollock (1982), whereby multiple trapping sessions take place within a season. Only individuals captured more than once within a season are retained for survival analysis, thereby removing most transients and individuals with low capture probabilities or permanent trap response, while still allowing for tests of differential survival or emigration of birds in their first year after ringing.

Even when the model assumptions are met, mark-recapture analyses are generally unable to differentiate mortality from emigration out of the study population, unless additional data are available from other study sites. Nevertheless, many studies based upon mark-recapture analyses tend to assume that emigration is negligible, and interpret their conclusions in terms of survival rates. For analyses comparing sexes, this assumption is very unlikely to be justifiable, as most animals are likely to show some sexual differences in dispersal or philopatry (Greenwood, 1980). Comparisons among species sampled at different times of year, or living in different environments must also be considered rather cautiously due to potential differences in emigration (e.g. Karr et al., 1990). Misleading conclusions may be reached even when emigration rates are low. As we have shown for female snow geese, changes in emigration rates over time completely obscured a long-term increase in true survival rates. Changing emigration rates are particularly likely in an expanding or contracting population--when changes in survival rate are of greatest interest (e.g. Croxall et al., 1990; Kanyamibwa et al., 1990). Even if all birds remain in the study area, it is important to realize that birds may permanently move to parts of the study area where they will not be captured, or where capture rates are much lower. Such movement will again be confounded with mortality.

Recovery data are generally assumed to be relatively unaffected by emigration, and Burnham (this volume) has proposed a model to measure emigration rates through simultaneous analysis of recapture and recovery data. However, as shown for the immature male snow geese, recovery analyses may still be biased by emigration if some birds move to areas with differing recovery rates. This effectively leads to heterogeneity of recovery probabilities, but unlike the case of recapture analyses, such emigration could lead either to over-estimating or under-estimating survival rates depending upon whether recovery rates are higher or lower in the new areas. The potential for such bias may be detectable if, for example, the geographic distribution of recoveries differs among age or sex classes. However, if some birds emigrate to areas where they are never recovered, such bias could be much harder to detect. For migratory birds ringed on the breeding grounds, some of the problems can be avoided by restricting analyses to recoveries in wintering areas where most populations are likely to mix. However, for many species with low recovery rates, this may lead to an unacceptable loss of data.

Overall, this study illustrates the value of multiple approaches and types of data for differentiating mortality from emigration, and detecting biases such as those due to heterogeneous site fidelity or capture probabilities. Unfortunately, for many species, recovery rates are too low to provide sufficient data for analysis, particularly in short-term studies. Where only mark-recapture (or resighting) data are feasible, using a robust design (Pollock 1982), collecting additional data away from the study area, or using supplemental methods such as radio-tracking should be considered to measure movements and avoid some of the problems discussed here.

Acknowledgments. Snow Goose research at La Pérouse Bay has been supported in part by the Natural Sciences and Engineering Research Council of Canada (NSERC), the Canadian Wildlife Service (CWS), the Manitoba Department of Renewable Resources, the Wildlife Management Institute, the Mississippi and Central Flyway Councils, Ducks Unlimited Canada, and Ducks Unlimited Incorporated. CMF was supported by an NSERC post-doctoral fellowship, the CWS university research support fund, and the Chapman Fund of the American Museum (Natural History). FC was on sabbatical at the Centre d'Ecologie Fonctionnelle et Evolutive, C.N.R.S. while preparing this contribution, and gratefully acknowledges assistance from many of their personnel. J.-D. Lebreton, G. C. White, J. D. Nichols, and J. Hines kindly provided computer programs, while Queen's University computer science department, the Duke University Zoology Department Morphometrics Laboratory and C.N.R.S. provided computing facilities.

References

Blondel, J., Pradel, R. & Lebreton, J.-D. (1992).- Low fecundity insular blue tits do not survive better as adults than high fecundity mainland ones. Journal of Animal Ecology, 61: 205-213.

Brownie, C., Anderson, D. R., Burnham, K. P. & Robson, D. S. (1985).- Statistical inference from band recovery data--a handbook. Second Edition. United States Fish and Wildlife Service Resource Publication 156.

Buckland, S. T. (1982).- A mark-recapture survival analysis. Journal of Animal Ecology, 36: 419-435.

Burnham, K. P., Anderson, D. R., White, G. C., Brownie, C. & Pollock, K. H. (1987).- Design and analysis methods for fish survival experiments based on release-recapture. American Fisheries Society Monograph 5.

Clobert, J., Lebreton, J.-D. & Allaine, D. (1987).- A general approach to survival rate estimation by recaptures or resightings of marked birds. Ardea, 75: 133-142.

Conroy, M. J., Hines, J. E. & Williams, B. K. (1989).- Procedures for the analysis of band-recovery data and user instructions for program MULT. United States Fish and Wildlife Service Resource Publication 175.

Cooch, E. G., Jefferies, R. L., Rockwell, R. F. & Cooke, F. (1992).- Environmental change and the cost of philopatry: an example in the lesser snow goose. Oecologia, in press.

Cooch, E. G., Lank, D. B., Rockwell, R. F. & Cooke, F. (1989).- Long term decline in fecundity in a snow goose population: evidence for density dependence? Journal of Animal Ecology, 58: 711-726.

Cooch, E. G., Lank, D. B., Rockwell, R. F. & Cooke, F. (1991).- Long term decline in body size in a snow goose population: evidence of environmental degradation? Journal of Animal Ecology, 60: 483-496.

Cooke, F., Findlay, C. S. & Rockwell, R. F. (1984).- Recruitment and the timing of reproduction in lesser snow geese (*Chen caerulescens caerulescens*). Auk, 101: 451-458.

Cooke, F., MacInnes, C. D. & Prevett, J. P. (1975).- Gene flow between breeding populations of Lesser Snow Geese. Auk, 92:493-510.

Cooke, F. & Sulzbach, D. S. (1978).- Mortality, emigration and separation of mated snow geese. Journal of Wildlife Management, 42: 271-280.

Croxall, J. P., Rothery, P., Pickering, S. P. C. & Prince, P. A. (1990).- Reproductive performance, recruitment and survival of wandering albatrosses *Diomedea exulans* at Bird Island, South Georgia. Journal of Animal Ecology, 59: 775-796.

Finney, G. H. & Cooke, F. (1978).- Reproductive habits in the snow goose: the influence of female age. Condor, 80: 147-158.

Francis, C. M. & Cooke, F. (1992a).- Sexual differences in survival and recovery rates of lesser snow geese. Journal of Wildlife Management, 56: 287-296.

Francis, C. M. & Cooke, F. (1992b).- Migration routes and recovery rates of lesser snow geese from southwestern Hudson Bay. Journal of Wildlife Management, 56: 279-286.

Francis, C. M., Richards, M. H., Cooke, F. & Rockwell, R. F. (1992a).- Long-term changes in survival rates of Lesser Snow Geese. Ecology, 73: 1346-1362.

Francis, C. M., Richards, M. H., Cooke, F. & Rockwell, R. F. (1992b).- Changes in survival rates of Lesser Snow Geese with age and breeding status. Auk in press.

Gibbs, H. L., Weatherhead, P. J., Boag, P. T., White, B. N., Tabak, L. M. & Hoysak, D. J. (1990).- Realized reproductive success of polygynous Red-winged Blackbirds revealed by DNA markers. Science, 250: 1394-1397.

Greenwood, P.J. (1980).- Mating systems, philopatry and dispersal in birds and mammals. Animal Behaviour 28: 1140-1162.

Hestbeck, J. B., Nichols, J. D. & Malecki, R. A. (1991). Estimates of movement and site fidelity using mark-resight data of wintering Canada Geese. Ecology, 72: 523-533.

Herrmann, C. & Holz, R. (1988).- On the mortality of the Ringed Plover, *Charadrius hiaticula*--a comparison of methods. Okologie der Vogel, 10: 13-26.

Karr, J. R., Nichols, J. D., Klimkiewicz, K. & Brawn, J. D. (1990).- Survival rates of tropical and temperate forest birds: will the dogma survive? American Naturalist, 136: 277-291.

Kanyamibwa, S., Schierer, A., Pradel, R. & Lebreton, J.-D. (1990).- Changes in adult survival rates in a western European population of the White Stork *Ciconia ciconia*. Ibis, 132: 27-35.

Lebreton, J.-D., Burnham, K. P., Clobert, J. & Anderson, D. R. (1992).- Modeling survival and testing biological hypotheses using marked animals: a unified approach with case studies. Ecological Monographs, 62: 67-118.

Perrins, C. M. (1971).- Age of first breeding and adult survival rates in the swift. Bird Study, 18: 61-70.

Pollock, K. H. (1982).- A capture-recapture design robust to unequal probability of capture. Journal of Wildlife Management, 46: 757-760.

Pollock, K. H., Nichols, J. D., Brownie, C. & Hines, J. E. (1990).- Statistical inference for capture-recapture experiments. Wildlife Monographs 107.

Rockwell, R. F., Findlay, C. S. & Cooke, F. (1983).- Life history studies of the lesser snow goose (*Anser caerulescens caerulescens*) I. The influence of age and time on fecundity. Oecologia, 56: 318-322.

Marked Individuals in the
Study of Bird Population
J.-D. Lebreton and P. M. North (eds.)

The Influence of Capture-Recapture Methodology on the Evolution of the North American Bird Banding Program

John Tautin

Bird Banding Laboratory, U.S. Fish and Wildlife Service, Laurel, Maryland 20708 USA

Summary: Capture-recapture methodology has advanced greatly in the last twenty years and is now a major factor driving the continuing evolution of the North American bird banding program. Bird banding studies are becoming more scientific with improved study designs and analytical procedures. Researchers and managers are gaining more reliable knowledge which in turn betters the conservation of migratory birds. The advances in capture-recapture methodology have benefited gamebird studies primarily, but nongame bird studies will benefit similarly as they expand greatly in the next decade. Further theoretical development of capture-recapture methodology should be encouraged, and, to maximize benefits of the methodology, work on practical applications should be increased.

1. Introduction

The great naturalist-artist John James Audubon was the first person known to have marked birds for study in North America. In what could be described as a simple capture-recapture exercise, he attached silver wires to the legs of Eastern Phoebes (Sayornis phoebe) to see if they returned to nest the following year (Audubon 1840). Dr. Paul Bartsch, however, is generally credited with initiating the first systematic banding of birds in North America. He used serially numbered bands to band Black-crowned Night Herons (Nycticorax violaceous) in 1902 at Washington, D.C., (Bartsch 1904). James Henry Fleming is recognized as having banded the first bird in Canada, an American Robin (Turdus migratorius) banded in 1905.

The American Bird Banding Association was founded in 1909 to organize banding activities. It had an auspicious start, but lost momentum during World War I as did many ringing programs in Europe (Lincoln 1960). In 1920 the Bureau of Biological Survey, forerunner of the U.S. Fish and Wildlife Service (FWS), assumed responsibility for bird banding in the U.S., and Canada's banding office was established in 1923. These events followed the signing of the Migratory Bird Convention between the two countries in 1916, and, in the U.S., enactment of enabling legislation in the form of the 1918 Migratory Bird Treaty Act. These instruments still serve as the foundation for federal pre-eminence and involvement with migratory birds.

2. The Evolution of the North American Bird Banding Program

2.1. Game Bird Banding

In tracing the subsequent evolution of the North American bird banding program, it is useful to contrast banding of game (hunted) birds with nongame birds. The 1918 Act clearly defined game birds by specifying those families that could be hunted. The Act also authorized and directed the Secretary of Agriculture (now Interior) "to adopt suitable regulations permitting and governing" the hunting of migratory game birds "having due regard to zones of temperature and to the distribution, abundance, ... and times and lines of migratory flight of such birds."

Giving due regard to distribution, lines of flight and other biological considerations in the formulation of hunting regulations would require data. Frederick Lincoln, the first director of the U.S. banding program, recognized the potential of banding and promoted it. As early as 1924 he published results of banding projects, presenting distributions of recoveries from Mallards (Anas platyrhynchos) banded in the Mississippi Valley (Lincoln 1924). Further analyses led to the development of the Flyways concept (Lincoln 1935) which still is a prominent factor in the regulation of waterfowl hunting in Canada and the U.S. (USDI 1988). Lincoln was an accomplished administrator, biologist and writer. He strongly influenced waterfowl management and the early development of banding programs.

From these beginnings, game bird banding quickly evolved away from nongame bird banding. By the 1940s game bird (essentially waterfowl) banding was largely a public sector activity dominated by state and federal wildlife agencies. Agency waterfowl managers recognized that banding was important to their research and management responsibilities, and waterfowl banding has generally been limited to agency personnel and their close collaborators in private conservation organizations and various academic institutions. It has been easy for agencies to control and direct waterfowl banding, because they control the issue of banding permits.

In addition to wildlife agencies being able to control and direct waterfowl banding, several other factors predisposed them to develop sound banding programs. They had an identifiable constituency (hunters) to both serve and seek support from. They had clearly defined statutory responsibilities and other mandates to guide their work. They had funding, albeit limited, usually in the form of license and stamp revenues, but in some cases (e.g. FWS) from the federal treasury. The agencies also had administrative mechanisms such as Flyway Councils (USDI 1988) for getting things done.

Strict financial and administrative constraints forced agencies to economize and establish priorities in banding programs. Random, opportunistic banding was abandoned in favor of planned cooperative banding programs having well-defined aims (Anderson and Henny 1972). Banding became an important fool for providing information upon which decisions would be made. Details on the evolution of waterfowl banding and management can be found in the historical accounts of Anderson and Henny (1972), Hawkins, et al. (1984) and USDI (1988).

2.2. Nongame Bird Banding

The development of nongame bird banding has lagged considerably behind game bird banding. The factors that stimulated progress in game bird programs – mandates, support bases, administrative mechanisms, etc. – were generally lacking for nongame species. As a result, nongame bird banding evolved largely as a private sector activity dominated by independent university groups, museum staff, and most significantly, amateur ornithologists. Lacking programmatic direction and coordination, much nongame bird banding has been random and opportunistic, yielding miscellaneous data which are not easy to analyze or interpret. Descriptive studies rather than hypothesis based studies have been the norm. Although 70 percent of all birds banded in North America have been nongame species (36 million in total), the benefits accrued to research, management and conservation of nongame birds have not been as significant as with game bird work.

However, the situation is now changing rapidly. The same factors that influenced the evolution of game bird banding programs are coming into play. A nongame bird constituency has emerged in the form of bird watchers whose involvement in a variety of ornithological activities (including political lobbying) has increased dramatically.

There are now some statutory requirements and mandates for migratory nongame bird conservation. The National Environmental Policy Act and the Endangered Species Act, while not specific to birds, are relevant. The Fish and Wildlife Conservation Act of 1980 and subsequent amendments are specific. More mandates are likely as Congress considers bills on biodiversity and other matters related to birds.

Funding sources for nongame bird programs are developing. The base budgets of some federal agencies now include funds for nongame bird work. Congressional "add-ons" occasionally augment these funds. Entities such as the National Fish and Wildlife Foundation provide some monies. Perhaps the largest source of funds comes from state programs through which people can donate portions of their income tax refunds to nongame programs.

Administrative mechanisms are developing. The Western Hemisphere Shorebird Reserve Network is coordinating banding and other work on that group of birds. The interagency Partners in Flight initiative is promoting much work on neotropical migrants. Federal and state wildlife agencies and private conservation groups are developing networks of nongame bird workers.

3. The Evolution of Capture-Recapture Methodology

3.1 Early Methodology

The need for reliable knowledge about waterfowl populations, their survival and the effect of hunting stimulated the development of methods for analyzing banding data. For an early example of developments in capture-recapture methodology (C-RM), we return to the work of Frederick Lincoln. In 1930 he described a method for estimating population size based on recaptures of banded birds (Lincoln 1930). Every student of C-RM is familiar with this estimator, the Lincoln-Petersen Index.[1] Lincoln's analyses of banding data also led him to conclude that band recovery rates could serve as indices of hunting pressure or harvest rates. The Lincoln-Petersen Index was used to a limited extent in waterfowl management, but band recovery rates, became important factors in studies of waterfowl survival and the effects of hunting (e.g. see Geis 1972, Brownie et al. 1978, and Martin et al. 1979).

If one can stretch the scope of C-RM to include life tables, then Hickey's (1952) comprehensive work on survival of birds accurately portrays the state of the art in

1 Nichols (1992) points out that others, notably C.G.J. Petersen of Sweden, independently developed essentially the same index. Bailey (1952) discusses the other developments and credits Lincoln with describing and publishing the population estimator.

the 1940's and 1950's. Hickey thoroughly reviewed and critiqued life tables, and used them to estimate survival rates in numerous species-specific case studies. The use of life tables to estimate survival from band recovery data persisted for another 20 years (e.g. see Henny 1972).

3.2 Developments in the 1960's

The 1960s saw major changes in waterfowl management in North America. Duck populations were depressed by drought, and traditional hunting regulations (bag limits, season lengths) were more restrictive than in earlier years. Demand for hunting opportunity remained high, though, making for an unsatisfactory situation.

New strategies for harvest management were developed by wildlife agencies with the aim of preserving or enhancing hunting opportunities. The High Plains Unit was established in the western U.S. to afford additional opportunity to harvest mallards, a favored species among hunters throughout North America. Then, as discussed in USDI(1988), "there began a drift toward exploiting there-to-fore lightly utilized (i.e. "under utilized" or "surplus") stocks and species through the use of special harvest strategies such as September teal seasons and bonus bags. Some of these new strategies were developed through experimental seasons and data gathering, while others were based more on a presumption that no known problems prevented using them."

By the late 1960s it was evident that more refined data on waterfowl populations and survival were needed to balance the demands of hunting with the capacity for populations to sustain themselves. Better results from banding were essential. Wildlife agencies were good at gathering banding data, but were limited in their ability to draw meaningful inferences from them. This was due, in large part, to limitations of unsound analytical methods such as the life tables approach (Eberhardt 1972). A breakthrough in C-RM was needed if waterfowl management in North America was to progress further.

Fortunately, in the mid 1960s major developments in C-RM were achieved with the works of Cormack (1964), Jolly (1965) and Seber (1965). For agency waterfowl managers the watershed breakthrough came with the band recovery models of Seber (1970), Robson and Youngs (1971) and Brownie and Robson (1976). Their models were special case C-RM models. They utilized band recovery data from dead birds, as opposed to utilizing recapture data which were not commonly available then. Their models were stochastic and theoretically sound, and they proved to be vastly superior to life tables. The models were adopted by agency waterfowl managers, especially those at FWS' Patuxent Wildlife Research Center, Laurel, Maryland. The involvement and support of FWS scientists stimulated further development in this exciting new area of C-RM.

3.3 Contemporary Methodology

The history of C-RM from the early 1970s to the present is well documented. Recent reviews with extensive references on the subject can be found in Lebreton et al. (1992), Nichols (1992), Pollock (1991) and Pollock et al. (1990).

4. The Impact of Capture-Recapture Methodology

4.1 Game Birds

Since the early 1970s C-RM has greatly influenced research and management of migratory game birds in North America. It has enabled definitive survival studies of important game species like: the Mallard (Anderson 1975), the number one duck in total harvest in both Canada and the U.S.; the Black Duck (Anas rubripes) (Krementz et al. 1987) whose populations are depressed; the Canada Goose (Branta canadensis) (Hestbeck and Malecki 1989a) whose populations have increased in many areas; the Mourning Dove (Zenaida macroura) (Tomlinson et al. 1988) whose population size and harvest exceed all other migratory game birds combined; and the American Woodcock (Scolopax minor) (Dwyer and Nichols 1982), popular with upland hunters in the eastern North America where its population has declined.

C-RM has shed light on the longstanding, fundamental question of whether hunting mortality is additive or compensatory to other forms of mortality (Anderson and Burnham 1976), (USDI 1988). As C-RM progressed from estimating survival and band recovery rates to testing for differences among them, it has facilitated comparative studies assessing the effects of different regulatory regimes (Trost 1987). C-RM was central to analyses of the important experiment in which duck hunting regulations were stabilized for five years. (Sparrowe and Patterson 1987). The knowledge gained from these studies has had a significant impact on the philosophies and policies of wildlife agencies in North America regarding hunting of migratory birds. This is evident in FWS' environmental impact statement (USDI 1988) which provides the framework within which annual regulations for hunting migratory birds are promulgated.

In recent years game bird managers have moved beyond analyzing survival and recovery rates to estimating population size based on observations of marked birds. For example, Hestbeck and Malecki (1989b) used a Jolly-Seber model to refine estimates of Canada goose populations using observations of neck-collared birds as input to the model.

There have also been major spinoff benefits from C-RM. Brownie et al's (1978) chapter on planning a banding study and determining sample sizes has been very useful to waterfowl managers, guiding their development of plans for specific studies and for ongoing operational banding programs (USDI 1989). Field banding operations have become more focused and effective as a result. Developments in other areas e.g., trend analyses of time series data, have benefitted from the quantitative environment and consciousness nurtured by interests in C-RM.

C-RM continues to influence studies of migratory game birds. Currently a number of large-scale cooperative waterfowl projects are under way or planned. Snow Geese (Chen careulescens) and White-Fronted Geese (Anser albifrons) are being neck-collared and banded in large numbers as part of the Arctic Goose Joint Venture of the North American Waterfowl Management Plan. Atlantic Flyway states and provinces have undertaken another Canada Goose study to differentiate population parameters between Canadian breeding flocks which may need additional protection, and local resident flocks which may withstand additional exploitation. The Atlantic Flyway states have also proposed what would be the largest study ever done on the population dynamics and movements of Tundra Swans (Cygnus columbianus). Capture-recapture models will be integral parts of the analytical phases of these studies.

4.2 Nongame Birds

Until recently C-RM has had comparatively little impact on the evolution of nongame bird banding programs in North America. However, as state, federal and private nongame bird programs develop, C-RM is slowly, but definitely being adopted as a means of gaining more reliable knowledge. More nongame bird banders are collecting recapture data. A substantial portion are using auxiliary markers and radio transmitters to obtain their data without actually having to recapture birds. Recent examples of species-specific studies incorporating C-RM include Spendelow's (1991) and Buehler et al.'s (1991) survival studies on endangered Roseate Terns (Sterna dougallii) and Bald Eagles (Haliaetus leucocephalus), and Arnason et al.'s (1991) Bald Eagle study employing a new estimator of a closed population size.

C-RM has shed light on some important questions in nongame bird management, e.g., does supplemental winter feeding enhance survival of birds (Brittingham and Temple 1988). C-RM is yielding results that challenge some tenets of avian survival. For example, Karr et al.'s (1990) results dispute conventional wisdom that tropical forest birds have higher survival rates than temperate forest birds. Krementz et al. (1989) effectively demonstrate the superiority of C-RM survival estimates over individual longevity records in life history analyses.

C-RM, particularly the SURGE program (Lebreton and Clobert 1986), will be used to analyze data collected in the Monitoring Avian Productivity and Survivorship (MAPS) program being established in North America. This program is patterned after the British Constant Effort Sites scheme (Baillie et al. 1986). MAPS is coordinated by the Institute for Bird Populations of Inverness, California, with financial support from FWS and others. MAPS is an important development in nongame bird management in North America. It will be the first coordinated attempt to use mist netting and banding to monitor survival, recruitment and population size on a wide scale.

5. Future Developments in Capture-Recapture Methodology

Clearly, C-RM has had a profound and positive impact on bird banding in North America. It is now a major force driving the continuing evolution of the North American bird banding program. Capture-recapture workers are to be commended for their outstanding contributions to migratory bird research and management, and further work is encouraged. Theoretical development of C-RM should continue, and to maximize the benefits of the methodology, work on practical applications should be increased. Nichols (1992) is correct in suggesting that the burden now falls on animal ecologists to incorporate C-RM in their work. However, this will happen faster if capture-recapture workers devote more effort to practical applications and communication of them.

Some improved C-RM products are needed. Revised and updated "how to" manuals would be helpful to bird workers. Study design as well as analysis of data should be discussed in these works as it was in Chapter 9 of Brownie et al. (1978), and in Chapter 2 of White and Garrott (1990). The latter's work is especially good in that it prompts the investigator to think about the scientific method and conceptual designs of studies. Too many banders ignore these considerations. They begin and end their studies with the data collection step.

More works similar to that of Hestbeck (in press) would help. Hestbeck's logical, easy to follow approach leads the reader through all the stages of his Canada goose study, beginning with a clear elucidation of the problem and questions being asked and progressing through the selection of models for analyzing data. His approach to the study virtually guarantees success, assuming that the data collection phase goes well which it appears to be doing (Hestbeck, personal communication).

Additional "user-friendly", menu driven computer software that guides users through data entry, model selection and interpretation of results would be helpful. Several programs already exist, e.g. CAPTURE (Otis et al. 1978), SURGE (Lebreton and Clobert 1986), SURVIV (White 1983), JOLLY and JOLLYAGE (Pollock et al. 1990) and MULT (Conroy et al. 1989). These and other programs are outlined in Lebreton et al. (1992).

More education about C-RM is needed in North America. At universities more courses should be taught emphasizing the scientific method, study design and the application of C-RM. Workshops such as those periodically offered at Colorado State University and North Carolina State University help maintain the graduate's knowledge and skills in C-RM.

Workshops held in conjunction with ornithological meetings would introduce more workers to applications of C-RM. Many ornithologists have by accident or design accumulated excellent sets of banding data that would lend themselves to analysis, but are unaware of contemporary methodology or are intimidated by it's sophisticated theory and statistics. Workshops combining introductory, overview presentations on the subject of C-RM with demonstrations of the associated software using real data sets would help to educate ornithologists and to allay their fears about using C-RM.

More species case studies published in ornithological journals are needed to demonstrate the possibilities of C-RM. This is especially important for nongame bird studies where researchers and managers have been slow to take advantage of the methodology.

These further practical developments would increase use of C-RM and would accelerate progress in the North American bird banding program. More reliable knowledge would be gained, and the conservation of birds would be advanced.

ACKNOWLEDGEMENTS: Danny Bystrak, Marshall Howe, William Howe, Kathleen Klimkiewicz, James Nichols, Robert Smith and two anonymous reviewers kindly provided helpful suggestions for improvement of the manuscript. Financial assistance was provided by Center d'Ecologie Fonctionnelle et Evolutive, C.N.R.S., Montpellier, France.

REFERENCES

Anderson, D.R. (1975).- Population ecology of the Mallard V. temporal and geographic estimates of survival, recovery and harvest rates. U. S. Fish and Wildl. Serv. Resour. Publ. 125. USGPO. Wash. D.C. 110 pp.

Anderson, D.R. and C.J. Henny. (1972).- Population ecology of the Mallard I. a review of previous studies and the distribution and migration from breeding areas. U.S. Fish and Wildl. Serv. Resour. Publ. 105. USGPO. Wash. D.C. 166 pp.

Anderson, D.R. and K.P. Burnham. (1976).- Population ecology of the Mallard. VI. the effect of exploitation on survival. U.S. Fish and Wildl. Serv. Resour. Publ. 128. USGPO. Wash. D.C. 66 pp.

Arnason, A.N., C.J. Schwarz and Jon M. Gerrard. (1991).- Estimating closed population size and number of marked animals from sighting data. J. Wildl. Manage., 55:716-730.

Audubon, J.J. (1840).- Birds of America. J.B. Chevalier. Philadelphia.

Bailey, N.T.J. (1952).- Improvements in the interpretation of recapture data. J. Anim. Ecol., 21:120-127.

Baillie, S.R., R.E. Green, M. Boddy, and S.T. Buckland. (1986).- An evaluation of the Constant Effort Sites Scheme. British Trust for Ornithology, Tring, U.K. 103 pp.

Bartsch, Paul. (1904).- Notes on the herons of The District of Columbia. Smiths. Miscl. Coll. Vol. XLV. Pub. No. 1419, pp. 104-111.

Beuhler, D.A., J.D. Fraser, J.K. Seegar, G.D. Therres and M.A. Byrd. (1991).- Survival rates and population dynamics of Bald Eagles on Chesapeake Bay. J. Wildl. Manage., 55: 608-613.

Brittingham, M.C. and S.A. Temple. (1988).- Impacts of supplemental feeding on survival rates of Black-capped Chickadees. Ecology, 69:581-589.

Brownie, C., D.R. Anderson, K.P. Burnham and D.S. Robson. (1978).- Statistical inference from band recovery data - a handbook. U.S. Fish and Wildl. Serv. Resour. Publ. No. 131. USGPO. Wash. D.C. 212 pp.

Brownie, C. and D.S. Robson (1976).- Models for allowing age - dependent survival rates for band return data. Biometrics, 32:305-323.

Conroy, M.J., J.E. Hines, and B.K. Williams. (1989).- Procedures for the analysis of band recovery data and user instructions for program MULT. U.S. Fish & Wildl. Serv. Resource Pub. 175. USGPO. Wash. D.C. 61 pp.

Cormack, R.M. (1964).- Estimates of survival from the sighting of marked animals. Biometrika, 51:429-438.

Dwyer, T.J. and J.D. Nichols. (1982).- Regional population inferences for the American Woodcock. pp. 12-21 in Woodcock ecology and management, T.J. Dwyer and G.L, Storm, (Eds). U.S. Fish and Wildl. Serv. Res. Rept 14, Wash. D.C. 191 pp.

Eberhardt, L.L. (1972).- Some problems with estimating survival from banding data. pp. 153-171 in Population Ecology of Migratory Birds - a Symposium. U.S. Fish & Wildl. Serv. Wildl. Res. Rep. 2. Wash. D.C. 278 pp

Geis, A.D. (1972).- Use of banding data in migratory game bird research and management. U.S. Fish and Wildl. Serv. Spec. Sci. Rept. Wildl. No. 154. 47 pp.

Hawkins, A.S., R.C. Hanson, H.K. Nelson and H.M. Reeves (Eds.). (1984).- Flyways: Pioneering Waterfowl Management in North America. USGPO. Wash. D.C. 517 pp.

Henny, C.J. (1972).- An analysis of the population dynamics of selected avian species - with special reference to changes during the modern pesticide era. U.S. Fish and Wildl. Serv. Wildl. Res. Rept. 1. Wash. D.C. 99 pp.

Hestbeck, J.B. In press. Estimating population parameters for northern and southern breeding populations of Canada geese in Inter. Canada Goose Symp., D.H. Rusch, M.D. Samuel and B.O. Sullivan (Eds.). Milwaukee, Wisconsin.

Hestbeck, J.B. and R.A. Malecki. (1989a).- Estimated survival rates of Canada geese within the Atlantic Flyway. J. Wildl. Manage., 53:91-96.

Hestbeck, J.B. and R.A. Malecki. (1989b).- Mark-resight estimate of Canada goose midwinter number. J. Wildl. Manage., 53:749-752.

Jolly, G.M. (1965).- Explicit estimates from capture-recapture data with both death and immigration - a stochastic model. Biometrika, 52:225-247.

Karr, J.R., J.D. Nichols, M.K. Klimkiewicz and J.D. Brawn. (1990).- Survival rates of birds of tropical and temperate forests: will the dogma survive? The Am. Naturalist, 136:277-291.

Krementz, D.G., M.J. Conroy, J.E. Hines, and H.F. Percival. (1987).- Sources of variation in survival and recovery rates of American Black Ducks. J. Wildl. Manage., 51:689-700.

Krementz, D.G., J.R. Sauer and J.D. Nichols. (1989).- Model-based estimates of annual survival rates are preferable to observed maximum life span statistics for use in comparative life-history studies. Oikos, 56:203-208.

Lebreton, J.D. and J. Clobert. (1986).- Users manual for program SURGE, Centre L. Emberger, Montpellier, France. 24 pp.

Lebreton, J.D., K.P. Burnham, J. Clobert and D.R. Anderson. (1992).- Modeling survival and testing biological hypotheses using marked animals: a unified approach with case studies. Ecological Monographs, 62(1):67-118.

Lincoln, F.C. (1924).- Returns from banded birds, 1920 to 1923. Bull. No. 1268. U.S. Dept. of Ag. Wash. D.C. 55 pp.

Lincoln, F.C. (1930).- Calculating waterfowl abundance on the basis of banding returns. U.S. Dept. of Ag. Circ. 118. Wash. D.C. 4 pp.

Lincoln, F.C. (1935).- The waterfowl Flyways of North America. U.S. Dept. of Ag. Circ. 342. Wash. D.C. 12 pp.

Lincoln, F.C. (1960).- Bird banding. pp. 623-626 in Encyclopedia Britannica, Vol.3. William Benton, Publ. Chicago, London, Toronto.

Martin, F.W., R.S. Pospahala and J.D. Nichols. (1979).- Assessment and population management of migratory birds. pp. 187-239 in Environmental and Biomonitoring Assessment, Prediction, and Management - Certain Case Studies and Related Quantitative Issues. Statistical Ecology Vol. S11, J. Cairns, G.P. Patil and W.B. Waters (Eds.). Int. Co-operative Publ. House. Fairland, Maryland.

Nichols, J.D. (1992).- Capture-recapture models - using marked animals to study population dynamics. Bioscience, 42:94-102.

Otis, D.L. K.P. Burnham, G.C. White and D.R. Anderson. (1978).- Statistical inference from capture data on closed populations. Wildl. Monogr. 62, 135 pp.

Pollock, K.H. (1991).- Modeling capture-recapture and removal statistics for estimation of demographic parameters for fish and wildlife population: past, present and future. J. Am. Statistical Assn., 86:225-238.

Pollock, K.H., J.D. Nichols, J.E. Hines and C. Brownie. (1990).- Statistical inference from capture-recapture experiments. Wildl. Mono. 107. 97 pp.

Robson, D.S. and W.D. Youngs. (1971).- Statistical analysis of reported tag-recaptures in the harvest from an exploited population. Rept. 369-M. Cornell Univ. Biometric Unit. Ithaca, New York.

Seber, G.A.F. (1965).- A note on the multiple recapture census. Biometrika, 52:249-259.

Seber, G.A.F. (1970).- Estimating time-specific survival and reporting rates for adult birds from band returns. Biometrika, 57:313-318.

Sparrowe, R. D. and J.H. Patterson. (1987).- Conclusions and recommendations from studies under stabilized duck hunting regulations: management implications and future directions. Trans. N.Am. Wildl. and Nat. Resour. Conf., 52: 320-326.

Spendelow, J. A. (1991).- Postfledgling survival and recruitment of know-origin Roseate Terns (Sterna dougalli) at Falkner Island, Connecticut. Colonial Waterbirds, 14:108-115.

Tomlinson, R.E., D.D. Dolton, H.M. Reeves, J.D. Nichols and L.A. Mckibben. (1988).- Migration, harvest, and population characteristics of mourning doves banded in the Western Management Unit,

1964-1977. U.S. Fish and Wildl. Serv. Tech. Rept. 13. Wash. D.C. 100 pp.
Trost, R.E. (1987).- Mallard survival and harvest rates: a re-examination of relationships. Trans, N. Am. Wildl. and Nat. Resour. Conf., 52: 264-284.
USDI. (1988).- Final Supplemental Environmental Impact Statement: Issuance of Annual Regulations Permitting the Sport Hunting of Migratory Birds. U.S. Dept. of the Interior, Fish and Wildl. Serv. Final Environmental Statement. Wash. D.C. 340 pp.
USDI-(1989).- The North American duck banding program: a revised approach. U. S. Fish and Wildl. Serv. MBMO. Wash. D.C.
White, G.C. (1983).- Numerical estimation of survival rates from band-recovery and biotelemetry data. J. Wildl. Manage., 47:716-728.
White, G.C. and R.A. Garrott. (1990).- Analysis of Wildlife Radio-tracking Data. Academic Press, Inc. New York. 383 pp.

Merging different sources of information

Marked Individuals in the
Study of Bird Population
J.-D. Lebreton and P. M. North (eds.)

A Theory for Combined Analysis of Ring Recovery and Recapture Data

Kenneth P. Burnham

U.S. Fish and Wildlife Service, Colorado Cooperative Fish and Wildlife Research Unit, Colorado State University, Fort Collins, Colorado 80523, USA

Summary. For capture-recapture studies the Jolly-Seber model theory for time-specific effects is well developed and established, as is the corresponding time-specific ring recovery model, M_1 (Brownie et al. 1985). Many capture-recapture studies on birds will generate both live recaptures and rings reported from birds shot, or found dead, throughout the year. The purpose of this paper is to give the basic generalization of the Jolly-Seber and ring recovery model M_1 for the simultaneous analysis of these recapture and recovery data. Under the generalized model, as under each model separately, the statistical distribution of the relevant data can be taken as a set of appropriately structured, independent, multinomial distributions but with response categories for two classes of events: (a) the first recapture after a known release of a live bird, and (b) ring recovery from a dead bird after a known release. These events can be considered a type of post-release stratification of the data. Moreover, these two types of events "interfere" (or compete) with each other, and it is that feature of the problem that causes the modeling difficulties here. Conditional on the known releases, these events are mutually exclusive and exhaustive (when considered together with the event, release and neither the bird nor its ring is ever encountered again). This paper gives the expanded multinomial representation of the correct statistical model for the combined data and gives alternative forms of the multinomial cell probabilities under two assumptions about emigration from the study area. Representation of some of these cell probabilities in the joint model is more complex than in either separate model. Finally, the abundance estimation part of the Jolly-Seber model derives from modeling the probability distribution of first captures; that model component is unaffected by the possibility of using recoveries as well as recaptures from the releases of ringed birds.

1. Introduction

Statistical theory for the analysis of data from bird ringing studies has been quite thoroughly developed for two seemingly distinct classes of studies: capture-recapture and ring recovery. In the former type of study, birds are repeatedly recaptured alive and then re-released, thus multiple encounters can occur after the initial ringing of a bird. The latter type of study typically applies to harvested or hunted species, such as game birds wherein re-encounters with rings come from dead birds, hence after ringing there is at most one re-encounter possible. However, in bird ringing studies we can expect both live recaptures (and re-releases) and reports of rings from dead birds. The purpose of this paper is to present basic statistical theory for the time-specific effects case when there are both live recaptures and dead recoveries resulting from a ringing study.

I assume the reader is familiar with the basic capture-recapture and ring-recovery statistical analysis literature. This literature includes such seminal publications as Cormack (1964), Jolly (1965) and Seber (1965, 1970). In particular, the reader should to be familiar with the more recent comprehensive publications of Brownie et al. (1985), Burnham et al. (1987), Pollock et al. (1990) and Lebreton et al. (1992). Of the models considered in these papers, I am only directly concerned here with the Cormack-Jolly-Seber capture-recapture model, the corresponding ring recovery model M_1, and their extension to joint recapture-recovery analysis.

The joint analysis of live recapture and ring-recovery data has not been much considered in the ringing analysis literature (cf, Clobert and Lebreton, 1991). When both types of data exist from one study, it has been common to just do the two separate analyses, see e.g., Anderson and Sterling (1974) and compare the results. Alternatives have been to force the live recapture data to "fit" into the dead recoveries framework (Mardekian and McDonald, 1981), or conversely, force the dead recoveries data component into the live capture-recapture framework (Buckland, 1980). Only recently have truly unified analysis methods been used for these type of "post-stratified" data (Szymczak and Rexstad, 1991; see also Schwarz et al., 1988) and the theory herein given for this analysis has not previously appeared.

One difficulty in producing a combined theory is the choice of notation. Theory for ring recovery and capture-recapture data have been developed separately. This has lead to the two theories using very different notation and indexing over capture-release occasions. A large part of this difference stems from an historic emphasis in capture-recapture on estimation of population size rather than survival rate. The only parameter in common for recovery and recapture data is survival rate; thus a joint theory must focus on survival rate. With respect to survival rate, there is a basic statistical theory for time-specific release-resampling models that applies to both ring-recoveries and capture-recapture. This common theory is presented in Burnham et al. (1987), Lebreton et al. (1992) and given in detail in Burnham (1991). Burnham (1988) presents a very thorough theory development including a general notation for release-resampling that I will use here. Essentially, I choose to use capture-recapture type notation and multinomial modeling as the basis for a joint theory.

I will use here terminology and ideas of bird ringing studies with ringing done annually; the theory given here is, of course, more general and time intervals between capture occasions need not be equal. Ring recovery models have generally been developed in terms of known releases (R_i) at the start of (study) year i with subsequent recoveries (m_{ijd}, the *d* denotes "dead") during years $j = i, i+1, \ldots, t$ as being multinomially distributed events. By putting capture-recapture data in this same modeling framework (Burnham, 1991) we represent either type of data in a common framework.

Under the joint analysis we have releases, R_i, and *first* recaptures, m_{ijl}, of live (denoted by the subscript *l*) birds at times $j = i+1, \ldots, t$, and ring recoveries, m_{ijd}, from dead birds during yearly time periods i to $i+1$, $i+1$ to $i+2, \ldots, t-1$ to t. Recaptures at year j are re-released as part of the R_j birds then known to be released (except losses on capture are allowed). Instead of only one type of re-sampling event under the separate theories, we have two types of events: if it is encountered again, the bird may be either alive (the m_{ijl}) or dead (the m_{ijd}). This can be thought of as post stratification of the encounter data, in the sense that at the time of release we cannot know which type of subsequent encounter will occur. (The subscripts *l* and *d* encoded the post stratification factor, hence are after the ij subscripts). In each cohort of releases (the R_i) the subsequent encounter events are mutually exclusive. If we assume independent fates of individuals then the basic data have a multinomial distribution, independent over each of the released cohorts.

Completion of model specification now focuses on the functional form of the cell probabilities in the multiple-multinomial probability model for the recapture-recovery data. This paper gives a probability form that correctly generalizes Cormack-Jolly-Seber and Model M_1. The types of rate parameters in this model are survival probability, S, site fidelity, F, live capture probability, p, and ring recovery probability, f (the classical Jolly-Seber model assumes any emigration from the study area is permanent, hence only the product $\phi = SF$ is then estimable). The multinomial cell probabilities for the joint model are expressed as functions of these four types of parameters. However, some of the cell functional forms are more complex than for corresponding events in either marginal model. This is because the two types of resampling events (live or dead) compete (interfere) with one another for removal of a bird from its last release cohort. In particular, if a ring is known to have been recovered, then that bird is known to be unavailable for future live recapture.

In what follows I first summarize the basic theory for the Cormack-Jolly-Seber model and banding model M_1 in a common notation (Section 2). Then Section 3 gives a theory of joint recapture-recovery data: Section 3.1 gives a multinomial model representation of the general probability model for the joint distribution of recapture and recovery data; Section 3.2 gives the specific form of the cell probabilities in this model under the usual permanent emigration assumption; Section 3.3 gives theory under the assumption that on any capture occasion it is an independent random event as to whether a live bird is in the subset of the population at risk of capture. To have a short way to refer to this later assumption, I will call it "random emigration."

2. Separate model theory summarized

2.1 Cormack-Jolly-Seber Probability Model

The basic notation needed for the survival component of the Cormack-Jolly-Seber model is defined below:

R_i The number of ringed birds released at capture occasion i; under the assumptions of the Jolly-Seber model these birds can be a mix of first captures and recaptures; $i = 1, \ldots, t-1$.

m_{ijl} The number of birds recaptured for the first time at occasion j, from those birds released at occasion i; $j = i+1, \ldots, t$, $i = 1, \ldots, t-1$.

I consider herein only the case of t capture and release occasions, hence $t-1$ recapture occasions, and $t-1$ time-intervals for recovery of rings from dead birds. Some derived statistics are also needed, in particular the total recaptures from the R_i releases,

$$r_{il} = \sum_{j=i+1}^{t} m_{ijl} \qquad , i = 1, \ldots, t-1.$$

As given in much more detail in, e.g., Burnham (1987, 1988 or 1991) the probability model for these data is the product of $t-1$ independent multinomial distributions:

$$\Pr\{\text{recaptures} \mid \text{releases}\} = \prod_{i=1}^{t-1} \Pr\{m_{i,i+1,l}, \ldots, m_{itl} \mid R_i\} =$$

$$\prod_{i=1}^{t-1} \binom{R_i}{m_{i,i+1,l}, \ldots, m_{itl}, R_i - r_{il}} \left(\prod_{j=i+1}^{t} (\pi_{ijl})^{m_{ijl}} \right) (1-\lambda_{il})^{(R_i - r_{il})} \tag{1}$$

with

$$\pi_{ijl} = \frac{E(m_{ijl} \mid R_i)}{R_i} = \begin{cases} \phi_i p_{i+1} & , j = i+1, \\ (\phi_i q_{i+1}) \cdots (\phi_{j-2} q_{j-1})(\phi_{j-1} p_j) & , j > i+1. \end{cases} \tag{2}$$

For any definition of π_{ijl} (not just (2) above), define

$$\lambda_{il} = \sum_{j=i+1}^{t} \pi_{ijl} \qquad , i = 1, \ldots, t-1.$$

When (2) holds then there is a convenient recursive formula,

$$\lambda_{il} = \phi_i(p_{i+1} + q_{i+1}\lambda_{i+1,l}) \qquad , i = 1, \ldots, t-1,$$

with boundary condition $\lambda_{tl} = 0$.

The parameters in the Cormack-Jolly-Seber model are defined below:

p_i The probability of capture on occasion i given that the bird is still in the population at risk of capture on occasion i, $q_i = 1 - p_i$; $i = 2, \ldots, t$.

ϕ_i The probability of surviving in the subset of the population at risk of capture from occasion i to occasion i+1 given that the bird is alive in the population at risk at the start of survival period i, $i = 1, \ldots, t-1$.

There are $2t-3$ identifiable parameters in this model: $\phi_1, \ldots, \phi_{t-2}, p_2, \ldots, p_{t-1}$, and the product $\phi_{t-1}p_t$. However, the parameter ϕ_i is a product of two biologically interpretable parameters: $\phi_i = S_iF_i$. The probability of surviving between capture occasion i and $i+1$ is S_i; thus $1 - S_i$ = probability of death. Given an animal is alive at occasion $i+1$, then F_i (fidelity) is the probability of it being in the study area at that time.

2.2 Ring Recovery Probability Model

Basic additional notation needed for the time-specific ring recovery model M_1 in a form that will be compatible with capture-recapture data is below (still for t occasions):

m_{ijd} The number of rings recovered between ringing occasions j and $j+1$, from birds released on occasion i, $j = i, \ldots, t$, $i = 1, \ldots, t$.

The total number of rings recovered from released cohort i is

$$r_{id} = \sum_{j=i}^{t} m_{ijd} \qquad , i = 1, \ldots, t.$$

As shown in detail in Burnham (1988 or 1991) (see also Brownie et al., 1985) the probability model for these data is the product of t independent multinomial distributions:

$$\Pr\{\text{recoveries} \mid \text{releases}\} = \prod_{i=1}^{t} \Pr\{m_{iid}, \ldots, m_{itd} \mid R_i\} =$$

$$\prod_{i=1}^{t} \binom{R_i}{m_{iid}, \ldots, m_{itd}, R_i - r_{id}} \left(\prod_{j=i}^{t} (\pi_{ijd})^{m_{ijd}} \right) (1-\lambda_{id})^{(R_i - r_{id})} \tag{3}$$

with

$$\pi_{ijd} = \frac{E(m_{ijd} \mid R_i)}{R_i} = \begin{cases} f_i & , j = i, \\ S_i \cdots S_{j-1} f_j & , j > i, \end{cases} \tag{4}$$

For any definition of π_{ijd}, define

$$\lambda_{id} = \sum_{j=i}^{t} \pi_{ijd} \qquad , i = 1, \dots, t.$$

If (4) applies then there is a convenient recursive formula,

$$\lambda_{id} = f_i + S_i \lambda_{i+1,d} \qquad , i = 1, \dots, t,$$

with boundary condition $\lambda_{t+1,d} = 0$.

The parameters in this ring recovery model are defined below:

f_i — The probability of a ring being recovered between occasions i and $i+1$ given that the bird is alive on occasion i, $i = 1, \dots, t$.

S_i — The probability of surviving from occasion i to occasion i+1 given that the bird is alive at occasion i, $i = 1, \dots, t-1$.

The above $2t-1$ parameters are all identifiable (i.e., $S_1, \dots, S_{t-1}, f_1, \dots, f_t$).

2.3 Comments on Separate vs. Joint Models

The data representation used above is not minimal sufficient for either model; rather, in both cases a simply expressed, minimal sufficient statistic exists with a probability distribution representable by a product of binomials. The dimensionality of the minimal sufficient statistic is the same as the number of identifiable parameters ($2t-3$ for the capture-recapture case as presented, and $2t-1$ for the ringing case with t ringing occasions). In both cases, closed-form (if not restricted to the unit interval) maximum likelihood estimators exist. When the models are "merged" for joint analysis of recapture and recovery data, this simple reduction to a minimal sufficient statistic no longer seems to hold under the permanent emigration assumption of the Jolly-Seber model, hence I do not give the reduced probability models for these two above cases.

At first thought one might think to simply combine the multinomials and use, as the joint model for the ith released cohort, the set of independent multinomials with cell probabilities as defined above in (2) and (4). Structurally this would be the set of cell probabilities

$$\pi_{iid},\ \pi_{i,i+1,l},\ \pi_{i,i+1,d},\ \pi_{i,i+2,l},\ \dots,\ \pi_{i,t-1,d},\ \pi_{itl},\ \pi_{itd},$$

and a final cell probability of $1 - \lambda_{id} - \lambda_{il}$. This is correct in the context of this modeling approach *only* if some of the cell probabilities are appropriately redefined. Once a recapture at occasion j ($>i$) occurs that bird is reassigned upon release to the jth release cohort and hence is no longer available to become a dead recovery from the ith release cohort after occasion j. Similarly, when a recovery is recorded in year j, that bird is known to be no longer available to become a live recapture in occasions beyond j. Thus, heuristically it should be clear that some modification to these sets of marginal cell probabilities (i.e., the π_{ijl}, or the π_{ijd}) is needed in the joint model to adjust for the fact that the two possible types of re-encounters, after a known release, compete (as in competing risks) to remove birds from that release cohort. In the next section I consider a correct form of the joint cell probabilities.

There is also an issue here about whether to include or drop the final year of recovery data, the m_{itd}, $i = 1, \dots, t$. If we think of having a classical capture-recapture study and adding possible dead recoveries to the data during the study years, then the study ends at occasion t and we would do the data analysis without the m_{itd} data. That is, we use ring recovery data m_{ijd} only for

$i=1, \ldots, t-1$, $j=i, \ldots, t-1$. However, if the ring recovery process is independent of the capture process, then we will ultimately have not only the m_{itd} data, but also recoveries beyond capture occasion t (as in the models of Brownie et al., 1985): m_{ijd}, $j=t, t+1, \ldots, k$, $i=1, \ldots, t$. It is beyond the scope of this paper to consider in detail the extension of theory to $k>t$.

3. A Probability Model for Joint Analysis of Recapture-Recovery Data

3.1 The general model structure

The organization of the joint recapture and recovery data under the multinomial modeling framework used here is displayed in Table 1 for the case of no dead recoveries tabulated after the end of the last live capture occasion. Each of the $t-1$ cohorts of releases, of size R_i, leads to a multinomial distribution for the counts of first recaptures and recoveries from that cohort. That this is so is a straightforward result of the usual assumptions of bird ringing studies (see e.g., Lebreton et al., 1992): the same parameters apply to each bird, independent fates of each bird and under the data summary form in Table 1, the possible fates (multinomial cells) within cohorts are mutually exclusive. Therefore, the probability model structure for these data is represented as below. Note that the cell probabilities (π_{ijd} and π_{ijl}) in this joint model are not all identical to the forms given in (2) and (4), which are the marginal cell probabilities of the two data types separately. Two possible cell forms for the joint model are given in Sections 3.2 and 3.3; since (2) and (4) are special cases I do not use new notation here to denote the cell probabilities. The joint model is below:

$$\Pr\{\text{recapture-recovery data up to occasion } t \mid \text{releases}\} =$$

$$\prod_{i=1}^{t-1} \Pr\{m_{iid}, m_{i,i+1,l}, \ldots, m_{i,t-1,d}, m_{itl} \mid R_i\} =$$

$$\left[\prod_{i=1}^{t-1} \binom{R_i}{m_{iid}, m_{i,i+1,l}, \ldots, m_{i,t-1,d}, m_{itl}, R_i - r_{id} - r_{il}}\right] \times$$

$$\left[\prod_{i=1}^{t-1}\left[\left\{\prod_{j=i}^{t-1} (\pi_{ijd})^{m_{ijd}} (\pi_{i,j+1,l})^{m_{i,j+1,l}}\right\} (1-\lambda_{id}-\lambda_{il})^{(R_i - r_{id} - r_{il})}\right]\right]. \tag{5}$$

Also

$$\lambda_{il} = \sum_{j=i+1}^{t} \pi_{ijl} \qquad , i=1, \ldots, t-1,$$

$$\lambda_{id} = \sum_{j=i}^{t-1} \pi_{ijd} \qquad , i=1, \ldots, t-1.$$

The number of releases, R_i, includes all newly ringed birds and all re-released live recaptures made on occasion i.

The probability model given in (5) is for the case where we do not have, or do not use, the recovery data m_{itd} obtained in time interval t to t+1 (i.e., in year t). If we include these data, then one more cell is added to each multinomial and a tth multinomial (binomial, actually) distribution is added to the probability model. This probability model is given below:

Table 1. Symbolic cohort representation of joint recovery-recapture data, m_{ijv}, for t occasions with effort ending at the last recapture occasion; i denotes release occasion, j denotes the time interval for recoveries or the recapture occasion; post stratification is $v = d$ (dead recoveries) or l (live recoveries).

i	Releases	j = 1	2		3		…	t − 1		t
1	R_1	m_{11d}	m_{12l}	m_{12d}	m_{13l}	m_{13d}	…	$m_{1,t\text{-}1,l}$	$m_{1,t\text{-}1,d}$	m_{1tl}
2	R_2			m_{22d}	m_{23l}	m_{23d}	…	$m_{2,t\text{-}1,l}$	$m_{2,t\text{-}1,d}$	m_{2tl}
3	R_3					m_{33d}	…	$m_{3,t\text{-}1,l}$	$m_{3,t\text{-}1,d}$	m_{3tl}
⋮	⋮						⋱		⋮	⋮
t-1	$R_{t\text{-}1}$								$m_{t\text{-}1,t\text{-}1,d}$	$m_{t\text{-}1,tl}$

Pr{recapture-recovery data through year t | releases} =

$$\left[\prod_{i=1}^{t-1} \Pr\{m_{iid},\, m_{i,i+1,l},\, \ldots,\, m_{i,t\text{-}1,d},\, m_{itl},\, m_{itd} \mid R_i\}\right]\left[\Pr\{m_{ttd} \mid R_t\}\right] =$$

$$\left[\prod_{i=1}^{t-1}\binom{R_i}{m_{iid},\, m_{i,i+1,l},\, \ldots,\, m_{i,t\text{-}1,d},\, m_{itl},\, m_{itd},\, R_i - r_{id} - r_{il}}\right]\binom{R_t}{m_{ttd}}\times$$

$$\left[\prod_{i=1}^{t-1}\left[\left\{\prod_{j=i}^{t-1}(\pi_{ijd})^{m_{ijd}}(\pi_{i,j+1,l})^{m_{i,j+1,l}}\right\}(\pi_{itd})^{m_{ijd}}(1-\lambda_{id}-\lambda_{il})^{(R_i - r_{id} - r_{il})}\right]\right]\times$$

$$\left[(\pi_{ttd})^{m_{ijd}}(1-\pi_{ttd})^{(R_t - m_{ttd})}\right]. \tag{6}$$

Also

$$\lambda_{il} = \sum_{j=i+1}^{t} \pi_{ijl} \qquad , i = 1, \ldots, t-1,$$

$$\lambda_{id} = \sum_{j=i}^{t} \pi_{ijd} \qquad , i = 1, \ldots, t.$$

Finally, note that cohort sums of re-encounters, r_{id} and r_{il}, mean the same thing for models (5) and (6) – separate sums of the m_{ijd} and m_{ijl} by cohort; I do not explicitly give their formulae under each model.

The correct cell probabilities for (5) and (6) can be given as formula (2) for the live (first) recapture-after-release probabilities, π_{ijl}. For the recoveries component we use (7) given below, which is a modified version of (4):

$$\pi_{ijd} = \frac{E(m_{ijd} \mid R_i)}{R_i} = \begin{cases} f_i & , j = i, \\ S_i \cdots S_{j\text{-}1}\gamma_{ij} f_j & , j > i, \end{cases} \tag{7}$$

where γ_{ij} is the probability that a bird released at capture occasion i and still alive at occasion j will never have been removed from the ith released cohort by being captured on one of occasions i+1 to, and including, occasion j. Such a removal from the cohort, and reassignment to a later released cohort, makes the bird (i.e., ring) not available to be recovered as a part of count m_{ijd}. Rather, if that ring is then subsequently recovered in year j, that count becomes part of a different recorded m_{hjd} for $i < h \leq j$. This bookkeeping device allows us to formulate our model in terms of familiar multinomial distributions, for which much theory is known. If no live capture data are used (or recorded) then, in effect, all $p_i = 0$ and as a result all $\gamma_{ij} = 1$. Similarly, if there are no recovery data collected or used, then all $f_i = 0$.

3.2 Joint cell probabilities under permanent emigration

Previously ringed birds may still be alive at a later capture occasion but may have moved out of the area (i.e., subset of the population) in which birds are at risk of capture. This sort of "movement" may be accompanied by an intervening round trip migration, but that is not relevant per se. Such movements out of the subset of the population at risk of capture are said to be emigration, especially if these are permanent, i.e., the bird never returns to the subset of the population at risk of capture, at least at any time capture is occurring. Permanence of emigration movements (if any occur) is the usual assumption in capture-recapture. By making this additional assumption we can determine an explicit form for γ_{ij}:

$$\gamma_{jj} = 1 ,$$

$$\gamma_{j-1,j} = 1 - F_{j-1}(1 - q_j) ,$$

$$\gamma_{j-2,j} = 1 - F_{j-2}(1 - q_{j-1}\gamma_{j-1,j}) ,$$

or, in general

$$\gamma_{ij} = 1 - F_i(1 - q_{i+1}\gamma_{i+1,j}) , \tag{8}$$

with boundary condition $\gamma_{jj} = 1$ and indexed for $i = 1, \ldots, j-1$ and $j = 1, \ldots, t$. To compute these parameters, do it for any given fixed j by iterating backwards from $i = j-1, j-2, \ldots,$ to 1. Written out in one long, awkward, form for $j > i$, we have

$$\gamma_{ij} = 1 - F_i(1 - q_{i+1}(1 - F_{i+1}(1 - q_{i+2}(1 - \cdots - F_{j-1}(1 - q_j)\cdots))))$$

and computation is from the inner-most component outward. Clearly, the probability γ_{ij} depends on only the parameters $F_i, \ldots, F_{j-1}, q_{i+1}, \ldots, q_j$; of course $q_i = 1 - p_i$. Thus the γ_{ij}'s do not introduce any new parameters into the model.

The γ_{ij} probabilities adjust the cohort-specific ring recovery probabilities, in the multinomial approach, for the expected live capture rate after known releases. For releases at occasion j there are no recaptures before recoveries in time interval j to j+1, hence $\gamma_{jj} = 1$. Under the data representation scheme of Table 1, in order to be counted as a recovery in $m_{j-1,jd}$ a bird must first survive year $j-1$ and then not be recaptured live at occasion j. This latter probability, $\gamma_{j-1,j}$, is the probability of two mutually exclusive events. Either the bird emigrates (note, $E_i = 1 - F_i$) in year $j-1$ or stays in the population at risk of capture but is not captured at occasion j. This probability is

$$\gamma_{j-1,j} = E_{j-1} + F_{j-1}q_j = 1 - F_{j-1}(1 - q_j).$$

In general for a bird released at occasion i, given that it is still alive at occasion j, to still be available to contribute a recovery *from release cohort i* in year j to the count m_{ijd}, the bird must either emigrate in year i or stay in the population at risk of capture but not be caught at occasion i+1 and then not be caught after occasion i+1. Translating this description to a mathematical formula we get the recursive definition of γ_{ij}:

$$\gamma_{ij} = E_i + F_i q_{i+1} \gamma_{i+1,j} \ ,$$

which is equal to form (8): $1 - F_i(1 - q_{i+1}\gamma_{i+1,j})$. This form for γ_{ij} is unique to the assumption of permanent emigration.

It is instructive to consider three special cases. If no emigration is possible, or if otherwise fidelity is perfect, then all $F_i = 1$ and an explicit expression for γ_{ij} exists:

$$\gamma_{ij} = q_{i+1} q_{i+2} \cdots q_j \, .$$

If no recapture occurs, say because recaptures are ignored (i.e., not recorded), then all $p_i = 0$, hence all $q_i = 1$ and this leads to all $\gamma_{ij} = 1$. This result is easily proven by induction on (8). If for a particular i and j, $\gamma_{ij} = 1$, then from (8), $\gamma_{i-1,j} = 1 - F_{i-1}(1 - q_i)$ and hence if all $q_i = 1$ then $\gamma_{i-1,j} = 1$. But we already know that all $\gamma_{jj} = 1$, thus it follows that if all $p_i = 0$, then all $\gamma_{ij} = 1$.

If the adjustment entailed by γ_{ij} in (7) to the marginal recovery probabilities is not made and distribution (5) or (6) is erroneously used, the cell probability totals $\lambda_{id} + \lambda_{il}$ can exceed 1. With the adjustment, these totals will never exceed 1. By way of example consider the case represented in Table 1 for t = 3, hence model (5), for $F_i = 1$ and the extreme (but mathematically possible) case of $f_i = 1 - S_i$ for i = 1, 2. For the proper model (multinomial cells defined by (2) and (7)), we have

$$\lambda_{1d} + \lambda_{1l} = 1 - (S_1 q_2 S_2 q_3) \ ,$$

which can never exceed 1. But with the cells as defined by (2) and (4) we have

$$\lambda_{1d} + \lambda_{1l} = 1 - S_1 S_2 + S_1(p_2 + q_2 S_2 p_3) \ .$$

Let $S_1 = S_2 = p_2 = p_3 = 0.9$ and the above is equal to 1.0729. This example is just to reinforce the point that in the joint analysis of recapture-recovery data the model must properly account for the fact that known recoveries cannot thereafter become live captures.

Under the case represented by model (5), permanent emigration and t occasions for capture and release with t − 1 ring recovery time intervals, there are 4t − 6 identifiable parameters: $S_1, \ldots, S_{t-2}, p_2, \ldots, p_{t-1}, f_1, \ldots, f_{t-1}, F_1, \ldots, F_{t-2}$, and the product $(S_{t-1}F_{t-1}p_t)$. If we add to model (5) a tth year of ring recovery data, thus we have model (6), there are 4t − 4 identifiable parameters: $S_1, \ldots, S_{t-1}, p_2, \ldots, p_{t-1}, f_1, \ldots, f_t, F_1, \ldots, F_{t-2}$, and the product $(F_{t-1}p_t)$.

3.3 Cell probabilities under random emigration

The permanent emigration assumption is not the only assumption under which the time-specific models remain structurally valid. As given in Burnham (1988) an alternative parameterization of the Jolly-Seber model cells given in (2) is possible. Specifically, the "location" of a live bird during a capture occasion could be random with regard to whether or not it is in the sub-population (area) at risk of capture. Now the parameter F_i is the probability at occasion i that the bird is at risk of capture given it is alive; p_i is the probability of capture given the bird is alive and at risk of capture. The structure of π_{ijl} remains as in (2) however, we must now interpret ϕ_i as in fact being true survival rate, S_i, and p_i is actually the product of F_i and the traditional conditional capture probability.

To avoid confusion I will explicitly give cell probabilities appropriately reparameterized under this "random emigration" assumption. For live captures we have

$$\pi_{ijl} = \frac{E(m_{ijl} \mid R_i)}{R_i} = \begin{cases} S_i p^*_{i+1} & , j = i+1, \\ (S_i q^*_{i+1}) \cdots (S_{j-2} q^*_{j-1})(S_{j-1} p^*_j) & , j > i+1. \end{cases} \qquad (9)$$

The model structure, under this assumption, for the recovery cell probabilities under the joint analysis model is given by (7) with

$$\gamma_{ij} = 1 \qquad , j = i,$$

$$\gamma_{ij} = q^*_{i+1} q^*_{i+2} \cdots q^*_j \qquad , j > i.$$

Here we have $p^*_i = F_{i-1} p_i$ and $q^*_i = 1 - F_{i-1} p_i$. One can think of this model as one wherein the birds have fidelity $F_i = 1$, capture probability p^*_i, and $\phi_i \equiv S_i$. The separate F_{i-1} and p_i are confounded, hence not estimable; only their product, $p^*_i = F_{i-1} p_i$ is estimable.

As a consequence of the multiplicative structure on the γ_{ij}, we have for this model

$$\pi_{ijd} = \frac{E(m_{ijd} \mid R_i)}{R_i} = \begin{cases} f_i & , j = i, \\ (S_i q^*_{i+1}) \cdots (S_{j-1} q^*_j) f_j & , j > i. \end{cases} \qquad (10)$$

Thus under the random emigration assumption, the model for joint recapture-recovery data is the probability distribution in (5), or (6), with cell probabilities as given in (9) and (10).

Under the case represented by model (5) with t occasions for capture and release, $t-1$ ring recovery time intervals, random emigration and with full time-specificity of parameters there are $3t-4$ identifiable parameters: $S_1, \ldots, S_{t-2}, f_1, \ldots, f_{t-1}$, and the products $p^*_2, \ldots, p^*_{t-1}$, $(S_{t-1} p^*_t)$ where $p^*_i = (F_{i-1} p_i)$. If we add to this model a tth year of ring recovery data, then there are $3t-2$ identifiable parameters: $S_1, \ldots, S_{t-1}, f_1, \ldots, f_t$, and the products $p^*_2, \ldots, p^*_t$.

4. Direct Construction of the Likelihood

The approach used here to give a probability model for capture-recapture and recovery data separately or jointly is not unique. This is because there are other approaches to the capture-recapture aspect of the problem, e.g., hypergeometric models (Pollock, 1975), Poisson models (Cormack and Jupp, 1991), log-linear models (Cormack, 1989) and capture history based models (Crosbie and Manly, 1985, Lebreton et al., 1992). All these approaches to capture-recapture give either the same or almost the same likelihood (the differences relate to what parameters are the focus of the model). The most flexible approach to creating likelihoods for bird ringing data, as regards inference on survival rates, is the encounter history approach. The data are represented as the ordered sequence of all possible encounters (captures, recoveries, etc.) coded 1 for encounter, 0 otherwise. For an encounter history, h, let X_h be the count of birds having this history and let $Y_h = 1$ if at the last recorded encounter the bird is re-released; $Y_h = 0$ if the bird is not re-released on last encounter. Trailing zeros in h are used for convenience after a ring recovery.

Under the encounter history formulation the likelihood contribution of each observed encounter history is first computed; some examples are given in Table 2 wherein $t = 5$ and $j =$ the capture occasion, or the between-occasion interval during which ring recoveries are made. The first encounter is a live capture at which time the bird is ringed. Thereafter potential dead and live encounters alternate. Assuming the usual Cormack-Jolly-Seber model with

Table 2. Examples of the probabilities of specific encounter histories (h) assuming the Cormack-Jolly-Seber model (with permanent emigration); $t=5$, the study ends at occasion 5, $j=$ the capture occasion or the between-occasion interval during which ring recoveries are made (*l*ive, or *d*ead); $X_h=$ number of birds with history, h; $Y_h=0$ if the bird is not re-released at its last encounter; probabilities are conditional on first capture.

j = 112233445 h = *ldldldldl*	X_h	Y_h	Pr(h | first capture)
100010001	3	1	$\phi_1 q_2 \phi_2 p_3 \phi_3 q_4 \phi_4 p_5$
100010010	2	1	$\phi_1 q_2 \phi_2 p_3 S_3 \gamma_{34} f_4$
100100000	9	0	$S_1 \gamma_{12} f_2$
100000010	9	0	$S_1 S_2 S_3 \gamma_{14} f_4$
100000000	44	1	$1-(\lambda_{1l}+\lambda_{1d})$
101010000	11	1	$\phi_1 p_2 \phi_2 p_3 (1-(\lambda_{3l}+\lambda_{3d}))$
101010000	1	0	$\phi_1 p_2 \phi_2 p_3$
001000100	6	1	$\phi_2 q_3 \phi_3 p_4 (1-(\lambda_{4l}+\lambda_{4d}))$
001010010	6	0	$\phi_2 p_3 S_3 \gamma_{34} f_4$
000010110	2	0	$\phi_3 p_4 f_4$

permanent emigration, each Pr(h | first capture) is easily written, as has been done in Table 2. The full likelihood is then

$$\mathcal{L} = \prod_h \left[\left\{ \Pr(h \mid \text{first capture}) \right\}^{X_h} \right].$$

Numerical-based likelihood inference is feasible as the "rules" for constructing Pr(h | first capture) can be encoded in a computer program. Technically, Y enters the model (or likelihood) via the form

$$\Pr(h \mid \text{first capture}) = \Pr(\text{first to last encounter events of h} \mid \text{first capture}) \times [\Pr(\text{never encountered again} \mid \text{final encounter of h})]^Y$$

This formalism is hardly needed as $Y=0$ denotes that the probability is 1 that the bird will never be encounted again.

If the last encounter is a live capture, the form of Pr(h | first capture) up to that last occasion, j (i.e., the first symbolic component above), is well known from capture-recapture theory. The remaining multiplicative component of the probability for that encounter history is $1-(\lambda_{jl}+\lambda_{jd})$ if the bird was released, and is 1 otherwise (bird not released). If the last encounter was a ring recovery (hence the bird is dead) the (first component) probability of h is more complex and we must partition the encounter history into two parts: from the first capture occasion (i) to last live recapture occasion (o); then from that last recapture to the next encounter (j) which is the ring recovery event. Symbolically the probability is $P_{io}P_{oj}$. If $i=o$, then $P_{io}=1$. For $o>i$ the probability terms in P_{io} are just the usual Cormack-Jolly-Seber product $\phi_i \cdots \phi_{o-1} p_o$, and the corresponding product of p_w or q_w for $w=i+1$ to $o-1$, according to whether capture occurred or not at occasion w. The second multiplicative component is

$$P_{oj} = f_o \qquad , j = o ,$$

$$= S_o \cdots S_{j-1} \gamma_{oj} f_j \qquad , j > o ,$$

for γ_{oj} as given in (8). Several examples of Pr(h | first capture) are in Table 2.

Using this encounter history approach the models for joint analysis of recapture-recovery data can be generalized to include factors such as age, control-treatment, sex, capture-history dependencies, and so forth, such as the models discussed in Lebreton et al. (1992). Also, it is clearly possible to use multinomial models (or explicit estimators) for inference and then construct estimated expected counts for any encounter histories; this is important for some types of model goodness of fit testing. This direct approach of getting Pr(h | first capture) is not as convenient for theory development because there are a very large number of histories to consider, much more than 2^t and $t > 10$ is not uncommon.

5. Discussion

The purpose of this paper is to give a form of the correct joint sampling probability distribution for recapture and recovery data arising from the same set of released ringed birds. Moreover, this is herein done mostly in terms of extending the Cormack-Jolly-Seber model for survival estimation to include the ring recoveries, assuming such recoveries would have a marginal probability distribution as in the Brownie et al. (1985) time-specific model M_1. There are easy extensions or special cases possible for this joint recapture-recovery model, in particular to the use of ring recoveries from years beyond year t.

If the same ring recovery process continues in years beyond capture occasion t, then there are also recovery data m_{ijd} for $j = t+1, \ldots, k$. These counts just extend the number of mutually exclusive cells in the underlying multinomial models. The corresponding cell probabilities in model (6) for these extended ring recovery process counts is a generalization of (7):

$$\pi_{ijd} = S_i \cdots S_{t-1} \gamma_{it} S_t \cdots S_{j-1} f_j \qquad , t+1 \le j \le k \, .$$

Because the live capture process ends at occasion t, birds are no longer subject to cohort reassignment after occasion t; therefore, the role of the γ_{ij}-parameters ends after occasion t.

These additional ring recovery data (i.e., having $k > t$) can increase the precision of some parameter estimators, such as $\hat{S}_i$; however, no additional parameters of real interest become identifiable by this extension. Rather, this model extension adds to the parameters the $k - t$ identifiable *products* $(S_t f_{t+1})$, $(S_t S_{t+1} f_{t+2})$, ..., $(S_t \cdots S_{k-1} f_k)$.

Other assumptions about the recovery process can be made which lead to different forms for the γ_{ij}, in particular, the assumption that birds are not hunted, but rather recoveries come from birds found dead. The nature of the model then depends on the spatial area and time interval over which rings are found and reported. One restricted case is that only the area in which captures occur is examined for dead birds. This case is further restricted if the searches for dead birds occur only during the time of live capture effort. For such a situation we take m_{iid} to be the number of rings recovered at occasion $i+1$ from birds ringed at occasion i and not live captured at occasion $i+1$. In general, now the time interval for recoveries is the limiting case of "time" being $j - \epsilon$ to j. Then the survival parameter in (7) is really ϕ_i, not S_i. Under permanent emigration and the recovery process limited to the capture area and to the capture occasions, (7) becomes

$$\pi_{ijd} = \frac{E(m_{ijd} \mid R_i)}{R_i} = \begin{cases} \phi_i f_i & , j = i, \\ \phi_i q_{i+1} \cdots \phi_{j-1} q_j \phi_j f_j & , j > i. \end{cases} \tag{11}$$

There are $3t-4$ identifiable parameters in this model: $\phi_1, \ldots, \phi_{t-2}, p_2, \ldots, p_{t-1}, f_1, \ldots, f_{t-2}$ and the products $(\phi_{t-1}p_t)$ and $(\phi_{t-1}f_{t-1})$.

The integrity of the notation is retained in this special case even though a different indexing ($j+1$ in place of j) would make sense and be easier to understand for this case because the counts m_{ijd} are now actually obtained on occasion $j+1$. This shows the difficulty of finding a general, logical notation that would cover all special cases. Note that fidelity is not an estimable parameter in this model.

If we reparameterized (11) to have $\phi_i f_i = f_i^*$ then (11) is *structurally* identical to (10). Hence, heuristically this case of all dead recoveries being from the capture area on the capture occasions is structurally like the joint model with all $F_i = 1$. A conclusion I draw here is that if one caught and released a bird on occasion j and then found it dead very soon afterwards, that bird could be treated as a loss on (live) capture. However, birds only found dead on occasion j should not be considered a loss on capture because that will bias the estimators of p_i. This is because the encounter probabilities are different for live recapture (or resighting) vs. for dead birds.

Closed-form maximum likelihood estimators and their variances do not seem possible for the joint recapture-recovery model under permanent emigration. Nor does it seem possible to get explicit probability distributions for a reduced dimension sufficient, or minimal sufficient, statistic under this assumption. However, under the assumption of random emigration the multiplicative nature of all the cell probabilities π_{ijd} and π_{ijl} allows determination of a full rank minimal sufficient statistic and its distribution and closed-form maximum likelihood estimators and their sampling variances. Those results require far too much space to be given here, but have been derived by the author.

Numerical likelihood-based estimation and inference for joint recapture-recovery data in the model given by (5) or (6) under cell structures given by (2), (7) and (8) is straightforward because this is a standard multinomial-based model. The model can be, for example, implemented in program SURVIV (White, 1983). This has been done recently by Szymczak and Rexstad (1991) for some gadwall (*Anas strepera*) data. Alternatively, the method presented in Burnham (1989) can be implemented using SAS PROC NLIN to achieve convenient and flexible data analysis of joint recapture and recovery data. Also, methods of parsimonious model construction and selection as presented in Lebreton et al. (1992) and Burnham and Anderson (1992) are directly applicable to this extended general model. In fact, Szymczak and Rexstad (1991) used these analysis methods, such as AIC (Akaike, 1985) based model selection, for the gadwall recapture-recovery data; their paper provides the interested reader with an example of this "joint" analysis theory.

The Jolly-Seber model includes components for the estimation of population abundance parameters, N_i's and B_i's. As shown in Burnham (1991) the additional model component needed is $Pr(u_1, \ldots, u_t)$, the probability distribution of the first captures of birds. At first-capture birds are not ringed; u_i denotes the number of unringed birds caught at occasion i. All or most of these birds would be ringed and released, hence contribute to R_i. The $Pr(u_1, \ldots, u_t)$ is not effected by whether or not there are dead recoveries augmenting the usual live recaptures of the Jolly-Seber model. Also, the sets of random variables $\{m_{ijd}, m_{ijl} \mid R_i\}$ and $\{u_i\}$ are independent, hence the models given here can be extended to include abundance estimation by just including the component for $Pr(u_1, \ldots, u_t)$ from Burnham (1991) with the recapture-recoveries likelihood. This added theory is too extensive to repeat here.

Basically, however, abundance estimation under the Jolly-Seber model is the same regardless of whether or not any ring recovery data are used.

There are additional theoretical models that should be developed for the joint analysis of recapture-recovery data, in particular extensions to age-specific models. However, I think the most pressing need now is to try the existing (above) theory on a wide variety of adult data to find out if this theory is useful: is the inclusion of recapture data with recovery data (in what are meant to be ring recovery studies) worthwhile, and conversely in recapture studies is the inclusion of any recovery data worthwhile? Calculation of theoretical efficiencies will also be a useful research direction and it could be done using Monte Carlo simulation.

Acknowledgments. The author's initial work on this problem (some years ago) produced the results for random emigration and was reviewed by Cavell Brownie who pointed out that those results did not apply under permanent emigration. I thank Dr. Brownie for first noting the form of (8) as being correct for the permanent emigration model and in general for discussions and help on some aspects of this work. Helpful review comments on drafts of this paper were supplied by David Anderson, whose questions caused me to discover and correct several errors, Gary White who reviewed the final draft, allowing me to improve it some more, and an anonymous reviewer, who made some appreciated, helpful comments.

References

Akaike, H. (1985).- Prediction and entropy. pp 1-24 *in* A Celebration of Statistics, A.C. Atkinson and S.E. Fienberg (Eds). Springer-Verlag, Berlin.

Anderson, D.R. and Sterling, R.T. (1974).- Population dynamics of molting pintail drakes banded in south-central Saskatchewan. Journal of Wildlife Management, 38:266-274.

Brownie, C., Anderson, D.R., Burnham, K.P. and Robson, D.S. (1985).- Statistical inference from band recovery data--a handbook. US Fish and Wildlife Service, Resource Publication 156, 2nd edition.

Buckland, S.T. (1980).- A modified analysis of the Jolly-Seber capture-recapture model. Biometrics, 36:419-435.

Burnham, K.P. (1988).- A unified approach to animal release-resampling studies of survival processes and population estimation. North Carolina State University, Institute of Statistics Technical Report Series, #1698. Raleigh, NC. USA.

Burnham, K.P. (1989).- Numerical survival rate estimation for capture-recapture models using SAS PROC NLIN. pp 416-435 *in* Estimation and Analysis of Insect Populations, L. McDonald, B. Manly, J. Lockwood and J. Logan (Eds). Springer-Verlag, New York.

Burnham, K.P. (1991).- On a unified theory for release-resampling of animal populations. pp 11-35 *in* Proceedings of the 1990 Taipei Symposium in Statistics, M.T. Chao and P.E. Cheng (Eds). Institute of Statistical Science, Academia Sinica. Taipei, Taiwan, R.O.C.

Burnham, K.P. and Anderson, D.R. (1992).- Data-based selection of an appropriate biological model: the key to modern data analysis. pp 16-30 *in* Wildlife 2001: Populations, D. McCullough and R. Barrett (Eds). Elsevier, New York.

Burnham, K.P., Anderson, D.R., White, G.C., Brownie, C. and Pollock, K.H. (1987).- Design and analysis of fish survival experiments based on release-recapture data. American Fisheries Society, Monograph 5. Bethesda, Maryland.

Clobert, J. and Lebreton, J.D. (1991).- Estimation of demographic parameters in bird population dynamics. pp. 75-104 *in* Bird Population Studies: Their Relevance to Conservation and Management, C.M. Perrins, J.D. Lebreton and G.J.M. Hirons (Eds). Oxford University Press, Oxford, England.

Cormack, R.M. (1964).- Estimates of survival from the sighting of marked animals. Biometrika, 51:429-438.

Cormack, R.M. (1989).- Loglinear models for capture-recapture. Biometrics, 45:395-413.

Cormack, R.M. and Jupp, P.E. (1991).- Inference for Poisson and multinomial models for capture-recapture experiments. Biometrika, 78:911-916.

Crosbie, S.F. and Manly, B.F.J. (1985).- Parsimonious modelling of capture-mark-recapture studies. Biometrics, 41:399-410.

Jolly, G.M. (1965).- Explicit estimates from capture-recapture data with both death and immigration - stochastic model. Biometrika, 51:225-247.

Lebreton, J.D., Burnham, K.P., Clobert, J. and Anderson, D.R. (1992).- Modeling survival and testing biological hypotheses using marked animals: case studies and recent advances. Ecological Monographs, 62:67-118.

Mardekian, S.Z. and McDonald, L. (1981).- Simultaneous analysis of band recovery and live-recapture data. Journal of Wildlife Management, 45:484-488.

Pollock, K.H. (1975).- A K-sample tag-recapture model allowing for unequal survival and catchability. Biometrika, 62:577-583.

Pollock, K.H., Nichols, J.D., Brownie, C. and Hines, J.E. (1990).- Statistical inference for capture-recapture experiments. Wildlife Monographs, 107.

Schwarz, C.J., Burnham, K.P. and Arnason, A.N. (1988).- Post-release stratification in band-recovery models. Biometrics, 44:765-785.

Seber, G.A.F. (1965).- A note on the multiple-recapture census. Biometrika, 52:249-259.

Seber, G.A.F. (1970).- Estimating time-specific survival and reporting rates for adult birds from band returns. Biometrika, 57:313-318.

Szymczak, M.R. and Rexstad, E.A. (1991).- Harvest distribution and survival of a gadwall population. Journal of Wildlife Management, 55:592-600.

White, G.C. (1983).- Numerical estimation of survival rates from band recovery and biotelemetry data. Journal of Wildlife Management, 47:7166-728.

Marked Individuals in the
Study of Bird Population
J.-D. Lebreton and P.M. North (eds.)

On Boundary Estimation in Ring Recovery Models and the Effect of Adding Recapture Information

E.A. Catchpole[1] , S.N. Freeman[2] and B.J.T. Morgan[3]

[1] *University College, University of New South Wales, Canberra, ACT 2600, Australia,* [2] *Renewable Resources Assessment Group, Centre for Environmental Technology, Imperial College of Science, Technology and Medicine, 8 Princes Gardens, London, SW7 1NA, England and* [3] *Institute of Mathematics and Statistics, University of Kent, Canterbury, Kent, CT2 7NF, England.*

Summary: When models for survival are fitted to data by maximum-likelihood, it is not unusual to find that some parameters are estimated as 0 or 1. This can reduce the utility of the modelling, introducing bias and posing questions of inference. The problem can be quite a serious one – see, e.g., North and Cormack (1981), and Heisey and Nordheim (1990). We start by describing explicit solutions for the Cormack/Seber model for ring-recovery data (Catchpole and Morgan, 1991), and go on to show how they may be used for determining when boundary estimation will occur under that model. The work is then extended to cover the situation in which recovery information is augmented by radio-tracking data (Catchpole and Morgan, 1992, Freeman, Morgan and Catchpole, 1992). We then focus on a combined recapture and recovery experiment, in which there is a single recapture exercise at the end of the recovery experiment. Similar results are available here, concerning explicit solutions and boundary estimates. A simulation study reveals the appreciable utility of the extra information, in increasing precision and reducing the frequency of boundary estimates under the extended models of Morgan and Freeman (1989). Avenues for further work are outlined.

1. Introduction

Recent work has shown that a rich family of models can now be fitted to data describing the recoveries of dead birds which were originally ringed in the nest (e.g. Freeman and Morgan, 1990, 1992). The excellent computer program SURVIV (White, 1992) provides a robust workhorse for fitting these models to data. In some cases we are able to go further, and describe in closed mathematical form what the estimates of the model parameters look like. In this paper, as well as summarising this work, we shall also present similar material for studies in which recovery information is supplemented by recapture data obtained for the same cohorts of birds. We shall see that there are clear advantages accruing from this supplementation, and this suggests that ornithologists should, where practicable, consider designing ring-recovery experiments to include recapture, or similar information.

The parameters of probability models for data which arise from animal recapture or recovery experiments are often themselves probabilities. Models are usually fitted to data using the method of maximum likelihood, which is well described in Lebreton et al. (1992), for example. Models only make sense for probabilities in the range [0, 1], and hence usually the likelihood is only defined for this range of parameters. As a result it is not uncommon to find that the likelihood is maximised with at least one probability estimated as zero or one. Instances are recorded, for example, by North and Cormack (1981) and by Heisey and Nordheim (1990). In the latter paper, the focus was on models for the survival of nests, and in some simulation studies boundary estimates resulted in more than 50% of the total cases. Boundary estimates are a nuisance; frequently they make no practical sense, and they interfere with standard inferential procedures, as in hypothesis testing using likelihood-ratio tests, and in forming standard errors. To some extent they may reflect a paucity of data, or an inappropriate model. A further unsatisfactory feature is that in a multi-parameter model there are so many potential boundaries that may be reached.

One of the simplest models for recovery data is the Cormack-Seber model (Cormack, 1970, Seber, 1971) for birds ringed in the nest. Here some progress has been made in identifying explicit solutions (Catchpole and Morgan, 1991), and then using these to show that only a small fraction of potential boundary optima can occur (Catchpole and Morgan, 1992). This work is useful also in evaluating when boundary estimates may arise if recovery data are augmented by field observations on the survival of radio-marked first-year birds (Freeman, Morgan and Catchpole, 1992), which provide information on only one of the model parameters.

There are three main objectives to this paper:

(i) To summarise the work done to date, mentioned above. This is done in Sections 2 and 3 of the paper.

(ii) To demonstrate how this work may be extended to the case of a simple model for recovery and recapture. This material is provided in Sections (4.1) and (4.2) of the paper.

(iii) To show through simulation studies how the data resulting from adding recapture information to a more realistic recovery model can result in models with greatly reduced incidences of boundary estimates. This is done in Sections (4.3) and 5.

Discussion and avenues for further work are given in Section 6.

2. The Likelihood Surface Ridge in the Cormack-Seber Model

We start by establishing the notation to be used for a recovery experiment. For simplicity we shall use the terminology of a bird-ringing experiment: the only birds ringed are pulli. We assume that R_i birds are ringed in the ith year year of study, for $1 \le i \le k$, and that m_{ij} of these are recovered (found dead and reported) in the jth year, $i \le j \le k$. Let $f_{k-i+1} = (R_i - \sum_{j=i}^{k} m_{ij})$ denote the numbers of birds ringed in year i which are not recovered by the end of the recovery experiment, $1 \le i \le k$. We shall let ϕ_i be the probability that a bird survives its ith year of life, conditional on its having survived $(i-1)$ years, and let λ be the probability that a bird, having died, is found and reported soon after death.

Further, let $t_j = 1 - \dfrac{d_j}{u_j}$, $1 \le j \le k$, where

$$d_j = \sum_{i=1}^{k-j+1} m_{i,i+j-1}, \quad \text{and}$$

$$u_j = \sum_{i=j}^{k} (d_i + f_i).$$

Note that d_j is the number of birds observed to have died in their jth year of life, and that u_j is the number of birds of age $(j-1)$ "available for recovery" during the time-span of the exercise (i.e. not recovered at an earlier age, and not censored by the end of the study). Then t_j is the estimate of ϕ_j if it were known that $\lambda = 1$.

As is well-known, the equations for the maximum likelihood estimates $(\hat{\boldsymbol{\phi}}, \hat{\lambda})$ are linearly dependent. Catchpole and Morgan (1992) show that the set of maximum likelihood estimates forms a ridge on which $\lambda, \phi_1, \ldots, \phi_k$ all increase from their minimum values:

$$\lambda = 1-(t_1 \ldots t_k)$$

$$\phi_i = \phi_{i,\min} = \frac{t_i-(t_i \ldots t_k)}{1-(t_i \ldots t_k)}, \quad 1 \le i \le (k-1)$$

$$\phi_k = 0$$

to their maximum values:

$$\lambda = 1; \quad \phi_i = t_i, \quad 1 \le i \le k$$

(apart from certain pathological forms for the recovery table). Seber (1971) suggested selecting a point on the ridge by setting $\phi_k = \phi_{k-1}$. This may result in boundary estimates, as demonstrated by North and Cormack (1981). Knowledge of the location of the ridge then leads to the following results (Catchpole and Morgan, 1991, 1992): under Seber's constraint, boundary estimates result if and only if:

(i) $d_k > 0$, $t_{k-1} \ge t_k$; the boundary estimate is then $\hat{\lambda} = 1$;

(ii) $d_k = 0$, $d_{k-1} > 0$; the boundary estimate is $\hat{\phi}_{k-1} = \hat{\phi}_k = 0$;

(iii) $d_i = 0$ for $j \le i \le k$, $d_j > 0$, for some $j \le (k-1)$; the possible boundary estimates are $\hat{\phi}_j = \ldots = \hat{\phi}_k = 1$, or $\hat{\phi}_{j-1} = 0$ (with $\hat{\phi}_j, \ldots, \hat{\phi}_k$ arbitrary).

In all other cases, there is an explicit solution, given by:

$$\hat{\lambda} = 1 - \frac{t_1 \ldots t_{k-1}\,(t_k - t_{k-1})}{1 + t_{k-1}\,t_k - 2t_{k-1}},$$

$$\hat{\phi}_1 = (t_1 + \hat{\lambda} - 1)/\hat{\lambda},$$

$$\hat{\phi}_i = (t_1 \cdots t_i + \hat{\lambda} - 1)/(t_1 \ldots t_{i-1} + \hat{\lambda} - 1), \quad 2 \le i \le k.$$

For particular values for the parameters it is possible to calculate the probability of the event $(t_{k-1} \ge t_k)$ which causes the boundary estimates $\hat{\lambda} = 1$. We have done this for the case $k = 4$, and obtained results which agree well with the simulation results in North and Cormack (1981). Despite the valid criticisms of the use of the Seber constraint in analysis (Lakhani and Newton, 1983), it is satisfying to see how complete knowledge of conditions under which boundary estimates can occur, may be obtained in this one simple model.

We shall now consider how knowledge of the above ridge is useful in determining the occurrence of boundary estimates when the recovery experiment is augmented by other information.

3. Augmenting Recovery Data with Radio-Tracking Information

Another way of overcoming the basic nonidentifiability problem with the Cormack-Seber model is to make use of an external estimate of one of the model parameters (Lakhani, 1987). One possibility is to radio-mark n pulli. If r of these are observed to survive their first year of life then that information alone provides an estimate of ϕ_1: $\phi_1^* = r/n$. Suppose the combined likelihood, obtained from the recovery and radio-marking experiments has maximum likelihood estimates of the parameters given by: $(\tilde{\lambda}, \tilde{\boldsymbol{\phi}})$. A theorem in Catchpole and Morgan (1992) then provides the following results (except for certain sparse forms of the recovery table):

$$\phi_1^* = \tilde{\phi}_1 \text{ if and only if } \phi_{1,\min} \leq \phi_1^* \leq t_1 . \tag{1}$$

If $\phi_1^* \leq \phi_{1,\min}$ then $\tilde{\phi}_k = 0$, while if $\phi_1^* \geq t_1$, then $\tilde{\lambda} = 1$. Otherwise boundary estimates do not occur. Thus once again we are able to be precise regarding which boundaries may be reached, and also when. The key condition here involves ϕ_1^*, which provides the new information, but the boundary reached is either $\phi_k = 0$ or $\lambda = 1$. In simulation studies we have found that the boundary $\lambda = 1$ is rarely attained, in contrast to the situation of Section 2, when the Seber constraint is applied to select a point on the ridge, and when $\lambda = 1$ was the only boundary to be reached, if various zero cells are excluded. The detail of simulation studies of augmentation by radio-tracking is given by Freeman, Morgan and Catchpole (1992). It was found that the condition of (1) was frequently violated, resulting in large biases for the estimator $\hat{\phi}_k$. However, especially if a large radio-marking experiment was carried out, the bias for the remaining parameters was not too large; values of the order of 14% or less were obtained for the simulations carried out.

These results apply however the external estimate of ϕ_1 was obtained. In a radio-tracking study one might envisage a number of experimental problems which for the most part we do not consider here. Should the external estimator, ϕ_1^*, be biased, e.g., due to survival being reduced as a consequence of a radio-marked bird having to carry an additional load, then condition (1) is violated more frequently than before. More serious bias can now be found for all of the parameters. Somewhat ironically, in this case it is a disadvantage to run a large radio-marking study, since that increases the precision of the estimate ϕ_1^*, but this is not an estimate of the annual first-year survival probability under the model for the recovery data. This clearly presents a serious dilemma for the experimenter.

We next consider an alternative approach to obtaining a unique maximum-likelihood estimate, which involves further experimental effort, but in this case of a recapture kind.

4. Combining Recovery and Recapture

4.1. A simple model: common recapture probabilities

Potentially, any ringing programme provides information both on the time of an individual's death and times at which it is resighted, alive; which component is dominant will depend on the species and the method in question. We suppose here that at the end of the recovery experiment there is a single planned recapture episode, which is used to gain additional information about the parameters. This differs from the scenario in Section 3 in two key ways:

(a) the new data provide information on more than a single parameter;

(b) The new data relate to the same birds, so that this is different from combining two independent sets of recapture and recovery data, as done by Buckland and Baillie (1987), for example. Therefore there is no need to make an assumption of identical survival rates between two groups of birds. We may note here that this recapture design, which excludes information on live birds except during the single recapture episode, results in data which Lebreton et al. (1990) describe as *transversal* data.

In this paper we shall not, because of restriction of space, present analyses of real data of the kind described here. As this paper goes to press we are analysing an extensive such data set on Shags, *Phalacrocorax aristotelis*, kindly provided by Dr. Nicholas Aebischer. This is one of three data sets which correspond to the recovery/recapture procedure described here, of which we became aware following discussions at EURING 92.

In many practical instances we might expect recapture at the conclusion of each year, and we shall discuss this extension later. The simplest assumption, which we shall now make, is that there is a single probability, c, of recapture, irrespective of age of bird/year of ringing. Many birds and mammals exhibit natal dispersal (Greenwood, 1980). The extent of this in relation to the study area is clearly crucial in determining the validity of this assumption in a particular case. However, there are birds, often passerines or colonial seabirds, in which a degree of philopatry is found (Greenwood, 1985). We shall later relax the assumption to allow for species which spend their early years away from the area of their birth. We may also note two further instances in which this model may be plausible: large mammals restricted by geography to a known area, and those wildfowl species in which entire populations winter in the same, relatively restricted area. In these cases the resighting probability can be high, see, e.g., Clutton-Brock, Price, Albon and Jewell (1991) and Owen and Black (1991).

For illustration, for a three-year study ($k = 3$), the recovery and recapture table will then be as shown below:

No. ringed	Recoveries			Recaptures	Birds lost
R_1	m_{11}	m_{12}	m_{13}	r_3	$f_3 - r_3$
R_2		m_{22}	m_{23}	r_2	$f_2 - r_2$
R_3			m_{33}	r_1	$f_1 - r_1$

The corresponding probability model is then given by the following values:

Recoveries			Recaptures
$(1-\phi_1)\lambda$	$\phi_1(1-\phi_2)\lambda$	$\phi_1\phi_2(1-\phi_3)\lambda$	$\phi_1\phi_2\phi_3 c$
	$(1-\phi_1)\lambda$	$\phi_1(1-\phi_2)\lambda$	$\phi_1\phi_2 c$
		$(1-\phi_1)\lambda$	$\phi_1 c$

The extra effort of the recapture experiment partitions up the $\{f_i\}$, providing extra information at the cost of introducing one new parameter, c. For a k-year experiment we have $(k+2)$ parameters and $\{k(k+3)/2\}$ degrees of freedom. Thus, for the case $k = 3$, under the model we have 9 degrees of freedom and 5 parameters. However, for the recovery data alone the equations for maximum-likelihood estimation are dependent. As we saw in Sections 2 and 3, we needed just one further equation for the parameters $(\lambda, \boldsymbol{\phi})$. The required information comes from the recapture experiment. In effect, there are just 6 degrees of freedom available from the likelihood, so that at most 6 identifiable parameters can be fitted in the model, and an extra 3 degrees of freedom in the data. Thus a "saturated" model with 6 parameters would still have 3 degrees of freedom for a χ^2 goodness-of-fit test. We shall see an example of this later in Section (4.3).

Each of the multinomial distributions contributing to the likelihood now contains an additional component. Mardekian and MacDonald (1981) made a similar extension to the recovery models developed by Brownie et al. (1985), in which adult survival depends on year rather than age. See also Szymczak and Rexstad (1991). Buckland (1980, 1982) describes a modification of the Jolly-Seber recapture model that can estimate age-dependent survival by including known deaths.

Under our model, the likelihood for the recovery and recapture data is given by:

$$L \propto \{\lambda(1-\phi_1)\}^{m_{11}+m_{22}+m_{33}}\{\lambda\phi_1(1-\phi_2)\}^{m_{12}+m_{23}} \times \{\phi_1\phi_2(1-\phi_3)\lambda\}^{m_{13}} \times$$
$$(\phi_1\phi_2\phi_3\, c)^{r_3}\,(\phi_1\phi_2\, c)^{r_2}\,(\phi_1\, c)^{r_1} \times \{1-\phi_1 c-\lambda(1-\phi_1)\}^{f_1-r_1} \times$$
$$\{1-\phi_1\phi_2\, c-\lambda(1-\phi_1\phi_2)\}^{f_2-r_2} \times \{1-\phi_1\phi_2\phi_3\, c-\lambda(1-\phi_1\phi_2\phi_3)\}^{f_3-r_3}\,. \quad (2)$$

This likelihood may now be factored, as $L \propto L_1 L_2$ where L_1 is the likelihood for the recovery data alone, under the Cormack-Seber model, and L_2 is a product binomial likelihood, where the probabilities are probabilities of recapture, conditional upon not having been recovered dead. (A similar factorisation, for a model for nest survival, appears in Pollock and Cornelius, 1988). Thus

$$L_2 = \frac{(\phi_1 c)^{r_1}\,\{1-\phi_1 c-\lambda(1-\phi_1)\}^{f_1-r_1}}{\{1-\lambda(1-\phi_1)\}^{f_1}} \times \frac{(\phi_1\phi_2 c)^{r_2}\,\{1-\phi_1\phi_2 c-\lambda(1-\phi_1\phi_2)\}^{f_2-r_2}}{\{1-\lambda(1-\phi_1\phi_2)\}^{f_2}} \times$$
$$\frac{(\phi_1\phi_2\phi_3 c)^{r_3}\,\{1-\phi_1\phi_2\phi_3 c-\lambda(1-\phi_1\phi_2\phi_3)\}^{f_3-r_3}}{\{1-\lambda(1-\phi_1\phi_2\phi_3)\}^{f_3}}\,. \quad (3)$$

The set of parameter values which maximise L_2 will form a surface, rather than a ridge, since in (3) we have 5 parameters and the optimisation will provide 3 equations. It is possible to show that for $k \geq 3$, except for a particular correspondence between the ringing and recapture data sets, it is not possible simultaneously to maximise L_1 and L_2 with a common set of parameter estimates. (In the practically uninteresting case of $k = 2$, however, an explicit solution maximising both L_1 and L_2 does exist.) It follows that, in general, for $k \geq 3$, the global maximum of L does not lie on the ridge of the likelihood surface under the Cormack-Seber model for the recovery data alone.

We have seen in Section 3 that with augmentation of the kind considered there, the boundary estimate of $\phi_k = 0$ was frequently obtained. Now, in general, ϕ_k is a multiplying factor, and so that automatically ensures that, unless $r_k = 0$, the boundary $\phi_k = 0$ is excluded in this case. We shall return to this particular point in Section 5.

A new possibility for boundary estimation now arises, viz. when $c = 1$. An illustrative 100 simulation run for the case $k = 3$, with 1000 birds ringed each year and $\phi_1 = 0.4$, $\phi_2 = 0.5$, $\phi_3 = 0.6$, $\lambda = 0.1$, $c = 0.2$ resulted in just one boundary value, with $\hat{c} = 1$. The averages of the parameter estimates were: $\hat{\phi}_1 = 0.39$ (0.045), $\hat{\phi}_2 = 0.49$ (0.059), $\hat{\phi}_3 = 0.59$ (0.110), $\hat{\lambda} = 0.10$ (0.009), $\hat{c} = 0.21$ (0.03) (sample standard deviations are given in parentheses). In this case we are not able to comment, as yet, on the possibilities for boundary estimation in more detail than we have already. However in the next section we allow c to vary with age of bird and, as we shall see, this returns us to a situation very similar to that described in Section 2.

4.2. Age-dependent recapture

We now suppose that the recapture probability for i-year-old birds is denoted by c_i. This adds k parameters to the Cormack-Seber model, while the recapture exercise also adds just k extra degrees of freedom. If we now consider the joint likelihood as a function of the parameters $\{\boldsymbol{\phi}, \lambda, \mathbf{c}\}$, then it is maximised along a further ridge, corresponding to a curve $\mathcal{C}$, in the parameter space, similar to that in the Cormack-Seber model. On $\mathcal{C}$, as λ increases, so do $\phi_1, \ldots, \phi_k$, but $c_1, \ldots, c_k$ all decrease. At one end of the ridge λ achieves its maximum value of 1, ϕ_i its maximum value of t_i and c_i its minimum value of $w_i = r_i / f_i$ $(1 \le i \le k)$. At the other end, λ and ϕ_i take on their minimum (non-zero) values and c_i their maximum values, which are < 1 for $1 \le i \le k-1$, but $c_k = 1$. This is a simplification of a general rather complicated result, but the behaviour described applies if, as is often the case,

$$\frac{1-t_1 \cdots t_i}{1-w_i t_1 \cdots t_i} \le \frac{1-t_1 \cdots t_k}{1-w_k t_1 \cdots t_k}, \quad \text{for } 1 \le i \le k . \tag{4}$$

If (4) fails to hold then the more complex general result (not given here) is needed. We see that in this case there are just two possible boundary estimates, viz. $\lambda = 1$, and $c_k = 1$. The boundary estimate $\phi_k = 0$ is now impossible as long as $r_k > 0$, which accords perfectly with commonsense, of course. (We can deal separately with the case $r_k = 0$, which may well arise from studies that are long, relative to life span, and with pathological forms of the recovery table.) Note that, on $\mathcal{C}$, the parameters are perfectly (but not linearly) related, so that the correlation coefficients within the group $(\hat{\phi}_1, \ldots, \hat{\phi}_k, \hat{\lambda})$ and within the group $(\hat{c}_1, \ldots, \hat{c}_k)$ will all be close to +1, and between these groups will be close to −1. If a constraint is added to obtain a unique solution, then we would expect the pattern of correlations between the estimators to adhere to this form, as far as the signs are concerned, and we shall see an example of this later.

An Example: Here we set $k = 3$ and make use of the following artificial data:

1000:	70	18	9	20
1000:		75	19	36
1000:			62	77

It is readily verified that the conditions (4) hold, and so the bounds for the 7 parameters are as shown below:

Bounds under $\{\boldsymbol{\phi}, \lambda, \mathbf{c}\}$	Bounds under $\{\boldsymbol{\phi}, \lambda\}$	% reduction of interval width due to recapture
$0.099 \le \lambda \le 1$	$0.097 \le \lambda \le 1$	0.22
$0.300 \le \phi_1 \le 0.931$	$0.286 \le \phi_1 \le 0.931$	2.17
$0.371 \le \phi_2 \le 0.980$	$0.327 \le \phi_2 \le 0.980$	6.74
$0.180 \le \phi_3 \le 0.990$	$0.000 \le \phi_3 \le 0.990$	18.18
$0.082 \le c_1 \le 0.255$		
$0.040 \le c_2 \le 0.326$		
$0.022 \le c_3 \le 1$		

These bounds correspond to the end points of the curve $\mathcal{C}$. Also shown above are the corresponding bounds which result if the recapture experiment had not taken place (from Section 2).

4.3. The case $c_{k-1} = c_k$

As with the original Cormack-Seber model, a single constraint is all that is required to return a unique maximum likelihood estimate for all of the parameters. This approach is open to the same criticism as the use of the Seber constraint, though it might be argued that to set $c_{k-1} = c_k$ is more plausible. In practice, for all but very small k as many as $(k-1)$ separate capture probabilities are unlikely to be necessary. We consider this arbitrary constraint not only as a practical device but also to tie in with the work of previous sections and as a prelude to a more realistic approach in Sections 5 and 6. This constraint is attractive in that it results in a simple explicit solution (not given here), with no boundary estimates, provided that

$$w_k < w_{k-1} < w_k + \frac{(1-t_k)(1-w_k)}{1-t_1 \cdots t_k} . \tag{5}$$

If (5) is not satisfied then either $w_{k-1} \le w_k$, and the boundary estimate is $\lambda = 1$, or $w_{k-1} \ge w_k + (1-t_k)(1-w_k)/(1-t_1 \ldots t_k)$, and the boundary estimate is $c_{k-1} = c_k = 1$.

For the data of the above example it is readily verified that (5) is satisfied, and our explicit solution agrees with the maximum likelihood estimates obtained from SURVIV (White, 1983, 1992) and shown in Table 1.

Table 1:
The maximum likelihood parameter estimates resulting from fitting the model $\{\boldsymbol{\phi}, \lambda, \mathbf{c}\}$ to the artificial data of the example of §4.2, with the constraint, $c_2 = c_3$. Estimated asymptotic standard errors are given along the diagonal in parentheses, with estimated correlations off the diagonal.

Parameter	Estimate	ϕ_1	ϕ_2	ϕ_3	λ	c_1	c_2
ϕ_1	0.36	(0.06)	0.77	0.65	0.77	-0.84	-0.88
ϕ_2	0.52		(0.12)	0.67	0.73	-0.65	-0.91
ϕ_3	0.55			(0.15)	0.62	-0.55	-0.77
λ	0.11				(0.01)	-0.66	-0.77
c_1	0.21					(0.04)	0.73
$c_2 = c_3$	0.19						(0.09)

The X^2 goodness-of-fit statistic is $X^2 = 1.37$, which is non-significant when referred to χ_3^2 tables. The maximum of the log-likelihood is $l_{\max} = -24.61$. Note that this barely changes when we set $c_1 = c_2 = c_3$, resulting in a value of $l_{\max} = -24.64$. We note finally that the pattern of signs for the correlations above is precisely as predicted.

5. Recovery and recapture with age-dependent first-year survival

It has been shown by Morgan and Freeman (1989) that the Cormack-Seber model can be extended by allowing ϕ_1 to be a function of calendar year, to give $\{\phi_{1i}\}$. This elaborated model is more biologically realistic and may be fitted by maximum-likelihood without constraint. Unless the data do not allow us to reject a model for which $\phi_{1i} = \phi_1$, for all i, ridge problems are no longer encountered. As pointed out by M. Crowder *(pers.comm.)*, under this model, if one establishes a null hypothesis that $\phi_{1i} = \phi_1$, for all i, then the situation is analogous to one where a nuisance parameter disappears under the null hypothesis, leading to a nonregular asymptotic procedure (see Davies, 1987). This is a feature which we plan to investigate further. The extended model has been successful as a tool for data analysis, and may also be further extended to allow λ to vary with calendar year of recovery (Freeman and Morgan, 1990, 1992), and in some cases with respect to age of bird. See here the paper by Rinne et al. in this volume. However there is always a possibility that following a likelihood-ratio test of $\phi_{1i} = \phi_1$, for all i, the model may reduce to the Cormack-Seber ridge, and also boundary optima may result. (The power of a test of equality of the $\{\phi_{1i}\}$ depends on the dimensions and parameters of any ring-recovery study. For further discussion, see Freeman and Morgan, 1992). We shall now consider the effect of adding an end-of-study recapture experiment to the recovery study, assuming that the basic Freeman-Morgan model holds. In this case the parameters of the model are:

$$(\lambda, c, \boldsymbol{\phi}_1, \phi_2, \ldots, \phi_k),$$

where $\boldsymbol{\phi}_1 = (\phi_{11}, \ldots, \phi_{1k})$, and we just consider a single constant recapture probability. So far we have only conducted an explorative simulation study, and the results of this are summarised in Table 2.

For this simulation example the picture is surprisingly clear and convincing: even a low value for c (for instance c as low as 5%) greatly improves overall precision, and much reduces the incidence of boundary estimates. In his study of Shags, Aebischer (pers.comm) estimated $c \approx 65\%$. It is interesting to see how little varying c over a wide range changes these conclusions, except that for large c (here $c = 0.9$) boundary estimates for c can arise. This is a feature that we plan to investigate thoroughly – for example through a detailed examination of the parameter correlation structure.

The kind of improvements that are seen here are roughly equivalent to carrying out a much larger ringing exercise (say 5- or 10-fold in extent – see Morgan and Freeman, 1989). This then raises the interesting question of relative costs of recovery and recapture exercises. However, apart from this, the added recapture exercise provides the safety-net that under the extended model one does not have to worry about the model reducing to the Cormack-Seber ridge.

Table 2:
A summary of a simulation study on the effect of the recapture probability c. The model simulated had parameters: $\{\lambda, \phi_{1i}, \phi_2, \phi_3, \phi_4, \phi_5, c\}$, with $\lambda = 0.1$, $\phi_{11} = 0.4$, $\phi_{12} = 0.5$, $\phi_{13} = 0.6$, $\phi_{14} = 0.5$, $\phi_{15} = 0.4$, $\phi_2 = 0.5$, $\phi_3 = 0.6$, $\phi_4 = 0.5$, $\phi_5 = 0.4$. 100 simulations were made using this parameter configuration, with 1000 birds ringed in each year. For each of the 100 recovery simulations, we made 7 simulations of a recapture experiment, corresponding to recapture probabilities $c = 0$, 0.05, 0.1, 0.3, 0.5, 0.7 and 0.9.
Shown below for each parameter is the frequency with which it was estimated in each interval. In all cases the frequency shown in boldface corresponds to the interval containing the value of the parameter used in the simulation.

			$\hat{\phi}_{11}$	$\hat{\phi}_{12}$	$\hat{\phi}_{13}$	$\hat{\phi}_{14}$	$\hat{\phi}_{15}$	$\hat{\phi}_2$	$\hat{\phi}_3$	$\hat{\phi}_4$	$\hat{\phi}_5$	$\hat{\lambda}$	c
	0 –	0.05								4	51		
	0.05	0.15								2		**84**	
	0.15	0.25					2			3	1	3	
	0.25	0.35	22				21	4	2	16	3	2	
	0.35	0.45	**43**	17		25	**31**	21	14	18	**4**	1	
$c = 0$	0.45	0.55	18	**50**	24	**40**	23	**38**	23	**16**	4	1	
	0.55	0.65	4	16	**42**	15	6	7	**24**	3	1	1	
	0.65	0.75	1	4	19	7	4	10	7	8	6		
	0.75	0.85	2	2	2	2	1	4	12	12	6	1	
	0.85	0.95	10	10	11	11	11	5	7	5	11		
	0.95	1.00		1	2			8	11	13	13	7	
	0 –	0.05									22		**72**
	0.05	0.15									1	**100**	28
	0.15	0.25					1			2	9		
	0.25	0.35	21			1	29	1		14	14		
	0.35	0.45	**64**	23		23	**45**	16	5	26	**16**		
$c = 0.05$	0.45	0.55	14	**61**	22	**65**	21	**74**	29	**26**	12		
	0.55	0.65	1	16	**67**	11	4	9	**43**	25	9		
	0.65	0.75			11				22	6	7		
	0.75	0.85							1	1	6		
	0.85	0.95											
	0.95	1.00									4		
	0 -	0.05									5		
	0.05	0.15									4	**100**	**99**
	0.15	0.25					2				16		1
	0.25	0.35	19			1	36			7	24		
	0.35	0.45	**67**	18		20	**27**	18	3	20	**13**		
$c = 0.1$	0.45	0.55	14	**67**	26	**68**	21	**73**	24	**39**	13		
	0.55	0.65		15	**63**	11	4	9	**57**	27	7		
	0.65	0.75			11				16	7	10		
	0.75	0.85									4		
	0.85	0.95											
	0.95	1.00									4		

Table 2 continued:

			$\hat{\phi}_{11}$	$\hat{\phi}_{12}$	$\hat{\phi}_{13}$	$\hat{\phi}_{14}$	$\hat{\phi}_{15}$	$\hat{\phi}_{2}$	$\hat{\phi}_{3}$	$\hat{\phi}_{4}$	$\hat{\phi}_{5}$	$\hat{\lambda}$	c
	0 –	0.05											
	0.05	0.15									3	**100**	
	0.15	0.25					1				12		16
	0.25	0.35	20			1	26			3	19		**67**
$c=0.3$	0.35	0.45	**69**	20		26	**54**	18	1	26	**34**		15
	0.45	0.55	11	**66**	20	**62**	16	**79**	27	**47**	13		2
	0.55	0.65		14	**70**	11	3	3	**57**	21	14		
	0.65	0.75			10				15	2	2		
	0.75	0.85								1			
	0.85	0.95									1		
	0.95	1.00									2		
	0 –	0.05											
	0.05	0.15									2	**100**	
	0.15	0.25									10		
	0.25	0.35	22			1	23				26		1
	0.35	0.45	**67**	23		25	**60**	17		28	**30**		21
$c=0.5$	0.45	0.55	11	**64**	20	**64**	15	**79**	25	**48**	22		**50**
	0.55	0.65		13	**71**	10	2	4	**62**	23	5		22
	0.65	0.75			9				13	1	3		4
	0.75	0.85									1		2
	0.85	0.95											
	0.95	1.00									1		
	0 –	0.05											
	0.05	0.15									1	**100**	
	0.15	0.25									15		
	0.25	0.35	23			1	26				24		
	0.35	0.45	**67**	22		23	**60**	17	1	27	**31**		
$c=0.7$	0.45	0.55	10	**63**	19	**64**	12	**79**	24	**53**	21		3
	0.55	0.65		15	**71**	12	2	4	**61**	19	5		29
	0.65	0.75			10				14	1	2		**34**
	0.75	0.85											22
	0.85	0.95											7
	0.95	1.00									1		5
	0 -	0.05											
	0.05	0.15										**100**	
	0.15	0.25									6		
	0.25	0.35	23				18				26		
	0.35	0.45	**67**	21		14	**67**	11		27	**45**		
$c=0.9$	0.45	0.55	10	**64**	18	**74**	13	**85**	24	**51**	15		
	0.55	0.65		15	**72**	12	2	4	**62**	19	5		
	0.65	0.75			10				14	3	2		11
	0.75	0.85									1		24
	0.85	0.95											**26**
	0.95	1.00											39

6. Discussion and further work

In this paper we have concentrated on very simple models in order to illustrate two basic features, viz.

(i) It is sometimes possible to make some progress with regard to obtaining explicit solutions for maximum-likelihood estimates of model parameters, and providing insight with regard to boundary estimation.

(ii) The simple expedient of terminating a recovery experiment with a single recapture exercise is useful in two ways. It adds extra information to the Cormack-Seber model without the need to bring in separate data, for which parameters may differ, and it greatly adds to the utility of fully age-specific Freeman-Morgan models, by reducing variability and also the incidence of boundary estimates.

So far explicit solutions have only been obtained for models with many parameters, the number of which one would hope to reduce in the interest of parsimony. However we feel it is important to realise that it is not always necessary to use numerical optimisation in order to fit models to data by maximum likelihood. Explicit solutions may be used to check the performance of numerical procedures, and may in turn be extended to correspond to other situations. We are currently investigating such extensions.

Much more work remains to be done. For instance we have not detailed a range of models for which explicit solutions exist. One example is provided by the Freeman-Morgan models, with c varying by age of birds and $k = 3$: $\{\lambda, \mathbf{c}, \boldsymbol{\phi}_1, \phi_2, \phi_3\}$. We have not yet considered the case of variable λ. In practice recaptures may be recorded each year, following the annual ringing exercise. Work has begun here on the construction and fitting of suitable models, with particular reference to data on Soay sheep (Clutton-Brock et al., 1991). In some cases we might even regard the recovery data as augmenting the recapture data, rather than the opposite, which has been the case here. As usual, multiple recaptures raise the possibility of having to account for capture effects. Precision and boundary estimation need to be fully evaluated here. Simulation studies of the $\{\lambda, \mathbf{c}, \boldsymbol{\phi}_1, \phi_2, \ldots, \phi_k\}$ and the annual recapture models have been highly encouraging. We need also to consider how resighting data may be incorporated.

An area which we have not discussed is how a Bayesian approach interacts with problems of ridges and so forth. Preliminary investigations are described by Freeman (1990, Chapter 6) and extensive work, making use of the Gibbs sampler, is described in Vounatsou and Smith (1992).

It is worth discussing here the need for elaborate probability models. In practice we can expect to reduce the numbers of parameters in models greatly, following the usual model-selection procedures, such as likelihood-ratio tests or using the Akaike Information Criterion, as now favoured by Lebreton et al. (1992). However, to initiate the model-selection procedure, it is attractive to be able to fit general models with many parameters. Undoubtedly boundary estimates can arise because sparseness of a data set leads to a flatness of the likelihood surface in certain directions. Catchpole and Morgan (1992) provide results which allow examination of how the incidence of boundary estimates, for the Cormack-Seber model under the Seber constraint, changes as sample sizes increase. However, boundary estimates may arise in other cases, for instance if it was found necessary to fit a model in which λ varied with respect to calendar year, in some cases one might still encounter a need for $\lambda_j = 1$ for some j. An alternative approach is to start with an acceptable *minimal* model, and then test to see how it should be elaborated to describe data using *score* tests (see, e.g., Atkinson, 1985, p.93). Preliminary work has been reported by Twell (1992) in which such tests have been shown to result in the same conclusions as likelihood-ratio tests in the context of the models of Freeman and Morgan (1990, 1992). In Twell (1992) the heavy algebra needed for the score tests was carried out by the symbolic algebra package, MAPLE. This approach might avoid many of the ridge/boundary estimation problems that can arise with elaborate multi-parameter models. Research into the relative merits of these different approaches is currently taking place.

We note finally that our demonstration of the effectiveness of adding recapture information to recovery information has led Ken Lakhani to suggest that, where possible, ring-recovery experiments should be *planned*, to include recaptures.

Acknowledgements: We are grateful to the editors and Kens Burnham and Lakhani for the helpful and careful review of this paper.

References

Atkinson, A.C. (1985).– Plots, Transformations and Regression. Oxford University Press, Oxford.

Brownie, C., Anderson, D.R., Burnham, K.P. and Robson, D.S. (1985).– Statistical inference from band recovery data – a handbook. United States Department of the Interior, Fish and Wildlife Service, Resource Publication No. 156. Washington, D.C., U.S.A.

Buckland, S.T. (1980).– A modified analysis of the Jolly-Seber capture-recapture model. Biometrics, **36**: 419-435.

Buckland, S.T. (1982).– A mark-recapture survival analysis. J.Anim.Ecol., **51**: 833-847.

Buckland, S.T. and Baillie, S.R. (1987).– Estimating bird survival rates from organized mist-netting programmes. Acta Ornithologica, **23**: 89-100.

Catchpole, E.A. and Morgan, B.J.T. (1991).– A note on Seber's model for ring-recovery data. Biometrika, **78**: 917-919.

Catchpole, E.A. and Morgan, B.J.T. (1992).– Boundary estimation in ring recovery models. Submitted for publication to J.R.Statist.Soc., B.

Clutton-Brock, T.H., Price, O.F., Albon, S.D. and Jewell, P.A. (1991).– Persistent instability and population regulation in Soay sheep. J.Anim.Ecol., **60**: 593-608.

Cormack, R.M. (1970).– Statistical appendix to Fordham's paper. J.Anim.Ecol., **39**: 24-27.

Davies, R.B. (1987).– Hypothesis testing when a nuisance parameter is present only under the alternative. Biometrika, **74**: 33-44.

Freeman, S.N. (1990).– Statistical analysis of avian breeding and survival. Unpublished Ph.D. thesis, University of Kent, Canterbury, England.

Freeman, S.N. and Morgan, B.J.T. (1990).– Studies in the analysis of ring-recovery data. The Ring, **13**: 271-287.

Freeman, S.N. and Morgan, B.J.T. (1992).– A modelling strategy for recovery data from birds ringed as nestlings. Biometrics , **48**: 217-236.

Freeman, S.N., Morgan, B.J.T. and Catchpole, E.A. (1992).– On the augmentation of ring-recovery data with field information. J.Anim.Ecol., **61**, 3, 649-657.

Greenwood, P.J. (1980).– Mating systems, philopatry, and dispersal in birds and mammals. Anim.Behav. **28**: 1140-1162.

Greenwood, P.J. (1985).– Article 'Dispersal'. pp.140-141 *in* Campbell, B. and Lack, E. (Eds) Dict.Birds. Calton and Vermillian.

Heisey, D.M. and Nordheim, E.V. (1990).– Biases in the Pollock and Cornelius method of estimating nest survival. Biometrics, **46**: 855-862.

Lakhani, K.H. (1987).– Efficient estimation of age-specific survival rates from ring-recovery data of birds ringed as young, augmented by field information. J.Anim.Ecol., **56**: 969-987.

Lakhani, K.H. and Newton, I. (1983).– Estimating age-specific bird survival rates from ring-recoveries – can it be done? J.Anim.Ecol., **52**: 83-91.

Lebreton, J-D., Hemery, G., Clobert, J. and Coquillart, H. (1990).– The estimation of age-specific breeding probabilities from recaptures or resightings in vertebrate populations. I.Transversal models. Biometrics, **46**: 609-622.

Lebreton, J-D., Burnham, K.P., Clobert, J. and Anderson, D.R. (1992). – Modeling survival and testing biological hypotheses using marked animals: a unified approach with case studies. Ecol.Monographs, **62**: 67-118.

Mardekian, S.Z. and McDonald, L. (1981).– Simultaneous analysis of band-recovery and live-recapture data. J.Wildl.Manage., **45**: 484-488.

Morgan, B.J.T. and Freeman, S.N. (1989).– A model with first-year variation for ring-recovery data. Biometrics, **45**: 1087-1101.

North, P.M. and Cormack, R.M. (1981).- On Seber's method for estimating age-specific bird survival rates from ringing recoveries. Biometrics, **37**: 103-113.

Owen, M. and Black, J.M. (1991).– The importance of migration mortality in non-passerine birds. pp.360-372 *in* Bird Population Studies – Relevance to Conservation and Management, C.M. Perrins, J.-D. Lebreton and G.J.M. Hirons (Eds). Oxford University Press, Oxford.

Pollock, K.H. and Cornelius, W.L. (1988).– A distribution-free nest survival model. Biometrics, **44**: 397-404.

Seber, G.A.F. (1971).– Estimating age-specific survival rates for birds from bird-band returns when the reporting rate is constant. Biometrika, **58**: 491-497.

Szymczak, M.R. and Rexstad, E.A. (1991).– Harvest distribution and survival of a Gadwall population. J.Wildl.Manage., **55**: 592-600.

Twell, J. (1992).- Score tests in models for bird survival. Unpublished Diploma project report, University of Kent, Canterbury, England.

Vounatsou, P. and Smith, A.F.M. (1992).– Bayesian analysis of ring-recovery data via Gibbs sampling. Tech. Report, Statistics Section, Imperial College, London.

White, G.C. (1983).– Numerical estimation of survival rates from band recovery and biotelemetry data. J.Wildl.Manage., **47**: 716-728.

White, G.C. (1992).– PC SURVIV User's Manual Version 1.4. Department of Fishery and Wildlife Biology, Colorado State University, Fort Collins, CO 80523, U.S.A.

Marked Individuals in the
Study of Bird Population
J.-D. Lebreton and P. M. North (eds.)

Survival Rate Estimation in the Presence of Tag Loss Using Joint Analysis of Capture-Recapture and Resighting Data

James D. Nichols and James E. Hines

U.S. Fish and Wildlife Service, Patuxent Wildlife Research Center, Laurel, MD 20708, USA

Summary: Studies using resightings of marked birds typically make use of readily-observable tags that are not retained as well as metal legbands. We review methods for estimating survival rate with open capture-recapture/resighting models when tag loss is not negligible. All methods rely on data from double-banding studies, usually carried out as part of the resighting study by application of metal legbands to all birds marked with alternative markers. When tag loss is homogeneous, the methods of Arnason and Mills (1981) and Pollock (1981) can be used. When rates of tag loss depend on time since marking, then a cohort approach can be used and is similar to the methods appropriate for homogeneous tag loss. In addition, Kremers (1987) and Nichols et al. (1992) developed models for the joint analysis of recapture and resighting data in the presence of tag loss. We emphasize the importance of obtaining recapture data in observation-based studies in which tag loss is likely to be a problem. We discuss the allocation of effort to recaptures and resightings for such studies.

1. Introduction

Many studies of birds rely on resightings, not recaptures, of marked birds. In such studies, birds are typically marked with metal legbands as well as with some other type of tag that is observable from a distance. Although many different types of observable tags have been used in bird studies, including neckbands, color legbands, patagial tags, and nasal saddles, these tags frequently do not have retention rates as high as those of metal legbands. If studies include recapture, as well as resighting, efforts, then data on presence or absence of observable tags on recaptured birds with metal legbands can be used to estimate rates of tag loss. Methods for estimating rates of tag loss from such double-tagging studies have been presented by Beverton and Holt (1957), Robson and Regier (1966), Caughley (1971), Seber (1982), and Hestbeck and Malecki (1989).

Loss of tags or bands produces negative bias in survival rates estimated from both band-recovery (Nelson et al. 1980) and capture-recapture/resighting (Arnason and Mills 1981, Pollock 1981) models. In this paper, we briefly review methods of obtaining unbiased (or less biased) estimates of survival rate in the presence of tag loss using capture-recapture/resighting models. We devote special attention to the case where rate of tag loss varies with time since initial tagging and discuss models using both recapture and resighting data to estimate survival rate in this case. We then discuss the relative contributions of recapture data and resighting data to the estimation of survival rate with these models.

2. General Study Design

The general study design that we consider is that of a capture-resighting study typical of those used to study goose and swan survival in North America (e.g., Hestbeck and Malecki 1989; Hestbeck et al. 1990; Samuel 1990*a*, 1990*b*). Birds are captured, marked with metal legbands and with readily-observable tags such as neckbands, and then released. Both the legbands and observable tags provide individual identifications (i.e., we do not consider batch marks). Observations of marked birds are sometimes obtained over a protracted period (e.g., October-March) each year permitting subdivision of the data into multiple sample periods per year (e.g., early winter and late winter). However, in this paper we will consider the situation of one sample period per year (e.g., a 6-week period in mid-winter) for a number of consecutive years. At each sample period, observers traverse a specified study area at least one time, recording tag numbers of all birds that are found. We consider only survival rate estimation, so observers need not obtain counts of unmarked birds.

In addition to data on observations of marked birds, birds are captured at each sample period. Captures include primarily new birds that are then marked and released. However, capture efforts also yield some recaptures of previously-marked birds. For each recaptured bird with a legband, the biologist records whether or not the bird has retained its observable tag.

3. Homogeneous Tag Loss

Arnason and Mills (1981) considered survival rate estimation under the Jolly-Seber model for the situation where all animals in the population at sampling period i exhibit the same probability of retaining their tags until $i+1$, θ_i. In situations where this rate can be estimated, as in a double-tagging study, the Jolly-Seber survival rate estimate, $\hat{\phi}_i$, is computed in the usual manner (Jolly 1965; Seber 1965, 1982) and then "corrected" for tag loss as:

$$\hat{\phi}_i^c = \hat{\phi}_i \;/\; \theta_i, \tag{1}$$

where $\hat{\phi}_i^c$ is the corrected survival estimate. The variance of the bias-corrected survival estimate can be estimated by:

$$v\hat{a}r(\hat{\phi}_i^c) = (\hat{\phi}_i^c)^2\left[\frac{v\hat{a}r(\hat{\phi}_i)}{\hat{\phi}_i^2}+\frac{v\hat{a}r\hat{\theta}_i}{\hat{\theta}_i^2}-\frac{2c\hat{o}v(\hat{\phi}_i,\hat{\theta}_i)}{\hat{\phi}_i\hat{\theta}_i}\right] , \tag{2}$$

where var and cov denote variances and covariances, respectively.

In many cases, tag retention rates are assumed to be constant over time (Pollock 1981, Samuel et al. 1990*a*). Zicus and Pace (1986) presented some evidence that this assumption was true for the Canada geese (*Branta canadensis*) that they studied. In such instances, the corrected survival estimate is obtained as in (1), but the tag retention estimate has no period-specific subscript (e.g., Pollock 1981).

4. Tag Loss Dependent On Time Since Marking

In many situations, rates of loss for neckbands and other readily-observable tags would be expected to vary with time since marking. Samuel et al. (1990*b*) and Nichols et al. (1992) provided evidence that annual neckband retention probabilities decreased with increasing time since application, whereas Zicus and Pace (1986), Hestbeck and Malecki (1989), and Raveling et al. (1992) found no such evidence. If some tags are improperly applied or if some birds exhibit behavioral responses (e.g., attempted tag removal) that result in tag loss, then the rate of loss for the period immediately following tagging might be higher than loss rates in subsequent periods. Regardless of the kind of tag or its composition, rates of loss would be expected to increase eventually as the tag wears over time. In general, we believe that rates of tag loss will often be a function of time since marking and will be much less likely to vary with calendar date or sampling period (also see Kremers 1987, 1988).

If rates of tag retention vary with time since marking, then the marked animals in the population at any period i will not represent a homogeneous group with respect to tag loss. Instead, the different tagged cohorts alive in period i will show different probabilities of losing a tag before period $i+1$. The approach of Arnason and Mills (1981) described above is not appropriate in such situations.

One way of estimating survival rate in the presence of this kind of time-dependent tag loss is to use a cohort approach to Jolly-Seber analysis of resighting data (Buckland 1980, 1982; Loery et al. 1987; Pollock et al. 1990), and then to use equations (1) and (2) to obtain estimates of survival rate unbiased by tag loss (Nichols et al. 1992).

The other general approach to estimating survival rate in the presence of tag loss rates that vary with time since marking is to develop a joint likelihood incorporating both recapture and resighting data. Kremers (1987) developed a generalization of the Jolly-Seber model that is parameterized not only with sighting and survival probabilities, but also with tag retention and capture probabilities. The models developed by Kremers (1987) probably represent the best approach to survival rate estimation in cases where time-specific resighting <u>and</u> capture probabilities are reasonably high.

Nichols et al. (1992) described a joint likelihood for recapture and resighting data that is similar to the binomial model of Kremers (1987), but that has fewer parameters and requires fewer recapture data. The model was developed for a data set that seems to be fairly typical of large-scale capture-resighting studies of neckbanded birds in North America, in that the number of resightings was very large relative to the number of recaptures. For example, the study of Canada Geese by Hestbeck and Malecki (1989) used 101,732 resightings and 447 recaptures, and the study of tundra swans (*Cygnus columbianus*) by Nichols et al. (1992) included 7,033 resightings and only 139 recaptures.

Because of the small numbers of recaptures and the need to use all recapture data even when resightings were restricted to a subset of study years, Nichols et al. (1992) chose not to model the recapture data in the usual manner with time-specific capture and survival probabilities. Instead, we viewed the recapture data not as an additional source of information on survival and population size, but strictly as a source of information about tag retention. This perspective led us to condition on the numbers of birds recaptured 1, 2, ... k periods after initial marking. The number of recaptured birds in each category that still had their neckbands were then modeled as binomial random variables, with the binomial probabilities modeled as functions of tag (neckband) retention rates. Resighting data of marked birds were modeled in terms of sighting probabilities, survival rates and tag retention rates. Details of this modeling approach are provided below.

4.1. Modeling recapture data

Let M_j be the number of birds recaptured j years after banding, where $j=1$, 2, ..., k. If a bird is recaptured on more than one occasion, we will allow it to contribute only to the M_j with the largest j (i.e., to the oldest recapture sample). In most cases it will be reasonable to assume that legbands represent permanent marks, but this is not necessary in our approach. The M_j must simply be birds that are recaptured with legbands. These M_j are random variables, and we condition on them in order to estimate tag retention. Let C_j represent the number of the M_j birds that have retained their neckbands. Then $P(C_j \mid M_j)$ is modeled as a binomial with parameter π_j.

If the individuals in each recapture sample (the M_j) are independent of each other and of individuals in the other samples (j) with respect to neckband loss, then we can write the joint probability function of the data as the product of the conditional binomials for $j = 1, 2, ... k$. If we allow neckband retention rates to vary by age, sex, or any other categorical variable, then we obtain a probability distribution for each class. The joint distribution for all classes can then be written as the product of these class-specific distributions. Let θ_l denote the probability that a bird alive and wearing a neckband at the beginning of the lth year after banding will retain its neckband until the end of the lth year after banding. Then we can write the π_j of the conditional binomials as:

$$\pi_j = \prod_{l=1}^{j} \theta_l \quad . \tag{3}$$

Hypotheses about class-specific variation in tag retention and variation associated with time since banding can be tested by placing various constraints on the θ_l. For example, Nichols et al. (1992) used 8 models incorporating different hypotheses about age-sex variation and variation with time since banding with tundra swan recapture data. These models were implemented using program SURVIV (White 1983), and a parsimonious model was selected using results of goodness-of-fit tests, likelihood ratio tests, and Akaike's Information Criterion, AIC (Akaike 1973, Lebreton et al. 1992, Burnham and Anderson 1992). Decisions about the appropriate model for tag retention rates can be based either on the recapture data alone or on the combined recapture and resighting data.

4.2. Modeling resighting data

Reobservation data can be modeled in a manner similar to that of the standard capture-recapture/resighting models of Jolly (1965) and Seber (1965) for a single age class and Pollock (1981) for multiple age classes. However, in addition to parameters for time-specific survival and resighting probabilities, the models include parameters for annual neckband retention rates, referenced to the number of years since banding rather than time or sampling period.

Define the following statistics:

R_i = number of adult birds caught for the first time and released in year i.

R_i' = number of young birds caught and released in year i.

$X_{abc..}$ = number of birds first captured as adults and exhibiting resighting history $abc..$, where $a=1$ if bird is initially caught and released in year i, 0 if it is not; $b=1$ if bird is initially caught or resighted in year $i+1$, 0 if it is not; etc.

$X'_{abc..}$ = number of birds first captured as young and exhibiting resighting history $abc..$, where $a=1$ if bird is initially caught and released in year i, 0 if it is not; $b=1$ if bird is initially caught or resighted in year $i+1$, 0 if it is not; etc.

Define the following model parameters:

ϕ_i = probability that an adult bird alive during the sampling period in year i is alive and in the population at the sampling period in year $i+1$.

ϕ'_i = probability that a young bird alive during the sampling period in year i is alive and in the population at the sampling period in year $i+1$.

p_i = probability that an adult bird alive and in the population during the sampling period in year i is resighted during that period.

θ_j = probability that an adult bird still wearing a neckband at the beginning of the jth year after banding retains its neckband until the end of the jth year after banding.

θ'_1 = probability that a young bird retains its neckband for the first year after banding.

Expected numbers of birds exhibiting different resighting histories under a general 2-age model with tag loss are presented for a 3-year, 1-area case in Table 1. The survival and sighting portion of this model is taken from Pollock (1981). Birds captured as young (first-year birds) are assumed to exhibit young survival and tag retention rates during their first year and adult rates thereafter. Note that each survival probability in the model is accompanied by a tag retention probability, although survival is subscripted by year (sampling period) and tag retention by years since banding. The number of birds from each R_i or R'_i exhibiting each possible capture history represents an observation from a conditional (on the releases) multinomial distribution. The probability structure associated with each of the 4 release cohorts of Table 1 is specified in the column of expected numbers of birds exhibiting each resighting history. The probability distribution for all of the resighting histories can be written as the product of these conditional multinomial distributions.

If the recapture and resighting data are independent of each other, then we can write the probability distribution for the combined data set as the product of the conditional binomials for the recapture data and the conditional multinomials for the resighting data. Maximum likelihood estimation under such models can then

Table 1. Expected numbers of resightings under a general 2-age model with collar loss.

Number of birds released[a]	Number of birds with specified capture/sighting history for periods $i, i+1, i+2$[b]	Expected number of sightings[c]
R_i	X_{111}	$R_i \phi_i \theta_1 p_{i+1} \phi_{i+1} \theta_2 p_{i+2}$
	X_{110}	$R_i \phi_i \theta_1 p_{i+1} (1-\phi_{i+1} \theta_2 p_{i+2})$
	X_{101}	$R_i \phi_i \theta_1 (1-p_{i+1}) \phi_{i+1} \theta_2 p_{i+2}$
	X_{100}	$R_i \left[1-\phi_i \theta_1 + \phi_i \theta_1 (1-p_{i+1})(1-\phi_{i+1} \theta_2 p_{i+2})\right]$
R_{i+1}	X_{011}	$R_{i+1} \phi_{i+1} \theta_1 p_{i+2}$
	X_{010}	$R_{i+1} (1-\phi_{i+1} \theta_1 p_{i+2})$
R'_i	X'_{111}	$R'_i \phi'_i \theta'_1 p_{i+1} \phi_{i+1} \theta_2 p_{i+2}$
	X'_{110}	$R'_i \phi'_i \theta'_1 p_{i+1} (1-\phi_{i+1} \theta_2 p_{i+2})$
	X'_{101}	$R'_i \phi'_i \theta'_1 (1-p_{i+1}) \phi_{i+1} \theta_2 p_{i+2}$
	X'_{100}	$R'_i \left[1-\phi'_i \theta'_1 + \phi'_i \theta'_1 (1-p_{i+1})(1-\phi_{i+1} \theta_2 p_{i+2})\right]$
R'_{i+1}	X'_{011}	$R'_{i+1} \phi'_{i+1} \theta'_1 p_{i+2}$
	X'_{010}	$R'_{i+1} (1-\phi'_{i+1} \theta'_1 p_{i+2})$

[a] R_i = the number of adult birds banded and released in year i; R'_i = the number of young birds banded and released in year i.

[b] $X_{a,b,c}$ = the number of birds with the capture history specified by the subscripts. Subscripts for the 3 sampling periods are either "0" indicating no capture or sighting or "1" indicating initial capture (the first "1") or sighting (subsequent "1's"). For example, X_{101} denotes a bird first captured in period i, not seen in period $i+1$, but sighted in period $i+2$. Prime denotes a capture history for birds initially captured and banded as young.

[c] ϕ_i = probability of surviving from period i to $i+1$. θ_j = probability that a bird with a neckband at the beginning of the jth year after banding will retain the neckband until the end of the jth year after banding; p_i = probability that a neckbanded bird alive and in the sampled area in period i is sighted.

be accomplished numerically using program SURVIV (White 1983). The model in Table 1 is fairly general, and it will frequently be useful to develop reduced-parameter models. For example, in the analysis of tundra swan data, we found evidence of variation in neckband retention rates with age, sex and time since banding. However, we were still able to model tag retention in this multi-year study with only 6 parameters (Nichols et al. 1992). We found no evidence of temporal or geographic variation in survival rates and no evidence of sex-specific variation among adults, leading to a large reduction in survival parameters (Nichols et al. 1992).

5. Bias Associated With Tag Loss: An Illustration

In order to illustrate the magnitude of bias that can result from a failure to adequately deal with tag loss, we computed survival rate estimates for a capture-recapture/resighting study of tundra swans using models that do and do not include tag retention parameters (Table 2). One set of estimates is based on the kind of model described here and by Nichols et al. (1992). The other model is identical except that it includes no tag retention parameters. The capture-resighting data are treated as though there was no tag loss (all tag retention rates = 1). Comparison of the estimates from these 2 models illustrates that a failure to consider tag loss can result in large negative biases in survival estimates (Table 2).

The results presented in Table 2 used models with identical survival parameterizations for comparative purposes. However, analyses that do not incorporate tag loss are also likely to yield erroneous conclusions about sources and patterns of variation in survival whenever tag retention rates vary with such factors age, sex, and time since marking. We believe that tag loss can be a serious problem and that it can yield misleading inferences if it is not considered in the estimation process.

6. On The Importance of Recapture Data

Data from some sort of double-tagging study are essential to the estimation of survival rate in the presence of tag loss. In all resighting studies of which we are aware, all birds are double-banded with both metal legband and whatever observable tag is used. However, efforts are not always made to obtain a sample of double-banded birds that can be examined for tag loss. Some large-scale resighting studies which were otherwise well thought out and carried out, have suffered from the absence of data on tag retention. For example, the inferences of Raveling et al. (1992) about sources of variation in survival rates were conditioned on assumptions about neckband retention rates that could not be tested.

Table 2. Estimated annual survival rates for tundra swans based on models that do and do not include tag retention parameters (θ_j) .

	Model with θ_j [a]		Model without θ_j [b]	
Age-sex class	$\hat{\phi}$	$\hat{SE}(\hat{\phi})$	$\hat{\phi}$	$\hat{SE}(\hat{\phi})$
Adult (male + female)	0.921	0.035	0.754	0.017
Young male	0.814	0.088	0.652	0.064
Young female	0.521	0.059	0.490	0.052

[a] Model is similar to that described in paper (e.g., equation 3 and Table 1) with specific constraints specified in Nichols et al. (1992).

[b] Model is exactly the same as that of Nichols et al. (1992), but with all tag retention parameters removed.

We believe that recaptures are the most appropriate source of data for double-tagging studies to estimate tag retention. Capture methods should not be selective for birds that do or do not retain tags, so it seems reasonable to condition on recaptured birds to estimate tag retention. Some workers have proposed the use of data from birds shot by hunters to estimate tag retention. The hunter is asked whether the reported bird still retains the neckband (or other observable tag), and numbers of reported birds with and without neckbands are accumulated. The danger in this approach is that hunters may selectively shoot birds with neckbands or, more likely, may report neckbands with higher probability (see Samuel et al. 1990a). In this case, the sample of birds to be examined is not unbiased with respect to the presence of the neckband, and neckband retention rate estimates will be biased high. Some methods of estimating tag retention rates do not require restrictive assumptions about reporting rates of neckbanded and legbanded birds (M.D. Samuel, personal communication) and hunter-shot birds would be useful with such methods.

Some workers have proposed the use of 2 different observable tags in double-tagging studies, so that actual capture is not necessary for estimating tag loss. Such studies depend on the assumption that the probability of sighting tags of one type (the type on which the conditional sample is to be based) is independent of the presence of the other type of tag.

It is our belief that insufficient thought has been devoted to the estimation of tag retention rates in many previous resighting studies. Even in most studies that have had sufficient recaptures to estimate tag retention, recapture samples were obtained incidentally, as part of the effort to mark new birds (e.g., Hestbeck and Malecki 1989, Nichols et al. 1992). In future resighting studies, we strongly recommend that recapture efforts for estimating tag retention be a part of the study design (also see Hestbeck et al. 1990). If functions can be developed to approximate the relationships between "effort" and both capture and resighting probabilities, then it is possible to optimize allocation of study effort to these 2 different activities.

We conducted a small simulation study to gain some insight into the relative contributions of capture and resighting probabilities to the precision of resulting survival estimates. Estimation was based on the described 2-age model incorporating parameters for resighting probability, survival rate and tag retention rate (Table 1), as implemented using SURVIV. The simulated data were generated for an initial population of 60,000 adults and 20,000 young, with 20,000 new young added each year. Annual survival rates for adults and young were 0.8 and 0.6, respectively. First-year tag retention rate for both young and adults was 0.95, and annual tag retention rate for all subsequent years was 0.85. Each approximation was based on 100 simulations. All results are based on 5-year studies, and we selected the average coefficient of variation of the second adult survival rate, CV($\hat{\phi}_2$), as the quantity of interest.

To obtain some idea of the sensitivity of estimate precision to changes in capture and resighting probabilities, we first fixed capture probability at 0.02 (p_c=0.02) and sighting probability at 0.50 (p_s =0.50). These values were selected as a baseline because they are similar to those that characterize some of the resighting data sets that we have encountered. This baseline run yielded averages of 191 captures, 2957 resightings and CV($\hat{\phi}_2$)=0.034. We then fixed p_c=0.02 and varied sighting probability as p_s =0.50(1+ α), with α varying on the interval (-1, 1). We then fixed p_s=0.50 and varied capture probability using p_c=0.02(1+ α). The resulting 2 plots of CV($\hat{\phi}_2$) against α appear very similar for α >0.5. For α <0.5, a specified decrease in capture probability brings about a larger loss in precision (larger CV[$\hat{\phi}_2$]) than the same proportional decrease in sighting probability (Figure 1).

The simple sensitivity analysis of Figure 1 contains no consideration of

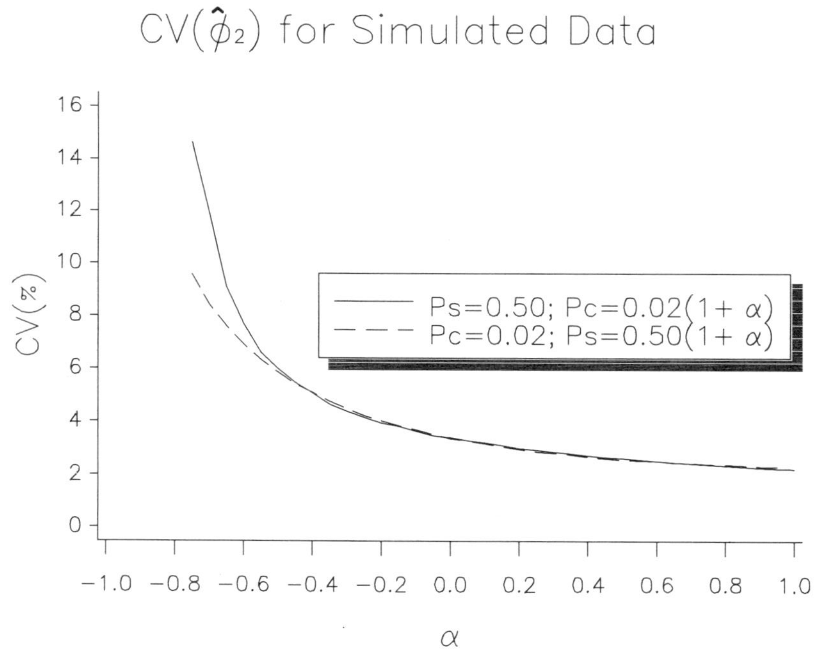

Figure 1. Precision of second survival estimate for fixed sighting probability (0.50) and variable capture probability (0.02 [$1+\alpha$]), and for fixed capture probability (0.02) and variable sighting probability (0.50 [$1+\alpha$]). See text for other study conditions of simulations.

investigator effort. Much more effort will typically be required to obtain recapture data than to obtain resighting data, and this difference will be an important consideration in study design. In order to illustrate this point, we considered optimal allocation of sampling effort for 2 simple (and certainly unrealistic) situations. We assumed that 20 "units" (denoted u) of sampling effort were available for a proposed study, and that each unit produced an increase of 0.05 in sighting probability ($p_s = 0.05u$) or an increase of either 0.005 ($p_c = 0.005u$) or 0.01 ($p_c = 0.01$) in capture probability. Another way of expressing these sampling constraints is p_s = 1-10p_c, or p_s = 1-5p_c. Although these simple functions reflect the general idea that catching a bird likely requires more effort (in these examples, 5 and 10 times more effort; in some studies the difference in required effort may be even larger, J.B. Hestbeck, personal communication) than resighting it, the linear nature of the relationships is unrealistic. The CV($\hat{\phi}_2$) is lowest near p_c = 0.04, p_s = 0.6 for the case where capture requires 10 times more effort than sighting and near p_c = 0.07, p_s = 0.65 for capture 5 times more expensive than sighting (Figure 2).

7. Discussion

Studies based on resighting marked birds offer much potential for the study of avian population dynamics. However, the utility of survival estimates that result from such studies is greatly limited if tag loss is not negligible. In most resighting studies, birds are marked with legbands as well as with more observable tags. Recapture samples of legbanded birds provide the data necessary to estimate tag retention rates.

If tag loss is homogeneous, then the methods of Arnason and Mills (1981) and Pollock (1981) can be used to correct survival estimates. If rate of tag loss varies with time since marking, then other methods are required. A cohort approach similar to that of Arnason and Mills (1981) can be used to correct survival rates in this situation. In addition, models by Kremers (1987) and Nichols et al. (1992) for joint analysis of recapture and resighting data can be used. The models of Kremers (1987) are fully efficient, whereas those of Nichols et al. (1992) are not (they do not use the information on survival contained in the recaptures). The models of Nichols et al. (1992) do not require the assumption that one tag type (e.g., legbands) is permanent. In addition, these models do not require that capture-recapture and resighting data correspond to exactly the same sets of years (Nichols et al. 1992 used all recaptures occurring over a 24-year period and resightings from much smaller sets of years).

Future statistical work might involve efforts to model the pattern of change in tag retention rate as a function of time since marking. Such infrastructural modeling could be accomplished using program SURVIV (White 1983). In

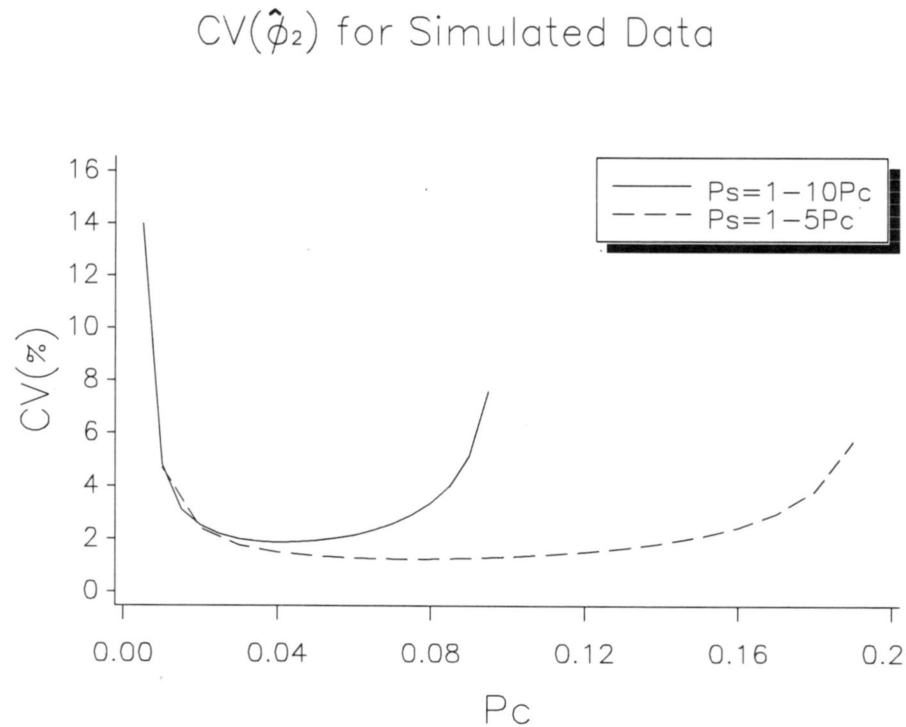

Figure 2. Precision of second survival estimate for different combinations of capture and sighting probabilities. Increases in capture probability require 5 and 10 times more effort than increases in sighting probability. See text for other study conditions of simulations.

addition, J.-D. Lebreton (personal communication) has pointed out that the $\phi_i\theta_j$ parameterization of our models (Table 1) can be viewed as SURGE model ϕ_{t+a} with a log link (Lebreton et al. 1992). This analogy may be useful both for understanding identifiability problems and for infrastructural modeling.

We recommend that the designs of future resighting studies include efforts to recapture birds. Knowledge of the relationship between effort and both resighting and recapture probabilities will permit the investigator to compute the optimal allocation of resources to recapture vs. resighting efforts in future studies.

Acknowledgements. We received helpful reviews of an earlier draft of this manuscript from C.M. Francis, J.B. Hestbeck, J.-D. Lebreton, M.D. Samuel, and J.A. Spendelow. We thank J. Bart, R.J. Limpert, and W.J.L. Sladen for useful discussions of the tag loss problem during our initial work on these models. Finally, we thank C. Hagen for typing the manuscript.

References

Akaike, H. (1973).- Information theory and an extension of the maximum likelihood principle. pp 267-281 *in* International Symposium on Information Theory, 2nd ed, B.N. Petran and F. Csaki (Eds) Akademiai Kiad., Budapest, Hungary.

Arnason, A.N. and Mills, K.H. (1981).- Bias and loss of precision due to tag loss in Jolly-Seber estimates for mark-recapture experiments. Can. J. Fish. Aquat. Sci., 38:1077-1095.

Beverton, R.J.H. and Holt, S.J. (1957).- On the dynamics of exploited fish populations. Her Majesty's Stationery Office, London.

Buckland, S.T. (1980).- A modified analysis of the Jolly-Seber capture-recapture model. Biometrics, 36:419-435.

Buckland, S.T. (1982).- A mark-recapture survival analysis. J. Anim. Ecol., 51:833-837.

Burnham, K.P. and Anderson, D.R. (1992).- Data-based selection of an appropriate biological model: the key to modern data analysis. pp 16-30 *in* Wildlife 2001: Populations, D.R. McCullough, and R.H. Barrett, (Eds) Elsevier Scientific Publ. Co., Amsterdam.

Caughley, G. (1971).- Correction for band loss. Bird-Banding, 42:220-221.

Hestbeck, J.B. and Malecki, R.A. (1989).- Estimated survival rates of Canada Geese within the Atlantic Flyway. J. Wildl. Manage., 53:91-96.

Hestbeck, J.B., Rusch, D.H. and Malecki, R.A. (1990).- Estimating population parameters for geese from band-recovery and mark-recapture data. Trans. N. Amer. Wildl. Nat. Resour. Conf., 55:350-373.

Jolly, G.M. (1965).- Explicit estimates from capture-recapture data with both death and immigration-stochastic model. Biometrika, 52:225-247.

Kremers, W.K. (1987).- Estimation of survival rates in the presence of tag loss: the binomial and hypergeometric model. Universitat Dortmund Fachbereich Statistik, Forschungsbericht Nr.87/6.

Kremers, W.K. (1988).- Estimation of survival rates from a mark-recapture study with tag loss. Biometrics, 44:117-130.

Lebreton, J.-D., Burnham, K.P., Clobert, J. and Anderson, D.R. (1992).- Modeling survival and testing biological hypotheses using marked animals: a unified approach with case studies. Ecol. Monogr., 62:67-118.

Loery, G., Pollock, K.H., Nichols, J.D. and Hines, J.E. (1987).- Age-specificity of avian survival rates: an analysis of capture-recapture data for a black-capped chickadee population, 1985-1983. Ecology, 68:1038-1044.

Nelson, L.J., Anderson, D.R. and Burnham, K.P. (1980).- The effect of band loss on estimates of annual survival. J. Field Ornithol., 51:30-38.

Nichols, J.D., Bart, J., Limpert, R.J., Sladen, W.J.L. and Hines, J.E. (1992).- Annual survival rates of adult and immature eastern population tundra swans. J. Wildl. Manage., 56:485-494.

Pollock, K.H. (1981).- Capture-recapture models allowing for age-dependent survival and capture rates. Biometrics, 37:521-529.

Pollock, K.H., Nichols, J.D., Brownie, C. and Hines, J.E. (1990).- Statistical inference for capture-recapture experiments. Wildl. Monogr., 107:1-97.

Raveling, D.G., Nichols, J.D., Hines, J.E., Zezulak, D.S., Silveira, J.G., Johnson, J.C., Aldrich, T.W. and Weldon, J.A. (1992).- Survival of Cackling Canada Geese, 1982-1988. J. Wildl. Manage., 56:63-73.

Robson, D.S. and Regier, H.A. (1966).- Estimates of tag loss from recoveries of fish tagged and permanently marked. Trans. Amer. Fish. Soc., 95:56-59.

Samuel, M.D., Rusch, D.H. and Craven, S. (1990*a*).- Influence of neckbands on recovery and survival rates. J. Wildl. Manage., 54:45-54.

Samuel, M.D., Weiss, N.T., Rusch, D.H., Craven, S.R., Trost, R.E. and Caswell, F.D. (1990*b*).- Neck band retention for Canada Geese in the Mississippi Flyway. J. Wildl. Manage., 54:612-621.

Seber, G.A.F. (1965).- A note on the multiple-recapture census. Biometrika, 52:249-259.

Seber, G.A.F. (1982).- The estimation of animal abundance and related parameters. Second edition. MacMillan, New York.

White, G.C. (1983).- Numerical estimation of survival rates from band-recovery and biotelemetry data. J. Wildl. Manage., 47:716-728.

Zicus, M.C. and Pace, R.M., III. (1986).- Neckband retention in Canada Geese. Wildl. Soc. Bull., 14:388-391.

Marked Individuals in the
Study of Bird Population
J.-D. Lebreton and P. M. North (eds.)

The "Robust" Capture-Recapture Design Allows Components of Recruitment to be Estimated

K. H. Pollock[1], W. L. Kendall[2], J. D. Nichols[3]

[1]*Department of Statistics, North Carolina State University, Box 8203, Raleigh, NC 27695, U.S.A.* [2]*Current Address: Office of Migratory Bird Management, U.S. Fish and Wildlife Service, Laurel, MD 20708, U.S.A.* [3]*Migratory Birds Section, Patuxent Wildlife Research Center, U.S. Fish and Wildlife Service, Laurel, MD 20708, U.S.A.*

Summary. The "robust" capture-recapture design (Pollock 1982) allows analyses which combine features of closed population model analyses (Otis et al., 1978, White et al., 1982) and open population model analyses (Pollock et al., 1990). Estimators obtained under these analyses are more robust to unequal catchability than traditional Jolly-Seber estimators (Pollock, 1982; Pollock et al., 1990; Kendall, 1992). The robust design also allows estimation of parameters for population size, survival rate and recruitment numbers for all periods of the study unlike under Jolly-Seber type models.

The major advantage of this design that we emphasize in this short review paper is that it allows separate estimation of immigration and in situ recruitment numbers for a two or more age class model (Nichols and Pollock, 1990). This is contrasted with the age-dependent Jolly-Seber model (Pollock, 1981; Stokes, 1984; Pollock et al., 1990) which provides separate estimates for immigration and in situ recruitment for all but the first two age classes where there is at least a three age class model. The ability to achieve this separation of recruitment components can be very important to population modelers and wildlife managers as many species can only be separated into two easily identified age classes in the field.

1. Introduction

Capture-recapture models are widely used to estimate demographic parameters (population sizes, survival rates, recruitment numbers) of many vertebrate populations (Seber, 1982) which are not easy to observe and count directly. The Lincoln-Petersen method for the simple two sample mark and recapture study has a long history and has been described in detail, for example by (Seber, 1982, p. 104).

Closed population models for multiple (more than two) sample capture-recapture studies have been researched extensively in recent years (Otis et al., 1978; White et al., 1982; Rexstad and Burnham, 1991). The primary emphasis of this research has been to relax assumptions of traditional approaches that all animals have equal capture probabilities in each sample.

Building on the pioneering papers of Jolly (1965) and Seber (1965) the open population capture-recapture models have also been extensively researched in recent years. Jolly (1982) and Brownie et al. (1986) have developed parsimonious models which allow for constant survival and/or for constant capture probabilities over time. Robson (1969), Pollock (1975) and Brownie and Robson (1983) develop temporary trap response models. Buckland (1982), Loery et al. (1987), and Pollock et al. (1990) develop cohort based Jolly-Seber models which allow age and time effects to be considered for known (or minimum) age animals. Pollock (1981) and Stokes (1984) develop an age dependent extension of the Jolly-Seber model. Burnham et al. (1987) provide an important monograph on another class of open population models where comparisons of survival rates for different populations are considered.

Special computer programs which provide maximum likelihood estimates of population parameters for various capture-recapture models are available. CAPTURE (White et al., 1982; Rexstad and Burnham, 1991) provides analyses of closed population models. Arnason and Schwarz (1986) have developed POPAN-3 for use with open populations. Pollock et al. (1990) have developed JOLLY and JOLLYAGE for general use with open populations. Burnham et al. (1987) have developed RELEASE for comparing populations. Lebreton et al. (1992) have developed SURGE for comprehensive modeling of survival and capture rates in open populations. SURVIV (White, 1983) can also be used for fitting a variety of models. It is very general and flexible but requires more knowledge by the user than other programs.

Pollock (1982) noted some limitations of the Jolly-Seber model and suggested an approach (or design) based on combining closed and open population model analyses. This so called robust design has been investigated further by Pollock et al. (1990), Nichols and Pollock (1990) and Kendall (1992). In Section 3 we discuss some important advantages of the design. In Section 4 we discuss one of these advantages in detail, which is the separation of in situ reproduction from immigration in a two or more age class model. This separation is very important to ecologists. We conclude with a general discussion section.

2. Description of the "Robust" Design

We consider the following representation of a capture-recapture sampling experiment.

Primary Periods	1		2		...	k		
Secondary Periods	1	$2 \ldots l_1$	1	$2 \ldots l_2$		1	$2 \ldots l_k$	

We have k primary sampling periods (e.g. months or years depending on the species) and within each of these we have $\ell_i (i = 1, \ldots, k)$ secondary sampling periods that are very close to each other in time (e.g. ℓ_i consecutive days of trapping). It is not necessary that the number of secondary periods be equal in all the primary periods but it is usually.

The design can be used to estimate the *population size* for each of the primary periods $(N_1, N_2, \ldots, N_k)$ assuming that the population is constant over the secondary sampling periods within each primary sampling period. We also can estimate *survival rates* $(\phi_1, \phi_2, \ldots, \phi_{k-1})$ between the primary sampling periods and *numbers of new individuals* entering the population between the primary sampling periods $(B_1, B_2, \ldots, B_{k-1})$.

Originally Pollock (1982) suggested an intuitive ad hoc approach to analysis which "would be to analyze each short-term study using the closed population models (to estimate population size) ... which allow unequal catchability. Then all the sampling periods in each short term study could be pooled and survival estimators between these short-term studies could be estimated using the Jolly-Seber Model. This approach allows population size estimation under models allowing unequal catchability while survival estimation, which is not so affected by unequal catchability, is under the Jolly-Seber Model. Estimators of the number of new animals would then be based on the population size and survival rate estimators described above ($\hat{B}_i = \hat{N}_{i+1} - \hat{\phi}_i\hat{N}_i$ assuming no losses on capture). From this discussion on estimation it is clear that the motivation of this robust approach is to find a design that has estimators robust to unequal catchability of animals. The unequal catchability might be due to heterogeneity and/or trap response (Otis et al., 1978).

Kendall (1992) compares the ad hoc estimation procedure suggested by Pollock (1982) with the Jolly-Seber estimators for a range of simulation conditions including heterogeneity and trap response of capture probabilities. Kendall (1992) also develops a more efficient maximum likelihood estimation procedure for models that allow for time and trap response effects on the capture probabilities. However, his maximum likelihood methods cannot be used for the important situation where there is heterogeneity of capture probabilities. For heterogeneity models the ad hoc approach described by Pollock (1982) is still the only real option unless the heterogeneity is very mild, in which case the Jolly-Seber estimators may have smaller mean squared error than the ad hoc estimators.

To aid the reader we now present a brief discussion of the simulation results of Kendall (1992). When there was a time effect the full-likelihood approach improved the efficiency of the population size and survival estimators by as much as 40% and 20%, respectively. The recruitment estimator for this approach was also as much as 40% more efficient, and less biased, but was not superior in all cases. When there was a temporary trap response, the results were similar to the time effect case, although not as pronounced. In the case of a permanent trap response, the full-likelihood approach failed to estimate population size and recruitment at a high rate when capture probabilities were low. However, in other cases, especially where there was a trap-shy response to capture, this method reduced the mean square error of the population size estimator by as much as 70%. In these cases the full-likelihood estimator for recruitment was usually less biased, but the results for mean square error were mixed.

When there was heterogeneity in capture probability the ad hoc approach was the only alternative to the Jolly-Seber method. Kendall (1992) found the Jolly-Seber estimator for survival to be somewhat superior, while for population size and recruitment the choice of estimator was scenario-dependent. The ad hoc estimators for population size tended to be lower in mean square error where heterogeneity was severe but inferior elsewhere. The ad hoc estimators of recruitment were almost uniformly less biased and less precise than the Jolly-Seber estimator. In addition, under the permanent trap response model the ad hoc (generalized removal) estimator of population size had a smaller mean square error than the Jolly-Seber estimator in most cases.

3. Advantages of the Robust Design

Here we very briefly list and discuss the important advantages of the design.

(*1*) *Population size and other parameters can be estimated without assuming equal catchability.* This advantage was alluded to in Section 2 and is due to the ability to use the closed population models (Otis et al., 1978; White et al., 1982;

Rexstad and Burnham, 1991) which do not require equal catchability. (However, we emphasize that the closed models cannot cope with extreme unequal catchability.)

(*2*) *All parameters can be estimated.* Under the Jolly-Seber model not all parameters are estimable. For example, N_1, N_k, ϕ_{k-1},B_1 and B_{k-1} in a *k* primary period study are not estimable whereas under the robust design all these parameters are estimable. Let us illustrate this further by considering the case of $K = 4$ primary periods. The parameters of the model are N_1, N_2, N_3, N_4; ϕ_1, ϕ_2, ϕ_3; and B_1, B_2, B_3; all of which are estimable under the robust design. The parameters of the model which are estimable under the Jolly-Seber model are N_2, N_3; ϕ_1, ϕ_2; and B_2 which is only 5 out of the 10 total parameters. Notice that some parameters can be estimated in more than one way which leads us to the next advantage.

(*3*) *Alternative estimators are available for most parameters.* Discrepancies between these alternative estimators might be useful in assessing assumption violations.

(*4*) *Abundance and Survival can be estimated with minimum dependence.* Under density dependent population models the focus of attention for ecologists is the true functional relationship between abundance (density) and survival. If the Jolly-Seber method provides estimators for abundance and survival the true functional relationship between density and survival will be confounded with the statistical dependence between their estimators. Kendall (1992) found using simulation that there is a very weak relationship between population size and survival when the ad hoc estimation procedure of the robust design is used. This means that the true functional relationship between density and survival is less likely to be obscured when the robust design estimators are used.

(*5*) *Designs can be customized more finely to meet specific needs.* Using the robust design, a given amount of sampling effort can be distributed in many ways, both within primary periods and across the study. Allocation of effort affects the precision of estimators differently. For example, the precision of the estimator of N_i from a closed model can be increased by increasing the number of trapping days in period i whereas one can increase the precision of the Jolly-Seber estimator of ϕ_i by increasing the sampling effort after period i. The secondary periods make the optimal sampling question complex but they do provide a lot of flexibility in allocation of sampling effort.

(*6*) *Practical advantages over the Jolly-Seber design* (i.e., only one secondary period per primary period) *exist.* For example, under the robust design a biologist might sample intensively for four consecutive days and then take three weeks off before repeating the process. Alternatively, a biologist may sample one day each week of the month and use the Jolly-Seber estimators. It seems to us that resources are much better utilized under the robust design scenario. For the other three weeks, the biologist might carry out other studies (habitat analyses etc.).

The final advantage of the robust design is the possibility of separation of *in-situ reproductive recruitment* and *immigration* in a two age class model. This aspect of the robust design is so important that we discuss it in detail in the next section.

4. Estimation of Separate Components of Recruitment

Nichols and Pollock (1990) state ... "Connor et al. (1983) asked why ecologists have not focused more attention on the relative contributions of immigration and in situ reproduction to recruitment. They stated that at least part of the answer involved the difficulty of assessing these contributions in the field. Several methods have been used for estimating rates of immigration and emigration in field studies (e.g., see Gaines and McClenaghan, 1980). However, many of these approaches require special elements of study design (e.g., radiotelemetry, "removal trapping", exclosures and enclosures, peripheral lines of traps surrounding grids), and none is without problems. There have also been several methods proposed for partitioning estimated rates of loss and gain into their respective components using data from standard capture-recapture studies (Jackson, 1939; Sheppard et al., 1969; Cameron and Williamson, 1977; Manly, 1985; Zeng and Brown, 1987). These methods generally assume that the maximum distances separating different portions of the sampled area are sufficiently large to include the longest distances likely to be moved by dispersing animals. For situations in which this assumption is reasonable, these methods should be used (especially see Manly, 1985). However, in many situations investigators are justifiably reluctant to make restrictive assumptions about dispersal distances."

Here we present very briefly a straightforward method of estimating in-situ reproduction numbers and immigration numbers using the robust design (Pollock et al., 1990; Nichols and Pollock, 1990). This method differs from others proposed for use with capture-recapture data in that it does not require any assumptions about the distances from which animals can immigrate.

Pollock (1981), Stokes (1984) and Pollock et al. (1990) consider a general age-dependent version of the Jolly-Seber model. Under their model it is possible to estimate separately immigration and in situ reproduction when there are at least 3 age classes (Pollock et al., 1990; p. 48). If there are two age classes, however, it is not possible to obtain the separate estimators unless the robust design is used. As there are many situations where only two age classes are easily identifiable in the field we concentrate on that case here.

Let us use the following notation for parameters.

$\phi_i^{(v)}$ = the probability that an animal of age v during primary sampling period i survives and is still in the sampled population at period $i + 1$. ($v = 0, 1; i = 1, \ldots, k$)

$N_i^{(v)}$ = the number of animals of age v in the sampled population at primary sampling period i. ($i = 1, \ldots, k$)

$B'^{(1)}_i$ = the number of adults in the population at primary sampling period $i + 1$ that were young animals in the population at time i ($i = 1, \ldots, k - 1$)

$B''^{(1)}_i$ = the number of adults in the population at primary sampling period $i + 1$ that entered the population as immigrants between periods i and $i + 1$. ($i = 1, \ldots, k - 1$)

The robust design method described briefly in Section 2 and for the two age class model by Pollock and Mann (1983) will yield estimates of survival rate (between primary periods) for young $\left[\hat{\phi}_i^{(0)}\right]$ and adult $\left[\hat{\phi}_i^{(1)}\right]$ animals, and of the number (for each primary period) of young $\left[\hat{N}_i^{(0)}\right]$ and adult $\left[\hat{N}_i^{(1)}\right]$ animals. The estimates can then be used in conjunction with the total sample size (n_i) and the total number released after the sample (R_i) to estimate the two components of recruitment as:

In Situ Reproduction Number

$$\widehat{B}_i^{\prime(1)} = \left[\widehat{N}_i^{(0)} - n_i^{(0)} + R_i^{(0)}\right]\widehat{\phi}_i^{(0)} \qquad i = 1, \ldots, k-1 \tag{1}$$

Immigration Number

$$\widehat{B}_i^{\prime\prime(1)} = \widehat{N}_{i+1}^{(1)} - \left[\widehat{N}_i^{(1)} - n_i^{(1)} + R_i^{(1)}\right]\widehat{\phi}_i^{(1)} - \left[\widehat{N}_i^{(0)} - n_i^{(0)} + R_i^{(0)}\right]\widehat{\phi}_i^{(0)}$$

$$i = 1, \ldots, k-1 \tag{2}$$

The estimate of in situ reproduction number given by (1) is simply the product of the estimated number of young animals alive just after primary period i and the estimated proportion of these young animals that survive until the next primary period. The estimate of immigration number is obtained by estimating the number of adults in primary period i + 1 and subtracting off the estimated number of adult and young survivors from primary period i.

We emphasize that this procedure makes several assumptions. (1) The time interval between successive primary periods must be at least as long as the interval required to make the transition from the young to the adult age class. (2) Young animals at period i that move to the adult class by period i + 1 are catchable at time period i (although their capture probability can differ from that of adults). A final assumption (3) is that all animals classified as young in any sampling period are products of in situ reproduction. We believe that these assumptions are likely to be met in many field situations.

Nichols and Pollock (1990) apply this method to capture-recapture data on meadow voles (*Microtus pennsylvanicus*) collected by Nichols at Patuxent Wildlife Research Center in 1981-82. While the example only provided estimates of the separated components of recruitment for two primary periods it clearly shows the potential of this estimation procedure.

5. Discussion

Recently Nichols et al. (1991) and Brownie et al. (1992) have developed general capture-recapture models that allow estimation of transition probabilities between states. Their models can be viewed as generalization of the Jolly-Seber model to multiple states (see also Arnason 1972, 1973). Nichols et al. (1991) note that the robust design could be used in this context if one were especially concerned with heterogeneity of capture probabilities. Once again the flexibility of the robust design can be exploited to get alternative estimators to Jolly-Seber type estimators.

Arnason (pers. comm.) has suggested another advantage of the robust design that we had never considered. "In the presence of a tag loss, we can often use different tagging methods for the primary and secondary tags (eg. a long lasting mark for the primary periods and a more losable mark within the primary periods). The primary tag provides the check needed for tag-loss of the secondary tag." We believe this is a very important point which needs further evaluation later.

The robust design approach has many advantages that we have described and has now been investigated sufficiently by Kendall (1992) using simulation and analytic methods to show clearly its practical utility. In this review we have emphasized the importance of being able to separate in situ reproduction and immigration under fairly reasonable assumptions, unlike earlier approaches.

References

Arnason, A. N. and Schwarz, C. J. (1987), *POPAN-3. Extended Analysis and Testing Features for POPAN-2*, Box 272, St. Norbert, Manitoba: Charles Babbage Research Center.

Arnason, A. N. (1972), Parameter estimates from mark-recapture experiments on two populations subject to migration and death. *Researches in Population Ecology 14*, 97-113.

Arnason, A. N. (1973), The estimation of population size, migration rates and survival in a stratified population. *Researches in Population Ecology 15*, 1-8.

Brownie, C., Hines, J. E., Nichols, J. D., Pollock, K. H. and Hestbeck, J. B. (1992), Capture-recapture studies for multiple strata including non-Markovian transitions. *Biometrics* (In Press).

Brownie, C. J., Hines, J. E. and Nichols, J. D. (1986), Constant-parameter capture-recapture models. *Biometrics*, 42, 561-574.

Brownie, C., and Robson, D. S. (1983), "Estimation of Time-Specific Survival Rates from Tag-Resighting Samples: A Generalization of the Jolly-Seber Model," *Biometrics*, 39-437-453.

Buckland, S. T. (1982), "A Mark-Recapture Survival Analysis," *Journal of Animal Ecology*, 51, 833-847.

Burnham, K. P., Anderson, D. R., White, G. C., Brownie, C., and Pollock, K. H. (1987), "Designs and Analysis Methods for Fish Survival Experiments Based on Release-Recapture," *American Fisheries Society Monographs* 5, Bethesda, MD.

Cameron, R. A. D., and Williamson, P. (1977), Estimating migration and the effect of disturbance in mark-recapture studies on the snail *Cepaea nemoralis* L. *Journal of Animal Ecology*, 46, 173-179.

Connor, E. F., Faeth, S. H., and Simerloff, D. (1983), Leafminers on oak: the role of immigration and in situ reproductive recruitment. *Ecology*, 64, 191-204.

Gaines, M. S. and McCleneghan, L. R. (1980), Dispersal in small mammals. *Annual Review of Ecology and Systematics*, 11, 163-196.

Jackson, C. H. N. (1939), The analysis of an animal population. *Journal of Animal Ecology*, 8, 238-246.

Jolly, G. M. (1965), Explicit estimates from capture-recapture data with both death and immigration--stochastic model. *Biometrika*, 52, 225-247.

Jolly, G. M. (1982), "Mark-Recapture Models With Parameters Constant in Time," *Biometrics*, 38, 301-321.

Kendall, W. L. (1992), "The Robust Design in Capture-Recapture Sampling: Modelling Approaches and Estimation Methods." Unpublished Ph.D. thesis, North Carolina State University, Biomathematics Program.

Lebreton, J. D., Burnham, K. P., Clobert, J. and Anderson, D. R. (1992), Modelling survival and testing biological hypotheses using marked animals: a unified approach with case studies. *Ecological Monographs*, 62, 67-118.

Loery, G., Pollock, K. H., Nichols, J. D., and Hines, J. E. (1987), "Age-Specificity of Avian Survival Rates: An Analysis of Capture-Recapture Data for a Black-Capped Chickadee Population, 1958-1983," *Ecology*, 68, 1038-1044.

Manly, B.F. J. (1985), A test of Jackson's method for separating death and emigration with mark-recapture data. *Researches in Population Ecology*, 27, 99-109.

Nichols, J. D. and Pollock, K. H. (1990), Estimation of recruitment from immigration versus in situ reproduction using Pollock's robust design. *Ecology*, 71, 21-26.

Nichols, J. D.. Sauer, J. R., Pollock, K. H. and Hestbeck, J. B. (1992), Estimating Transition Probabilities for Stage Based population projection matrices using capture-recapture data. *Ecology*, 73, 306-312.

Otis, D. L., Burnham, K. P., White, g. C., and Anderson, D. R. (1978), Statistical inference from capture data on closed animal populations, *Wildlife Monographs* 62.

Pollock, K. H. (1975), "A K-Sample Tag-Recapture Model Allowing for Unequal Survival and Catchability," *Biometrika*, 37, 521-529.

Pollock, K. H. (1981), Capture-recapture models allowing for age-dependent survival and capture rates. *Biometrics*, 37, 251-259.

Pollock, K. H. (1982), A capture-recapture design robust to unequal probability of capture. *Journal of Wildlife Management*, 37, 757-760.

Pollock, K. H. and Mann, R. H. K. (1983), Use of an age-dependent mark-recapture model in fisheries research. *Canadian Journal of Fisheries and Aquatic Sciences*, 40, 1449-1455.

Pollock, K. H., Nichols, J. D., Hines, J. E., and Brownie, C. (1990), "Statistical Inference for Capture-Recapture Experiments," *Wildlife Monographs*, 107, 1-97.

Rexstad, E. and Burnham, K. P. (1991), User's guide for interactive program CAPTURE. Colorado Cooperative Fish and Wildlife Research Unit, Fort Collins, Colorado.

Robson, D. S. (1969), Mark-Recapture Methods of Population Estimation in *New Developments in Survey Sampling*, eds. N. L. Johnson and H. Smith, New York: John Wiley, pp. 120-140.

Seber, G. A. F. (1965), A note on the multiple-recapture census, *Biometrika*, 52, 249-259.

Seber, G. A. F. (1982), The estimation of animal abundance and related parameters. Second edition. MacMillan, New York, New York, USA.

Sheppard, P. M., MacDonald, W. W., Tonn, R. J., and Grab, B. (1969), The dynamics of an adult population of *Aedes aegypti* in relation to dengue haemorrhagic fever in Bangkok, *Journal of Animal Ecology*, 38, 661-702.

Stokes, S. L. (1984), The Jolly-Seber method applied to age-stratified populations, *Journal of Wildlife Management*, 48, 1053-1059.

White, G. C. (1983), "Numerical Estimation of Survival Rates from Band-Recovery and Biotelemetry Data," *Journal of Wildlife Management*, 47, 716-728.

White, G. C., Anderson, D. R., Burnham, K. P. and Otis, D. L. (1982), *Capture-Recapture and Removal Methods for Sampling Closed Populations*, Los Alamos, NM: Los Alamos National Laboratory.

Zeng, Z., and Brown, J. H. (1987), A method for distinguishing dispersal from death in mark-recapture studies. *Journal of Mammalogy*, 68, 656-665.

Recruitment and dispersal

Marked Individuals in the
Study of Bird Population
J.-D. Lebreton and P. M. North (eds.)

Estimating migration rates using tag-recovery data

Carl J. Schwarz

Department of Statistics, University of Manitoba, Winnipeg, Manitoba, Canada R3T 2N2

Summary. The use of tag-recovery information to describe the movement or migration of birds has been traditionally limited to qualitative descriptions. The problem in most studies has been that the unequal and unknown tag-recovery rates in the different areas imply that migration rates cannot be estimated separately from tag-recovery rates. An experimental protocol that resolves this confounding will be presented. The statistical model for data obtained from this protocol is a matrix extension of the formulation of a tag-recovery experiment discussed by Brownie *et al.* (1985). Estimation is more difficult because of the convolution of parameters between release and recovery and numerical methods must be used. The model is applied to estimate the interchange of North American mallards (*Anas platyrynchos*) among wintering areas in the southern United States.

1. Introduction

In many species of birds, regular migrations occur and these migration patterns have often been qualitatively outlined through the use of tagging data. Birds are tagged in a number of areas (strata), and the locations of the recoveries of these tags in the future provide an indication of the migration routes and migration intensities. In some cases, the tagging and recovery strata are the same and interest focuses upon the rates of interchange among the strata. For example, birds may be tagged in their wintering areas, recoveries may also occur in the wintering areas, and interest focuses upon the rates of interchange among the wintering areas from year to year (e.g. Hestbeck, Nichols, and Malecki, 1991).

Models for these types of experiments in a capture-recapture context have been considered by previous authors. Chapman and Junge (1956) and Darroch (1961) considered the problem of migration from strata determined at the time of release to strata determined at the time of recovery with only two sampling times in closed populations. Arnason (1972, 1973) and Seber (1982, p. 555) extended these models to the case of more than two sampling times in open populations.

In tag-recovery contexts, recoveries typically are from dead birds and there is little (if any) chance of recapturing and releasing a bird. A bird is usually seen at most twice, at release and recovery. More importantly, tag recoveries may occur over an extended period of time between tagging times, not just at a point in time as assumed in the capture-recapture models.

The model formulations in this paper were developed in Schwarz, Schweigert, and Arnason (1992) and are matrix extensions to the model formulations of simple tag-recovery experiments developed by Brownie *et al.* (1985). Estimation is more difficult because of the convolution of parameters between release and recovery and because of the typically low tag-recovery rates. These may cause estimates of the survival/migration parameters to have low precision.

Following the definition of the notation, the assumptions of the model will be outlined. Next, the stochastic model is developed. An example of the migration of mallards (*Anas platyrynchos*) among wintering areas in the southern United States is presented. Finally, a discussion of how inferences in this model can be improved through modifications of the experimental protocol is given.

2. Notation

Bold face letters will be used to represent matrices and vectors. Elements within a matrix or vector will be identified by superscripts.

Parameters

- a the number of strata in which releases and recoveries take place
- k the number of years of releases
- l the number of years of recoveries
- $\boldsymbol{f}_i$ an $a \times a$ matrix whose $(s,t)^{\text{th}}$ element (f_i^{st}) represents the probability that a bird present in stratum s at the time of tagging in year i will be recovered in stratum t between the time of tagging in year i and year $i+1$ and its tag reported.
- $\boldsymbol{S}_i$ an $a \times a$ matrix whose $(s,t)^{\text{th}}$ element (S_i^{st}) represents the probability that a bird alive in stratum s at the time of tagging in year i will survive to and be present in stratum t at the time of tagging in year $i+1$.

Statistics

- $\boldsymbol{N}_i$ an $a \times 1$ vector whose s^{th} element (N_i^s) represents the number of birds tagged and released in stratum s in year i.
- $\boldsymbol{R}_{ij}$ an $a \times a$ matrix whose $(s,t)^{\text{th}}$ element (R_{ij}^{st}) is the number of birds released in stratum s in year i that are recovered in stratum t in year j.

3. Assumptions

The usual assumptions for tag-recovery models as outlined by Brownie *et al.* (1985, Section 1.3) and discussed by Nichols *et al.* (1982) and Pollock and Raveling (1982) are applicable. As well, it is assumed that

- releases occur in all strata over a number of years and effort is expended in all strata to recover birds in all years. Strata with no releases or recovery effort in some years will imply that certain parameters are not identifiable.

- no birds migrate temporarily to a stratum where recoveries and releases do not occur, returning after one or more years of absence. Birds may migrate out of the sampled strata permanently; such losses are indistinguishable from mortality.
- all birds behave independently with respect to migration and recovery. It is assumed that flocking of birds after release does not occur, or does not influence the migration pattern of the bird.
- birds behave in a Markovian fashion in each year, i.e., the current migration route does not depend upon previous migration choices.

4. Model Development

The number of birds that are released in stratum s in year i and recovered in stratum t in year j can be displayed using matrices as shown in Table 1. Note the resemblance to the display of a simple tag-recovery experiment (Brownie *et al.,* 1985, pp. 13-17), except that now each element of the table is a matrix or a vector.

Table 1. Symbolic representation of the number of birds released in stratum s in year i and recovered in stratum t in year j in the case of k=3 years of releases and l=4 years of recoveries

Year Released	Number Released	Number recovered by year 1	2	3	4
1	N_1	$\boldsymbol{R}_{11}$	$\boldsymbol{R}_{12}$	$\boldsymbol{R}_{13}$	$\boldsymbol{R}_{14}$
2	N_2		$\boldsymbol{R}_{22}$	$\boldsymbol{R}_{23}$	$\boldsymbol{R}_{24}$
3	N_3			$\boldsymbol{R}_{33}$	$\boldsymbol{R}_{34}$

N_i are $a \times 1$ vectors and $\boldsymbol{R}_{ij}$ are $a \times a$ matrices.

The probability that a bird released in stratum s in year i is recovered in stratum t in year j is a function of the parameters and can be computed using matrices as illustrated in Table 2. Again note the similarity of the form of the cell probabilities to those from simple tag-recovery experiments (Brownie *et al.*, 1985, pp. 13-17). The expected number of recoveries can then be computed from the elements of Table 2.

Table 2. Symbolic representation of the probability that a bird released in stratum s in year i is recovered in stratum t in year j in the case of k=3 years of releases and l=4 years of recoveries

Year Released	Number Released	Probabilities of recovery by year 1	2	3	4
1	N_1	f_1	S_1f_2	$S_1S_2f_3$	$S_1S_2S_3f_4$
2	N_2		f_2	S_2f_3	$S_2S_3f_4$
3	N_3			f_3	S_3f_4

f_i and S_i are $a \times a$ matrices..

For example, of the N_1^2 birds released in stratum 2 in year 1, the expected number of birds recovered in stratum 1 in year 1 is $E(R_{11}^{21}) = N_1^2 \times f_1^{21}$. In general, $E(\boldsymbol{R}_{11}) = \boldsymbol{D}(\boldsymbol{N}_1)$ f_1. where $\mathbf{D}(\bullet)$ transforms a column vector into a diagonal matrix.

The situation is more complicated for those birds recovered in year 2 because there are two possible paths that birds could have followed for each cell of $\boldsymbol{R}_{12}$. For example $E(R_{12}^{21}) = N_1^2 \times S_1^{22} \times f_2^{21} + N_1^2 \times S_1^{21} \times f_2^{11}$. In general, $E(\boldsymbol{R}_{12}) = \boldsymbol{D}(\boldsymbol{N}_1)\, \boldsymbol{S}_1 \boldsymbol{f}_2$.

The tag-recovery parameters are interpreted as the probability that a bird alive in stratum s at the time of tagging in year i will be recovered in stratum t between the time of tagging in year i and the time of tagging in year $i+1$ and the tag reported. This parameter includes components for migration between stratum s and stratum t and harvest, tag-retrieval, and tag-reporting. These parameters are indexed by both stratum s and stratum t since it is likely that the migration components to stratum t will differ depending upon the stratum where the bird was present at the time of tagging in year i even if subsequent harvest, tag-retrieval, and tag-reporting rates in stratum t are homogeneous. These parameters are assumed not to depend upon the migration patterns prior to year i.

An important implication is that the tag-recovery rates are assumed to be equal for all birds that leave stratum s in year i and are recovered in stratum t in the next year, regardless of the route they choose. If recoveries are obtained by hunting, all birds that migrate between two strata should arrive simultaneously in the hunting season. If one subset arrives at the beginning of the season and another subset arrives halfway through the season (because of a longer migration route), the tag-recovery rates will not be the same. If the rates are heterogeneous among subsets of birds migrating between strata s and t, then f_i^{st} can be thought of as an average rate. The model will no longer be valid; but, if the degree of heterogeneity is small, the results will still be approximately valid.

The survival/migration parameters are interpreted as the probability that a bird alive in stratum s at the time of tagging in year i will be alive and present in stratum t at the time of tagging in year $i+1$. These parameters include components for mortality and migration. All permanent migration to strata where no recoveries or releases occur is indistinguishable from, and treated as, another form of mortality. It is assumed that these rates do not depend upon migration patterns prior to year i. The row totals of the S_i^{st} (i.e., the $S_i^{s\cdot}$) represent the overall survival rates (regardless of where the bird migrated). The relative transition rates can be found by standardizing each row to sum to 1.

As in the tag-recovery rates, the survival/migration rates are assumed to be applicable to all birds regardless of the route chosen between strata. If there is considerable heterogeneity among subsets, then S_i^{st} can be thought of as an average survival/migration rate.

5. Model Fitting and Testing

The number of birds released in stratum s in year i and recovered in stratum t in year j (R_{ij}^{st}; $j=i,\ldots,l$, $t=1,\ldots,a$) are modelled using a multinomial distribution with index N_i^s and cell probabilities defined by Table 2. Birds that are released in different strata in the same year are assumed to be independent of each other. Releases in successive years are also assumed to be independent. Consequently, the likelihood can be written as a product of multinomial distributions.

There are a total of $a^2(k\text{-}1)$ survival/migration parameters (S_1,..., S_{k-1}), a^2k recovery rate parameters (f_1,..., f_k), and $a^2(l\text{-}k)$ confounded parameters representing the product of survival and recovery in years $k+1,\ldots,l$ (S_kf_{k+1},..., $S_kS_{k+1}S_{k+2}\ldots S_{l-1}f_l$) for a total of $a^2(l+k\text{-}1)$ parameters. Again note the resemblance to simple tag-recovery experiments where there are $(l+k\text{-}1)$ parameters.

A closed form moment-estimator of the parameters was derived in Schwarz, Schweigert, and Arnason (1992). These estimators are matrix extensions of the results of Brownie *et al.* (1985, pp.13-17). Numerical methods must be used to find the maximum likelihood estimates starting from these moment estimates. A goodness-of-fit test can be constructed using a likelihood ratio test comparing the full model to a saturated model where every cell in the recovery matrices has its own parameter.

Since the variances and covariances are found numerically, it is difficult to give quantitative statements about the precision of the estimates. However, extensive simulations (Schwarz 1988) have shown that the precision of the estimated tag-recovery rates is comparable to that in simple tag-recovery experiments, but the precision of the estimated survival/migration rates may be poor. This loss of precision is caused by the convolution of the intermediate survival/migration terms between release and recovery and the generally low tag-recovery rates in practice. There is a high correlation among the estimated survival/migration rates which implies that these parameters may be nearly non-identifiable.

There are many types of restrictions that can be placed upon the parameters of the general model to obtain reduced models of biological interest examples of which are presented in Schwarz, Schweigert, and Arnason (1992). For example, covariates could be used to relate the parameters to external factors such as the environment. Estimates of the parameters from these reduced models can be obtained numerically using the methods outlined in Don (1985). Likelihood ratio tests can be used to differentiate between models.

6. Example

The North American mallard makes yearly migrations from breeding areas in the northern United States and Canada to wintering areas in the mid- to southern United States. Nichols and Hines (1987) examined the winter distribution pattern of winter banded mallards and found evidence of a southward movement in the centre of the recovery distributions. However, they could not distinguish (using their methods) between the alternative possibilities of (1) increased harvest rates (and f_i) in the south and (2) a greater tendency of birds to winter in the south. The models developed in this paper permit the investigator to distinguish between these alternatives.

Data were obtained from the Patuxent Wildlife Research Centre on the band recoveries of adult male winter-banded (Jan.-Feb.) mallards from birds shot or found dead (Dec.-Feb.) from 1966-1977. Releases and recoveries were partitioned into the 15 Major Reference Areas (Figure 1) of Nichols and Hines (1987). A cross-tabulation of the recoveries by area of release and recovery (Table 3) indicates some evidence of a southward movement from areas 25, 27, 29 and 31 to areas 28, and 30. The raw data should not be over interpreted because of the differing band-recovery rates across years and reference areas.

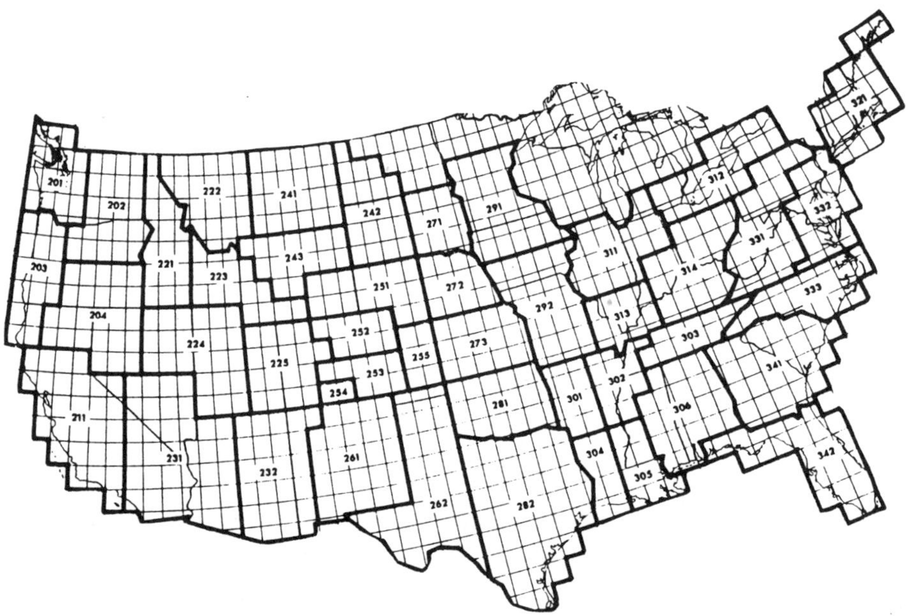

Figure 1. Minor reference areas in the United States. Major Reference Areas are formed by combining Minor Reference Areas with the same first two characters, e.g., Major Reference Area 30 consists of Minor Reference Areas 301-306.

Table 3. Cross-tabulation of the total bands recovered by Major Reference Area of release and recovery for 1966-1977. Birds were banded Jan-Feb. Recoveries were from birds shot or found dead Dec-Feb.

Area	Major Reference Area of Recovery														
Banded	20	21	22	23	24	25	26	27	28	29	30	31	32	33	34
20	1432	39	49	0	3	2	2	0	1	0	5	1	0	0	0
21	55	505	5	0	0	0	0	0	0	0	0	0	0	0	0
22	94	11	1410	13	33	29	3	2	4	0	19	2	0	0	0
23	0	1	18	31	1	0	0	0	0	0	1	0	0	0	0
24	17	1	51	0	804	76	4	48	37	12	73	3	0	2	1
25	29	8	67	7	45	2302	72	66	128	6	89	6	0	1	1
26	1	0	26	12	4	59	380	11	33	0	10	0	0	0	0
27	5	1	9	0	5	75	10	406	283	39	327	12	0	1	0
28	1	0	6	1	4	23	19	73	461	9	79	1	0	0	1
29	3	1	2	0	2	5	1	46	87	41	221	26	0	0	0
30	2	1	2	0	3	7	0	22	53	30	1636	194	0	34	77
31	2	0	6	0	0	2	3	21	28	26	1197	474	1	88	60
32	0	0	0	0	0	0	0	0	0	0	4	3	128	12	0
33	0	0	0	0	0	1	0	0	1	2	46	31	14	547	32
34	1	0	0	0	0	0	0	2	2	0	144	64	1	85	560

Table 4. Number of birds banded in Jan.-Feb and recovered in Dec.-Feb. 1966-1977 in Major Reference Areas 30 and 31. Totals do not match those in Table 3 because releases in 1971-1976 are omitted.

Year	Ref.	Number	Number of tags recovered														Totals	
			1967		1968		1969		1970		1971		1972		1973-77			
	Area	Released	30	31	30	31	30	31	30	31	30	31	30	31	30	31		
1966	30	10577	144	29	83	12	70	1	42	3	53	7	20	3	55	11	467	66
	31	5115	50	20	24	13	17	7	19	1	23	7	8	7	21	11	162	66
1967	30	7006			81	10	63	1	40	1	57	9	33	2	67	6	341	29
	31	7779			50	39	51	11	41	4	51	13	26	6	44	16	263	89
1968	30	2619					29	3	21	3	17	3	14	4	27	2	108	15
	31	3749					13	5	11	0	26	2	17	5	33	15	100	27
1969	30	2578							19	2	37	8	18	2	43	7	117	19
	31	3718							17	4	21	11	19	12	41	18	98	45
1970	30	1664									36	7	19	5	35	7	90	19
	31	4764									60	20	48	12	87	32	195	64
1971	30	2939											42	6	98	14	140	20
	31	3058											42	23	59	34	101	57
Total	30		144	29	164	22	162	5	122	9	200	34	146	22	325	47		
	31		50	20	74	52	81	23	88	9	181	53	160	65	285	126		

The yearly data are too sparse to investigate the interchange in more detail except for releases in reference areas 30 and 31 in 1966-1971 with recoveries in 1967-1977 (Table 4).

The model allowing yearly variation in recovery and survival rates was fitted to the data. A likelihood goodness-of-fit test against a model that allows cell-specific parameters did not provide evidence of a lack-of-fit (X^2=70.8, 60 df, p-value=0.16). Parameter estimates and estimated standard errors of the migration rates are shown in Table 5. The tag-recovery estimates are not shown because they are of little interest.

The estimates of overall survival (the sum of the rows of $\boldsymbol{S}_i$) are comparable to those reported by Nichols and Hines (1987). There appears to be some evidence of a southward movement from areas 30 to 31 in 1967, 1968 and 1970 and a northwards movement in 1970, but the precision of the estimates is poor. This is caused by the small number of recoveries in Area 31.

A restricted model allowing for differing yearly overall survival rates but constant yearly relative emigration rates, i.e., $S_i^{st} / S_i^{s\bullet}$ is constant from year to year for each stratum s, was then fit to the data (X^2=12.8, 8 df, p-value=0.12). This model estimated that a 0% yearly interchange of surviving birds occurred from Area 31 to 30 and a 14% yearly interchange from area 30 to 31.

Table 5. Parameter estimates and estimated standard errors of the migration rates

Year	Ref Area of Origin	Estimates Destination Ref. Area 30	31	Estimated s.e.[a] Destination Ref. Area 30	31
1966-67	30	0.6445	0.0000	0.0477	-
	31	0.0000	0.6826	-	0.0654
1967-68	30	0.8379	0.0000	0.0891	-
	31	0.3128	0.5720	0.1067	0.1530
1968-69	30	0.6646	0.0000	0.0840	-
	31	0.0959	0.5966	0.1245	0.1870
1969-70	30	0.6339	0.0000	0.0825	-
	31	0.0000	0.6057	-	0.0653
1970-71	30	0.6954	0.0427	0.1262	0.0939
	31	0.2148	0.5123	0.1052	0.1215

[a] If an estimate occurred on the bounds of the parameter space, its estimated standard error is not reported.

Table 6. Parameter estimates and estimated standard errors of the migration rates for a restricted model with S_i constant over time.

Years	Ref Area of Origin	Estimates Destination Ref. Area 30	31	Estimated s.e.[a] Destination Ref. Area 30	31
1966-71	30	0.7053	0.0000	0.0133	-
	31	0.1171	0.5986	0.0349	0.0417

[a] If an estimate occurred on the bounds of the parameter space, its estimated standard error is not reported.

Finally, a further restricted model where S_i is constant over time, i.e., both yearly overall survival and relative emigration rates are constant over time, was also fitted to the data (Table 6, X^2=23.6, 16 df, p-value=0.09 relative to initial model). The yearly movement from area 31 to area 30 is estimated to be 16% of birds that survive and return to areas 30 and 31. The power of the tests to detect differing survival rates over time is likely low because of the small number of bands recovered. This model provides a parsimonious, albeit, gross summary of the data.

7. Discussion

Ordinary tag-recovery data are theoretically adequate to estimate migration in these models but inference is not completely satisfactory. The parameters of most biological interest, the survival/migration rates (S_i) may have poor precision because of the convolution of these parameters between the time of release and recovery in the model formulation and the generally low tag-recovery rates present in practice.

Data requirements are stringent; releases must occur in all strata over a number of years along with recoveries in the same period of time. The practice of tagging in different areas in different years is not recommended since it is impossible to estimate most parameters unless very strong assumptions of equality among parameters over time and among areas are made.

In many cases, birds do not migrate in subsequent years independently of previous years. Schwarz (1988) explored a set of models allowing for fidelity to one or both of wintering and breeding sites over years. The assumption of Markovian movement cannot be tested using tag-recovery data unless ancillary information as to where the bird has migrated between release and recovery (e.g., resightings) are obtained. Schwarz (1988) showed how to combine resighting and recapture date in three types of migration models and Brownie *et al.* (1992) and Hestbeck *et al.* (1991) discuss the use of resighting data to estimate transition probabilities among strata and to test the Markovian assumption of movement.

If the bird could be sighted every year between release and recovery, then the complete migration history of the bird will be known, and the relatively simple method of inference in Markov chains can be used. This may be possible if the birds are fitted with a radio-telemetry device. Furthermore, if the "sightings" take place in a small period of time (rather than throughout the year), the methods of Arnason (1972, 1973) can be used since the "sighting" of a bird from its radio-telemetry device will be similar to a recapture and release in his capture-recapture formulation. With modern technology, inexpensive and reliable devices that permit the monitoring of a bird as it migrates among strata between release and recovery may be the most cost-effective method of obtaining good estimates of the survival/migration rates.

Acknowledgements. This work was supported in part by an Operating Grant from the Natural Science and Engineering Research Council of Canada. The author would like to thank Jim Nichols and Jim Hines for their help in extracting and interpreting the mallard database and the referees Jim Nichols and Jean Clobert for their comments.

References

Arnason, A. N. (1972). Parameter estimation from mark-recapture experiments on two populations subject to migration and death. *Researches in Population Ecology* **13**, 97-113.

Arnason, A. N. (1973). The estimation of population size, migration rates, and survival in a stratified population. *Researches in Population Ecology* **15,** 1-8.

Brownie, C., Anderson, D. R., Burnham, K. P., Robson, D. S. (1985). *Statistical Inference from band-recovery data - a handbook* , 2nd edition. U.S. Fish and Wildlife Service, Resource Publication No. 156. Washington, D.C. : U.S. Department of the Interior.

Brownie, C., Hines, J. E., Nichols, J. D., Pollock, K. H., and Hestbeck, J. B. (1992). Capture recapture studies for multiple strata including non-Markovian transition probabilities. *Biometrics* (in press).

Burnham, K. P. (1990). A unified theory for release-resampling studies of animal populations. Proceedings of the 1990 International Statistical Symposium , Taipei, R. O. C., 28-30 June 1990.

Chapman, D. G. and Junge, C. O. (1956). The estimation of the size of a stratified animal population. *Annals of Mathematical Statistics* **27**, 375-389.

Darroch, J. N. (1961). The two-sample capture-recapture census when tagging and sampling are stratified. *Biometrika* **48**, 241-260.

Don, F. J. H. (1985). The use of generalized inverses in restricted maximum likelihood estimation. *Linear Algebra and its Applications* **70**, 225-240.

Hestbeck, J. B., Nichols, J. D. and Malecki, R. A. (1991). Estimates of movement and site-fidelity using mark-resight data of wintering Canada geese. *Ecology* **72**, 523-533..

Nichols, J.D., and Hines, J.E. (1987). *Population ecology of the mallard. VIII. Winter distribution patterns and survival rates of winter-banded mallards.* U.S. Fish and Wildlife Service Resource Publication 162. Washington, D.C. : U.S. Department of the Interior.

Nichols, J. D., Stokes, S. L., Hines, J. E., and Conroy, M. J. (1982). Additional comments on the analysis of homogeneous survival rates in modern bird banding estimation models. *Journal of Wildlife Management* **46**, 953-962.

Pollock, K. H. and Raveling, D. G. (1982). Assumptions of modern band-recovery models with emphasis on heterogeneous survival rates. *Journal of Wildlife Management* **46**, 88-98.

Schwarz, C. J. (1988). Post-release stratification and migration models in band-recovery and capture-recapture models. Unpublished Ph. D. Thesis, University of Manitoba.

Schwarz, C. J., Burnham, K. P., and Arnason, A. N. (1988). Post-release stratification in band-recovery models. *Biometrics* 44,765-785.

Schwarz, C.J., Schweigert, J.F. and Arnason, A.N. (1992). Estimating migration rates using tag recoveries. Biometrics ••: •••-••• (in press).

Seber, G. A. F. (1982). *The estimation of animal abundance and related parameters,* 2nd edition. London: Griffin.

Marked Individuals in the
Study of Bird Population
J.-D. Lebreton and P. M. North (eds.)

The Estimation of Exchanges Among Populations or Subpopulations

James D. Nichols[1], Cavell Brownie [2], James E. Hines[1], Kenneth H. Pollock[2] and Jay B. Hestbeck[3].

[1]*U.S. Fish & Wildlife Service, Patuxent Wildlife Research Center, Laurel, MD 20708,*
[2]*Department of Statistics, North Carolina State University, Raleigh, NC 27695-8203, and*
[3]*U.S. Fish & Wildlife Service, Massachusetts Cooperative Fish & Wildlife Research Unit, University of Massachusetts, Amherst, MA 010031*

Summary. Capture-recapture methods for estimating rates of immigration, emigration, and movement among geographic strata are reviewed. We discuss likelihood-based estimation methods under models incorporating Markovian and non-Markovian movement. We briefly describe a computer program developed by Brownie et al. (1992) to carry out the necessary estimation and testing.

1. Introduction

The movement of an animal from one place to another is an event that can be of great significance to both evolutionary and population ecology. Population geneticists have long recognized the potential evolutionary importance of movement of animals among patches subjected to different selective pressures (Wright 1931). Recent experimental work on populations with local demic structure has corroborated the importance of movement to the results of interdemic selection (Wade 1982, Wade and McCauley 1988, Wade and Goodnight 1991).

All changes in population size result from the action of birth rates, death rates, and rates of migration in and out of the population. However, early work on modeling animal population dynamics focused on reproductive and survival rates within single populations (review in Hutchinson 1978). Experiments and observations during the last 25 years have led to increased recognition of the potential importance of movement parameters in the fundamental equation defining population change.

Fencing experiments in which Krebs et al. (1969) enclosed populations of voles, *Microtus pennsylvanicus* and *M. ochrogaster*, preventing movement of animals in and out, produced densities 3 times higher than those of unfenced control populations. The fenced populations overgrazed their habitat, resulting in starvation and declines in numbers. This and later experimental and descriptive work provided strong evidence of the involvement of emigration in population regulation of vertebrates (Krebs et al. 1969, Lidicker 1975, Hestbeck 1986).

Connor et al. (1983) found that populations of leafmining insects on oak trees were maintained by immigration rather than by in situ reproductive recruitment. Observations such as these have led to the development of models in which the migration of animals between "source" and "sink" habitats is essential to population regulation (Holt 1985, Pulliam 1988).

Levins (1969, 1970) and Boorman and Levitt (1973) proposed models to describe the demographic and evolutionary dynamics of metapopulations. The increasing fragmentation of habitats throughout the world has generated a great deal of interest in metapopulation dynamics in recent years (Hanski 1991, Hanski and Gilpin 1991). Movement among patches or subpopulations is a very important component of metapopulation models and is thought to have important effects on metapopulation size and persistence (Hanski 1991, Hansson 1991, Hastings 1991). Consequently, there is a need to develop statistical models for the estimation of movement among subpopulations.

2. Estimation of Immigration and Emigration

One of the earliest attempts to estimate quantities related to animal movement in capture-recapture studies involved the efforts of Jackson (1939) to decompose rates of gain and loss into components associated with death, reproduction and movement. Although Jackson's methods for drawing inferences about movement involved subdividing a large study area into smaller areas, his interest was clearly on the single study population rather than on rates of exchange among neighboring subpopulations. Other efforts to estimate movement in and out of a single study population using capture-recapture methods include Sheppard et al. (1969), Cameron and Williamson (1977), Manly (1985), Zeng and Brown (1987), Nichols and Pollock (1990), and Pollock et al. (1990, this volume).

3. Estimation With Geographic Stratification: Historical Development

Historically, many studies of animals have produced observations of marked animals caught or seen in one location at some time, i, and then caught/resighted at another location in some later period, $i + \Delta$. Such observations demonstrate that movements do occur, but they tell us little about rates or probabilities of movement. In some cases it may be possible to simply compare locations of animals that are recaptured, resighted or recovered to draw inferences about comparative rates of movement. For example, waterfowl biologists have addressed questions about sex-specificity of wintering ground locations using simple comparisons of band recovery locations (Nichols and Haramis 1980, Perdeck and Clason 1983, Nichols and Hines 1987). However, this approach is not appropriate for all comparative questions and does not lead to estimates of rates of movement.

The initial work on capture-recapture studies of populations with geographic stratification focused on the 2-sample Lincoln-Petersen estimator (Chapman and Junge 1956, Beverton and Holt 1957, Darroch 1961; also see Overton and Davis 1969, Seber 1982, Dorazio and Rago 1991). Animals are captured and marked at period 1 in each of a locations or strata and then released back into the population. A short period elapses during which animals may move among strata, and then a second sample is taken and recaptures recorded. If the overall population is closed (no births, deaths, or movements in or out of the total study area) between the 2 sample periods, then estimation of migration rates and stratum-specific numbers of animals is possible (Chapman and Junge 1956, Darroch 1961, Seber 1982). If stratum-specific mortality occurs between the 2 sampling periods, then the migration and capture probabilities are not identifiable. Darroch (1961) imposed the constraint of equal survival for all strata, permitting estimation of rates of movement.

Iwao (1963) extended the work on capture-recapture with 2 geographic strata to 3 sampling periods. His approach was deterministic and assumed that movement was possible between sampling periods 1 and 2 but not between 2 and 3. He derived estimators for rates of movement and survival. Arnason (1972) used a stochastic approach and derived moment estimators for the case of 2 strata with 3 sampling periods, removing Iwao's (1963) constraint of no movement between periods 2 and 3.

Arnason (1973) generalized his earlier work to more than 2 areas and derived moment estimators for rates of survival and movement for 3-sample capture-recapture studies. Seber (1982) showed that Arnason's (1973) methods could be generalized to more than 3 sample periods.

4. Estimation With Geographic Stratification: Recent Work

Despite the growing recognition during the 1970's and 1980's that movement was an important component of animal population dynamics, animal population ecologists seem to have made very little use of the estimation methods described above. Perhaps this was a consequence of the lack of available software, although program POPAN-2 had some capability to handle interstrata movement

(Arnason and Baniuk 1978). During the last 5 years, however, there has been a substantial increase in both interest and work on the estimation of rates of movement among geographic strata. This recent work has involved both band-recovery models and capture-recapture/resighting models.

4.1. Band recovery models

Band-recovery models were developed for studies in which animals are marked and released, and then recovered (generally as a dead animal, e.g., by hunters or fishermen) at some future time. The relevant data for models including migration are thus the time and location of both banding and recovery. Kania and Busse (1987) presented a method for computing point estimates of the number of birds in a banded sample migrating to different banding areas. Schwarz et al. (1988) developed post-stratification models in which animals are released at one location and recovered in 1 of several possible geographic strata. These models permit estimation of stratum-specific survival and recovery rates, under the assumption that animals exhibit complete fidelity to 1 recovery stratum (Schwarz et al. 1988).

Schwarz and Arnason (1990), Hilborn (1990), and Schwarz et al. (1992) considered band-recovery models with various assumptions about the fidelity of animals to banding and recovery locations. Some of these models are very general and permit movement among areas to vary with time, under the Markovian assumption that movement between times i and $i+1$ depends only on location at i. These general models permit estimation of time-specific rates of movement and survival. These models are the subject of Schwarz (this volume) and will not be discussed further here.

4.2. Capture-recapture/resighting models

Capture-recapture/resighting models include possible multiple captures or sightings of the same animal. These data can be summarized in capture histories specifying the time (sampling period) and location or stratum of each capture and/or sighting. Hestbeck et al. (1991) analyzed data from a large capture-resighting study of Canada Geese, *Branta canadensis*, and were interested in year-to-year movement among 3 Atlantic Flyway wintering areas. They used the basic model of Arnason (1973, the likelihood for this model was subsequently presented by Schwarz et al. 1992; hence we will refer to it as the Arnason-Schwarz model) and conditioned on the numbers of geese released in each area during 1 winter. Hestbeck et al. (1991) then considered the probabilities associated with all possible resighting histories for the subsequent 2 winters. Instead of using the moment estimators of Arnason (1973), Hestbeck et al.

(1991) used program SURVIV (White 1983) to compute maximum likelihood estimates numerically. Goodness-of-fit tests led to the rejection of the Arnason-Schwarz model for all 3 tested cohorts.

Hestbeck et al. (1991) then proposed a model in which movement between years i and $i+1$ depends not only on location at i, but also on location at i-1. This model fit the data for 2 of the 3 cohorts adequately, and likelihood ratio tests provided strong evidence that the new "Memory" model explained the data better than the model with Markovian movement.

Recently, we noted that the approach of Hestbeck et al. (1991) was neither fully efficient nor easily generalized (Brownie et al. 1992). We presented methods and an associated computer program to estimate parameters for geographically-stratified populations under some models in which movement can be considered a first-order Markov process and others in which movements depend on locations in previous periods (i.e., animals exhibit memory). Here, we will briefly outline these models.

4.2.1. Arnason-Schwarz model

First, we consider the model of Arnason (1973; also see Appendix B of Schwarz et al. 1992) which assumes Markovian transitions. Define the following parameters:

ϕ_i^{rs} = probability of being alive and in stratum s at time $i+1$, for an animal alive and in stratum r at time i,

p_i^s = probability of recapture or resighting at time i, for an animal in stratum s at time i.

Assume a study in which $a=3$ strata and $k=4$ sampling periods and define the following matrices of transition and recapture/resighting probabilities:

$$\underset{\sim}{\phi_i} = \begin{bmatrix} \phi_i^{11} & \phi_i^{12} & \phi_i^{13} \\ \phi_i^{21} & \phi_i^{22} & \phi_i^{23} \\ \phi_i^{31} & \phi_i^{32} & \phi_i^{33} \end{bmatrix}, \; i = 1, 2, 3, \text{ and } \; \underset{\sim}{p_i} = \begin{bmatrix} p_i^1 \\ p_i^2 \\ p_i^3 \end{bmatrix}, \; i = 2, 3, 4.$$

Define $D(\underset{\sim}{p_i})$ to be the diagonal matrix with diagonal elements equal to the elements p_i^s of the parameter vector $\underset{\sim}{p_i}$. Let $q_i^s = 1 - p_i^s$, and define $D(\underset{\sim}{q_i})$ as the corresponding diagonal matrix.

Define the following statistics:

R_i^r = the number of marked animals released in stratum r at time i (this includes both newly-banded birds and recaptures),

m_{ij}^{rs} = the number of marked animals recaptured/resighted in stratum s at time j that were last captured/sighted in stratum r at time i.

Then, define the following data matrices:

$$\underset{\sim}{R}_i = \begin{bmatrix} R_i^1 \\ R_i^2 \\ R_i^3 \end{bmatrix}, i=1,2,3, \text{ and } \underset{\sim}{m}_{ij} = \begin{bmatrix} m_{ij}^{11} & m_{ij}^{12} & m_{ij}^{13} \\ m_{ij}^{21} & m_{ij}^{22} & m_{ij}^{23} \\ m_{ij}^{31} & m_{ij}^{32} & m_{ij}^{33} \end{bmatrix}, i=1,2,3,\ j=i+1,...,4.$$

Conditional on the number of releases, R_i^r, the associated recaptures/resightings $(m_{i,\,i+1}^{r1},\ m_{i,\,i+1}^{r2},...,\ m_{i,\,i+1}^{ra};\ m_{i,\,i+2}^{r1}\ m_{i,\,i+2}^{r2},\ ...,\ m_{i,\,i+2}^{ra};...;\ m_{i,\,k}^{r1}\ m_{i,\,k}^{r2}\ ...,\ m_{i,\,k}^{ra})$, together with the number of R_i^r never seen or caught again, represent a multinomial random variable. Recapture/resighting statistics and corresponding cell probabilities are presented in compact matrix notation (Brownie et al. 1992, Schwarz et al. 1992) in Table 1.

Each 0 in a capture history (indicating that the animal is not captured or seen during a period), followed by a recapture/resighting in a later period, must include the possibility that the animal was in any stratum during the period in which it was not seen or caught. For example, the probability associated with m_{13}^{12} is $p_3^2(\phi_1^{11}q_2^1\phi_2^{12}+\phi_1^{12}q_2^2\phi_2^{22}+\phi_1^{13}q_2^3\phi_2^{32})$, and the 3 terms inside the brackets reflect the unknown location of the animal during period 2.

Brownie et al. (1992) described a computer program, MSSURVIV, to be used in conjunction with SURVIV (White 1983), which implements the Arnason-Schwarz model and computes maximum likelihood estimates. This program also computes estimates under preselected reduced-parameter models and permits the user to implement other reduced-parameter models using the CONSTRAINTS option.

The transition probabilities of the Arnason-Schwarz and related reduced-parameter models will be the parameters of interest in many applications. However, these parameters can be thought of as representing the product of probabilities of surviving and making a particular transition (moving). Interest in decomposing the transition probabilities, ϕ_i^{rs}, leads to the following derived parameters:

$\phi_i^{r\cdot}=\sum_{s=1}^{a}\phi_i^{rs}$, the total survival rate for animals in stratum r at time i, and

Table 1. Symbolic representation of recapture data and corresponding multinomial cell probabilities, both using compact matrix notation. Release and recapture matrices $\underset{\sim}{R}_i$ and $\underset{\sim}{m}_{ij}$, and parameter matrices $\underset{\sim}{\phi}_i$, $D(\underset{\sim}{p}_i)$, and $D(\underset{\sim}{q}_i)$, are defined in the text.

Recapture array

Year of release	Number released	Year of first recapture 2	3	4
1	$\underset{\sim}{R}_1$	$\underset{\sim}{m}_{12}$	$\underset{\sim}{m}_{13}$	$\underset{\sim}{m}_{14}$
2	$\underset{\sim}{R}_2$		$\underset{\sim}{m}_{23}$	$\underset{\sim}{m}_{24}$
3	$\underset{\sim}{R}_3$			$\underset{\sim}{m}_{34}$

Multinomial cell probabilities

Year of release	Year of first recapture 2	3	4
1	$\underset{\sim}{\phi}_1 D(\underset{\sim}{p}_2)$	$\underset{\sim}{\phi}_1 D(\underset{\sim}{q}_2)\underset{\sim}{\phi}_2 D(\underset{\sim}{p}_3)$	$\underset{\sim}{\phi}_1 D(\underset{\sim}{q}_2)\underset{\sim}{\phi}_2 D(\underset{\sim}{q}_3) D(\underset{\sim}{p}_4)$
2		$\underset{\sim}{\phi}_2 D(\underset{\sim}{p}_3)$	$\underset{\sim}{\phi}_2 D(\underset{\sim}{q}_3)\underset{\sim}{\phi}_3 D(\underset{\sim}{p}_4)$
3			$\underset{\sim}{\phi}_3 D(\underset{\sim}{p}_4)$

$\psi_i^{rs} = \dfrac{\phi_i^{rs}}{\phi_i^{r\cdot}}$, the probability of being in stratum s at time $i+1$ for animals that were in stratum r at i and survived until $i + 1$.

If survival probability depends only on location at i and is independent of the transition made between i and $i+1$, then the ψ_i^{rs} represent rates of interstratum movement (Hestbeck et al. 1991, Brownie et al. 1992). In some situations these derived parameters may be of primary interest, and an option of MSSURVIV reparameterizes the Arnason-Schwarz model in terms of $\phi_i^{r\cdot}$ and ψ_i^{rs}. Parameters are still identifiable under this reparameterization because of the constraint that $\sum_{s=1}^{a} \psi_i^{rs} = 1$.

4.2.2. Models with memory

Hestbeck et al. (1991) and Brownie et al. (1992) provided evidence that in some cases, movement between times i and $i+1$ depends not only on location at i but also on location at i -1. Hestbeck et al. (1991) used a model in which new releases exhibit transition probabilities that depend only on location at the time of release, i (because location in the previous time period, i -1, is not known for these animals), whereas the transition between $i+1$ and $i+2$ depends on location at both $i+1$ and i. Hestbeck et al. (1991) specified all cell probabilities and obtained estimates using SURVIV. Brownie et al. (1992) noted that this approach was time-consuming and not easily generalized to larger numbers of time periods and strata. Thus, they recommended a partial or conditional likelihood approach, conditioning on releases at time i for which location at i -1 is known.

Define the following statistics:

$R_{i-1,i}^{rs}$ = the number of releases in stratum s at i, that were also seen in stratum r at i -1, i=2,..., k-1; r, s =1,..., a, and

$m_{i-1,i,j}^{rst}$ = the number of animals in release cohort $R_{i-1,i}^{rs}$ that are next captured in stratum t at time j, i =2,..., k-1; j =i+1, .., k; r, s, t=1..., a.

The corresponding matrices of releases and recaptures/resightings for $a=2$ strata can be represented as:

$$\underset{\sim}{R}_{i-1,i} = \begin{bmatrix} R^{11}_{i-1,i} \\ R^{12}_{i-1,i} \\ R^{21}_{i-1,i} \\ R^{22}_{i-1,i} \end{bmatrix}, \quad i=2, ..., k\text{-}1, \text{ and}$$

$$\underset{\sim}{m}_{i-1,i,j} = \begin{bmatrix} m^{111}_{i-1,i,j} & m^{112}_{i-1,i,j} \\ m^{121}_{i-1,i,j} & m^{122}_{i-1,i,j} \\ m^{211}_{i-1,i,j} & m^{212}_{i-1,i,j} \\ m^{221}_{i-1,i,j} & m^{222}_{i-1,i,j} \end{bmatrix}, \quad i=2, ..., k\text{-}1, \; j=i+1, ..., k.$$

Define the following parameters of the Memory model:

$\phi^{rst}_{i-1,i}$ = probability of being present at stratum t at time $i+1$ for an animal that was in r at time i-1 and is in s at time i.

p^s_i = probability of capture at time i for an animal present in stratum s at i.

The corresponding matrices of transition and recapture/resighting probabilities (again for $a=2$ strata) are:

$$\underset{\sim}{\phi}_{i-1,i} = \begin{bmatrix} \phi^{111}_{i-1,i} & \phi^{112}_{i-1,i} & 0 & 0 \\ 0 & 0 & \phi^{121}_{i-1,i} & \phi^{122}_{i-1,i} \\ \phi^{211}_{i-1,i} & \phi^{212}_{i-1,i} & 0 & 0 \\ 0 & 0 & \phi^{221}_{i-1,i} & \phi^{222}_{i-1,i} \end{bmatrix}, \quad i=2, ..., k\text{-}1,$$

$$\underset{\sim}{p}_i = \begin{bmatrix} p^1_i & 0 \\ 0 & p^2_i \\ p^1_i & 0 \\ 0 & p^2_i \end{bmatrix}, \text{ and } D(\underset{\sim}{q}_i) = \begin{bmatrix} q^1_i & 0 & 0 & 0 \\ 0 & q^2_i & 0 & 0 \\ 0 & 0 & q^1_i & 0 \\ 0 & 0 & 0 & q^2_i \end{bmatrix}, \quad i=3,...,k,$$

where $q_i^s = 1 - p_i^s$. The data for the Memory model and the corresponding cell probabilities are represented for $k = 6$ sample periods in Table 2.

Because of the structural similarity of the Arnason-Schwarz model (Table 1) and the conditional likelihood of the Memory model (Table 2), estimates for the Memory model can be computed using MSSURVIV (Brownie et al. 1992). Reduced-parameter models can also be implemented using MSSURVIV. For example, in the case of 3 strata, the Arnason-Schwarz model can be obtained by setting $\phi_{i-1,i}^{1st} = \phi_{i-1,i}^{2st} = \phi_{i-1,i}^{3st}$ for all s and t. The test between the Arnason-Schwarz and Memory models provides a biologically-interesting test of the null hypothesis of Markovian transitions.

5. Discussion

We believe that the movement of animals among subpopulations is a topic that historically did not receive the attention it deserved in animal population ecology. The recent interest of ecologists and conservation biologists in landscape ecology, fragmentation and metapopulation structures and dynamics has appropriately begun to focus attention on animal movement. The model of Arnason (1973) and the recent models and software of Hilborn (1990), Schwarz et al. (1992) and Brownie et al. (1992) provide a conceptual framework and computational methods for estimating movement parameters and testing hypotheses about the movement process.

We believe that there are opportunities for future statistical work with these models. For example, it will be very interesting to build infrastructural models in which movement parameters are expressed as functions of environmental covariates (e.g., weather variables) or aspects of metapopulation structure (e.g., distance between patches, patch size, habitat characteristics of patches, independent measures of fitness within patches, etc.). Although such infrastructural modeling of movement probabilities can be conducted using MSSURVIV, it would be desirable to develop software to facilitate such modeling, as has been done for survival and capture probabilities (SURGE 4.1, Lebreton et al. 1992).

A robust design approach can be used to estimate movement probabilities as in Nichols et al. (1992). This approach requires substantial data but may be useful for very large data sets for which failures of certain assumptions (e.g., homogeneous capture probabilities among individuals within a patch) result in problems with model fit. When estimates of patch-specific population size are available (e.g., from capture-recapture models for closed populations or from other estimation methods), then numbers of immigrants and emigrants can be estimated (Hestbeck et al. 1991).

Table 2. Symbolic representation of recapture data and corresponding multinomial cell probabilities, both using compact matrix notation. Release and recapture matrices $\underset{\sim}{R}_{i-1,i}$ and $\underset{\sim}{m}_{i-1,i,j}$, and parameter matrices $\underset{\sim}{\phi}_{i-1,i}$, $\underset{\sim}{p}_i$, and $D(\underset{\sim}{q}_i)$, are defined in the text.

Year of release	Number released	Recapture array Year of recapture 3	4	5	6
2	$\underset{\sim}{R}_{12}$	$\underset{\sim}{m}_{123}$	$\underset{\sim}{m}_{124}$	$\underset{\sim}{m}_{125}$	$\underset{\sim}{m}_{126}$
3	$\underset{\sim}{R}_{23}$		$\underset{\sim}{m}_{234}$	$\underset{\sim}{m}_{235}$	$\underset{\sim}{m}_{236}$
4	$\underset{\sim}{R}_{34}$			$\underset{\sim}{m}_{345}$	$\underset{\sim}{m}_{346}$
5	$\underset{\sim}{R}_{45}$				$\underset{\sim}{m}_{456}$

Year of release	Multinomial cell probabilities Year of first recapture 3	4	5	6
2	$\underset{\sim}{\phi}_{12}\underset{\sim}{p}_3$	$\underset{\sim}{\phi}_{12}D(\underset{\sim}{q}_3)\underset{\sim}{\phi}_{23}\underset{\sim}{p}_4$	$\underset{\sim}{\phi}_{12}D(\underset{\sim}{q}_3)\underset{\sim}{\phi}_{23}D(\underset{\sim}{q}_4)\underset{\sim}{\phi}_{34}\underset{\sim}{p}_5$	$\underset{\sim}{\phi}_{12}D(\underset{\sim}{q}_3)\underset{\sim}{\phi}_{23}D(\underset{\sim}{q}_4)\underset{\sim}{\phi}_{34}D(\underset{\sim}{q}_5)\underset{\sim}{\phi}_{45}\underset{\sim}{p}_6$
3		$\underset{\sim}{\phi}_{23}\underset{\sim}{p}_4$	$\underset{\sim}{\phi}_{23}D(\underset{\sim}{q}_4)\underset{\sim}{\phi}_{34}\underset{\sim}{p}_5$	$\underset{\sim}{\phi}_{23}D(\underset{\sim}{q}_4)\underset{\sim}{\phi}_{34}D(\underset{\sim}{q}_5)\underset{\sim}{\phi}_{45}\underset{\sim}{p}_6$
4			$\underset{\sim}{\phi}_{34}\underset{\sim}{p}_5$	$\underset{\sim}{\phi}_{34}D(\underset{\sim}{q}_5)\underset{\sim}{\phi}_{45}\underset{\sim}{p}_6$
5				$\underset{\sim}{\phi}_{45}\underset{\sim}{p}_6$

Extension of the robust design approach of Nichols and Pollock (1990) should also permit separation of estimated new recruits in a patch into components associated with (1) within-patch reproduction, (2) immigration from other patches within the metapopulation system, and (3) immigration from outside the studied system.

Radiotelemetry has been used extensively to study animal movement in recent years. It would be possible to use capture-recapture/resighting methods together with radiotelemetry to study movements within a metapopulation system. A single likelihood could be developed in which the 2 types of data would share certain parameters (e.g., movement or transition probabilities). For the models described in this chapter, the decomposition of transition probabilities into survival and movement probabilities requires the assumption that survival is independent of the specific transition made. Additional data from telemetered animals could be used to test this assumption. More importantly, such data could be used to relax the assumption by permitting estimation of one survival probability for animals that remain in a patch and another for animals that leave the patch. Again, this estimation could be efficiently handled by developing a multinomial likelihood incorporating both types of data.

Even in the absence of new statistical development, we believe that the models described in this chapter can be used to address a number of interesting biological questions about animal movement. Island biogeographic theory (MacArthur and Wilson 1967) is based on certain assumptions about animal movement that are based on only limited empirical evidence. For example, it would be instructive to test hypotheses about the functional relationships between rates of movement and such factors as patch (or island) size, inter-patch distance, and patch position (e.g., with respect to the "stepping-stone" model).

Similarly, a large study area could be divided into areas corresponding to different habitat types or to different food types or abundances, and movement estimates could be used to test hypotheses about habitat selection (e.g., Fretwell 1972) or optimal foraging (e.g., Pyke 1984), although radio-telemetry studies may be more appropriate for such tests (Conroy this volume). The memory models could be used to test for hypothesized tendencies to return to (e.g., because of familiarity with specific locations of food or cover) or avoid (because of "knowledge" of factors that prompted patch departure in the first place, e.g., predator numbers, depleted food, etc.) previously-used patches. In summary, we look forward both to the results of applying these existing models to studies of animal movement and to the development of new extensions of these models and methods.

Acknowledgments. We thank J. Clobert, W.A. Link, J.-D. Lebreton, and J.R. Sauer for helpful comments on an early draft of the manuscript, and C. Hagen for her patience in typing the paper.

References

Arnason, A.N. (1972).- Parameter estimates from mark-recapture experiments on two populations subject to migration and death. Res. Pop. Ecol., 13:97-113.

Arnason, A.N. (1973).- The estimation of population size, migration rates and survival in a stratified population. Res. Pop. Ecol., 15:1-8.

Arnason, A.N. and Baniuk, L. (1978).- POPAN-2: a data maintenance and analysis system for mark-recapture data. The Charles Babbage Research Centre, St. Pierre, Manitoba, Canada.

Beverton, R.J.H. and Holt, S.J. (1957).- On the dynamics of exploited fish populations. Her Majesty's Stationery Office, London.

Boorman, S.A. and Levitt, P.R. (1973).- Group selection on the boundary of a stable population. Theor. Pop. Biol., 4:85-128.

Brownie, C., Hines, J.E., Nichols, J.D., Pollock, K.H. and Hestbeck, J.B. (1992).- Capture-recapture studies for multiple strata including non-Markovian transition probabilities. Biometrics, 48 (in press)

Cameron, R.A.D. and Williamson, P. (1977).- Estimating migration and the effects of disturbance in mark-recapture studies on the snail *Cepaea nemoralis* L.J. Anim. Ecol., 46:173-179.

Chapman, D.G. and Junge, C.O. (1956).- The estimation of the size of a stratified animal population. Ann. Math. Stat., 27:375-389.

Connor, E.F., Faeth, S.H. and Simberloff, D. (1983).- Leafminers on oak: the role of immigration and in situ reproductive recruitment. Ecology, 64:191-204.

Conroy, M.J.- Testing hypotheses about the relationship of habitat to animal survivorship. *In* The Use of Marked Individuals in the Study of Bird Population Dynamics: Models, Methods, and Software, J.-D Lebreton and P.M. North (Eds) Birkhauser Verlag, Berlin.

Darroch, J.N. (1961).- The two-sample capture-recapture census when tagging and sampling are stratified. Biometrika, 48:241-260.

Dorazio, R.M. and Rago, P.J. (1991).- Evaluation of a mark-recapture method for estimating mortality and migration rates of stratified populations. Can J. Fish. Aquat. Sci., 48:254-260.

Fretwell, S.D. (1972).- Populations in a seasonal environment. Princeton Univ. Press, Princeton, New Jersey.

Hanski, I. (1991).- Single-species metapopulation dynamics: concepts, models and observations. Biol. J. Linn. Soc., 42:17-38.

Hanski, I. and Gilpin, M. (1991).- Metapopulation dynamics: brief history and conceptual domain. Biol. J. Linn. Soc., 42:3-16.

Hansson, L. (1991).- Dispersal and connectivity in metapopulations. Biol. J. Linn. Soc., 42:89-103.

Hastings, A. (1991).- Structured models of metapopulation dynamics. Biol. J. Linn. Soc., 42:57-71.

Hestbeck, J.B. (1986).- Multiple regulation states in populations of the California vole, Microtus *californicus*. Ecol. Monogr., 56:161-181.

Hestbeck, J.B., Nichols, J.D. and Malecki, R.A. (1991).- Estimates of movement and site fidelity using mark-resight data of wintering Canada Geese. Ecology, 72:523-533.

Hilborn, R. (1990).- Determination of fish movement patterns from tag recoveries using maximum likelihood estimators. Can. J. Fish. Aquat. Sci., 47:635-643.

Holt, R.D. (1985).- Population dynamics in two-patch environments: some anomalous consequences of an optimal habitat distribution. Theor. Pop. Biol., 28: 181-208.

Hutchinson, G.E. (1978).- An introduction to population ecology. Yale Univ. Press, New Haven, Connecticut.

Iwao, S. (1963).- On a method for estimating the rate of population interchange between two areas. Res. Pop. Ecol., 5:44-50.

Jackson, C.H.N. (1939).- The analysis of an animal population. J. Anim. Ecol., 8: 238-246.

Kania, W. and P. Busse. 1987.- An analysis of the recovery distribution based on finding probabilities. Acta Ornithologica, 23:121-127.

Krebs, C.J., Keller, B.L. and Tamarin, R.H. (1969).- *Microtus* population biology: demographic changes in fluctuating populations of *M. ochrogaster* and *M. pennsylvanicus* in southern Indiana. Ecology, 50:587-607.

Levins, R. (1969).- Some demographic and genetic consequences of environmental heterogeneity for biological control. Bull. Entomol. Soc. Amer., 15:237-240.

Levins, R. (1970).- Extinction. pp77-107 *in* Some Mathematical Problems in Biology, M. Gesternhaber, (Ed) Amer. Math. Soc., Providence, Rhode Island.

Lidicker, W.Z., Jr. (1975).- The role of dispersal in the demography of small mammals. pp.102-128 *in* Small Mammals, Their Productivity and Population Dynamics, F.B. Golley, K. Petrusewicz, and L. Ryszkowski (Eds) Cambridge Univ. Press, Cambridge, England.

MacArthur, R.H. and Wilson, E.O. 1967.- The theory of island biogeography. Princeton Univ. Press, Princeton, New Jersey.

Manly, B.F.J. (1985).- A test of Jackson's method for separating death and emigration with mark-recapture data. Res. Pop. Ecol., 27:99-109.

Nichols, J.D. and Haramis, G.M. (1980).- Sex-specific differences in winter distribution patterns of canvasbacks. Condor, 82:406-416.

Nichols, J.D. and Hines, J.E. (1987).- Population ecology of the mallard. VIII. Winter distribution patterns and survival rates of winter-banded mallards. US Fish and Wildlife Service Resource Publication no. 162.

Nichols, J.D. and Pollock, K.H. (1990).- Estimation of recruitment from immigration versus in situ reproduction using Pollock's robust design. Ecology, 71: 21-26.

Nichols, J.D., Sauer, J.R., Pollock, K.H. and Hestbeck, J.B. (1992).- Estimating transition probabilities for stage-based population projection matrices using capture-recapture data. Ecology, 73:306-312.

Overton, W.S. and Davis, D.E. (1969).- Estimating the numbers of animals in wildlife populations. pp 403-455 *in* Wildlife Management Techniques, 3rd ed. revised, R.H. Giles (Ed) The Wildlife Society, Washington, D.C.

Perdeck, A.C. and Clason, C. (1983).- Sexual differences in migration and winter quarters of ducks ringed in the Netherlands. Wildfowl, 34:137-143.

Pollock, K.H., Nichols, J.D., Brownie, C. and Hines, J.E. (1990).- Statistical inference for capture-recapture experiments. Wildl. Monogr. 107.

Pollock, K.H., Kendall, W.L. and Nichols, J.D.- The "robust" capture-recapture design allows components of recruitment to be estimated. *In* The Use of Marked Individuals in the Study of Bird Population Dynamics: Models, Methods, and Software, J.-D Lebreton and P.M. North (Eds) Birkhauser Verlag, Berlin.

Pulliam, H.R. (1988).- Sources, sinks, and population regulation. Amer. Natur., 132:652-661.

Pyke, G.H. 1984.- Optional foraging theory: a critical review. Ann. Rev. Ecol. Syst., 15:523-575.

Schwarz, C.J.- Estimating migration rates using ring recoveries. *In* The Use of Marked Individuals in the Study of Bird Population Dynamics: Models, Methods, and Software, J.-D Lebreton, and P.M. North (Eds) Birkhauser Verlag, Berlin.

Schwarz, C.J. and Arnason, A.N. (1990).- Use of tag-recovery information in migration and movement studies. Amer. Fish. Soc. Symp., 7:588-603.

Schwarz, C.J., Burnham, K.P. and Arnason, A.N. (1988).- Post-release stratification in band-recovery models. Biometrics, 44:765-785.

Schwarz, C.J., Schweigert, J.F. and Arnason, A.N. (1992).- Estimating migration rates using tag-recovery data. Biometrics, 48 (in press)

Sheppard, P.M., MacDonald, W.W., Tonn, R.J. and Grab, B. (1969).- The dynamics of an adult population of *Aedes aegypti* in relation to dengue haemorrhagic fever in Bangkok. J. Anim. Ecol., 38:661-702.

Wade, M.J. (1982).- Group selection: migration and the differentiation of small populations. Evolution, 36:949-961.

Wade, M.J. and Goodnight, C.J. (1991).- Wright's shifting balance theory: an experimental study. Science, 253:1015-1018.

Wade, M.J. and McCauley, D.E. (1988).- Extinction and recolonization: their effects on the genetic differentiation of local populations. Evolution, 42:995-1005.

White, G.C. (1983).- Numerical estimation of survival rates from band-recovery and biotelemetry data. J. Wildl. Manage., 47:716-728.

Wright, S. (1931).- Evolution in Mendelian populations. Genetics, 16:93-159.

Zeng, Z. and Brown, J.H. (1987).- A method for distinguishing dispersal from death in mark-recapture studies. J. Mammal., 68:656-665.

Marked Individuals in the
Study of Bird Population
J.-D. Lebreton and P.M. North (eds.)

The Components of Local Recruitment

J. Clobert[1], R. Julliard[1] and R.H. McCleery[2]

[1]*Laboratoire d'Ecologie, Ecole Normale Supérieure, 46 Rue d'Ulm 75230 Paris Cedex 05 France, and* [2]*Edward Grey Institute, Department of Zoology, Oxford University, South Parks Road, OX1 3PS Oxford, England.*

Summary: Local recruitment can play an important role in population dynamics and in evolutionary processes. The local recruitment rate is the product of several components (immature survival rate, return rate, age-specific proportions of breeders). Recently, methods were proposed to estimate some of these components separately. We show how these methods could be imbedded into the framework of existing goodness of fit tests and standard CMR methodology. In particular, we focus on ways of testing model assumptions. We illustrate this using an example on the great tit (*Parus major*) and some simulated data.

1. Introduction

Local recruitment is a complex phenomenon in animal populations, playing a key role in both population dynamics (Lebreton, 1987; Porter and Coulson, 1987; Massot et al., 1992) and evolutionary processes (Clutton-Brock, 1988; Newton, 1989). It is usually estimated as the number of young found breeding in their natal population divided by the total number of young marked, although many steps are involved in determining whether a young individual will breed at its site of origin. In fact, three major components of the local recruitment can be recognized: 1) survival to the age of first breeding, 2) fidelity to the natal area and 3) age of first return to the breeding ground. In the absence of relevant data (Hestbeck et al., 1991; Nichols et al., this volume), the first two components cannot usually be identified separately, and their product is simply the local immature survival rate (Clobert et al., 1990). However, using data on birds marked as nestlings, it is possible to get a separate estimate of both local immature survival rate and age-specific proportion of breeders in the case of a batch marking (Lebreton et al., 1990) or of individual marking (Clobert et al., 1990 and in press). These methods are based on the fact that individuals marked as nestlings do not breed successfully for the first time at the same age. This implies that, up to a certain age, their capture probability is lower than that of adult breeders. Placing this problem into the context of capture-recapture models for several data sets (Lebreton et al., 1992), Clobert et al. (in press) presented various models where local immature survival rates and age-specific proportions of breeders can be estimated separately.

However, the behaviour of such models is poorly known. Here, we provide ways for checking the hypotheses underlying the use of these models (goodness of fit tests, sensitivity to heterogeneity) with an application to a study of the Great Tit (*Parus major*).

2. The model and its assumptions

During n occasions (years), individuals are marked at birth. Individuals born in year i (batch i) will not start to breed before age k and some will delay their breeding until age f. Only breeders at a given stage in their reproduction are captured. Hence, from year i to year i+k, the animal will not be catchable. The probability that it will survive up to age k, i.e. the local immature survival probability, is given as $\phi'_{i,1} = \phi'_{i,i}\phi'_{i,i+1}\cdots\phi'_{i,i+k-1}$ (Clobert et al., in press) where i indices the year of birth. The probability that an individual, surviving at the beginning of year i+k, will be captured or observed at age k depends both on the probability of starting to breed at age k ($\alpha_{i,k}$) and, for breeding individuals, on the capture probability $p_{i,i+k}$. The product (increasingly complex as age increases) of these two probabilities is noted $p'_{i,i+k}$.

Aside from the classical assumptions of independence between individuals and identity of rates among individuals, three assumptions are needed: 1) a first-time breeder will breed in all subsequent years at the same rate as previous breeders. Then, all individuals will have the same probability of capture $p_{i,j}$ (j=i+k+1,...,n); 2) from age k, both breeders and non-breeders of the same age will have the same survival probability $\phi_{i,j}$ (j=i+k,...,n-1). This assumption implies: i) there is no survival cost associated with a first reproduction; ii) the probability that they return to the natal area to breed for the first time is independent of age; 3) there is some age f at which all surviving individuals born in year i have started to breed ($p'_{ij}=p_{ij}$ for $j \geq i+f$).

Under these three assumptions, Clobert et al. (in press) give a formal description of the relation between $a_{i,j-i}$, $p'_{i,j}$ and $p_{i,j}$ and give explicit estimates for a time-dependent model.

3. Parameter estimation

The main difficulty in estimating the parameters arises from the fact that capture probabilities vary between the first recapture and the subsequent ones. Although standard models do not handle this situation, Clobert et al. (1990, in press) showed how to solve the problem. They cut the capture-recapture histories according to the first recapture in two pieces (Table I) and analyzed them as if they were distinct groups. The first group (group I, Table I) contains the part of the capture-recapture history from marking to the first recapture. The second group (group B) contains the capture-recapture history from first recapture to the end of the study. p'_{ij} and ϕ'_{i1} are estimated from group I; p_{ij} and ϕ_{ij} are estimated from group B. The first recapture can occur at different ages. Two individuals marked in year i and i+1 respectively, and breeding for the first time in i+2 will generate the same capture-recapture history in group B. This can be avoided by creating one group of individuals per observed age at first breeding (group B1, B2...; Table I).

Table 1. Splitting capture-recapture histories of individuals marked at birth into pseudo groups. A recapture is symbolyzed by 1, a no capture by 0. 6 indivuduals were marked at occasion 1, 5 at occasion 2. Group I: capture-recapture history from birth to the first breeding observation. Group B: part of the capture-recapture history from the first breeding observation to the end (6). Group B can be further split into group B1 (animals first observed as breeder at age 1), and into group B2 (animals first observed as breeder later than age 1).

individuals	Overall capture-recapture history	From birth to first breeding (Group I)	From first breeding (Group B)
1	111101	110000	011101
2	111010	110000	011010
3	100101	100100	000101
4	101101	101000	001101
5	110000	110000	010000
6	100000	100000	
7	010100	010100	000100
8	011010	011000	001010
9	010110	010100	000110
10	011101	011000	001101
11	010011	010010	000011

From first breeding at age 1 (Group B1)	individuals	From first breeding at age≥2 (Group B2)
011101	1	
010100	2	
	3	000101
	4	001101
010000	5	
	6	
	7	000100
001010	8	
	9	000110
001101	10	
	11	000011

The groups have some parameters in common such as the survival probabilities after age k and the recapture probabilities after age f ($p'_{ij}=p_{ij}$ for $j-i \geq f$). This requires to use computer software which enables: 1) to analyze multi-groups data, 2) to set some parameters equal across groups (for example: option "user defined model" in SURGE; Pradel et al., 1990).

This procedure is flexible and relies on existing computer programs. In addition, model selection can be done in the same theoretical framework as standard survival analysis (Burnham et al., 1987; Lebreton et al., 1992). From p'_{ij} and p_{ij} estimates, explicit formulae provide estimations of the age-specific proportions of breeders α_{ij} (Clobert et al., in press). Their variances and covariances are estimated from variances and covariances of the p'_{ij} and p_{ij} by the delta method (e.g. Seber, 1982).

In general, we start by first modelling ϕ_{ij} and p_{ij} in group B. After selection of an appropriate model for this group, we add group I to the analysis by searching for the age f at which $p'_{ij}=p_{ij}$. With this age f, the pattern of ϕ'_{i1} and p'_{ij} variation can be analyzed. In turn, the appropriateness of the selected f can be evaluated again.

4. Goodness of fit

When estimating demographic parameters, population ecologists have not often considered testing model hypotheses (Begon, 1983). Trap response or heterogeneity in survival or capture probabilities can have pernicious effects on parameter estimation or on model selection (Johnson et al., 1986). Contingency table tests for survival models were introduced by Brownie and Robson (1976) and Pollock et al. (1985) (see also Burnham et al., 1987). By using these procedures on data in group B, one can test for the adequacy of the Cormack-Jolly-Seber model (Cormack, 1964; Jolly, 1965; Seber, 1965; see Clobert and Lebreton, 1987) for breeders.

The contingency table approach often results in a test equivalent to a likelihood ratio test between two models given that these two models have full-rank minimal sufficient statistics (Burnham et al., 1987). This is the case in our general time-dependent model (Clobert et al., in press) for each f. Therefore, a contingency table test is likely to exist and the computer program RELEASE (Burnham et al., 1987) likely to provide its ingredients.

Table 2. Expected values for the 2I elements of Test 2 (see Burnham et al., 1987 for more details on Test 2) for a single batch. An example is given with simulated data (all $\phi_{ij}=0.8$, all $p_{ij}=0.6$, $\phi'_{11}=0.5$, $p'_{12}=0.2i$, $p'_{13}=0.4$ and f=3). R_{1j} numbers (in bold) of individuals first captured at occasion 1 and recaptured and released at occasion j (cohort).

cohorts		numbers of first recapture from release R_{1j} — occasions 2	3	4	5	6
R_{11}	**1000**	101	135	91	33	3
	2I.2					
R_{12}		**101**	47	15	5	3
R_{13}			**182**	83	33	8
	2I.3					
R_{14}				**189**	90	30
						2I.4

expected numbers in contingency tables for j=i+2 to n

$R_{1,1}\phi'_{i,1}(\prod_{m=2}^{j-1}\delta'_{i,m})p'_{ij}$	$\sum_{r=j+1}^{n} R_{1,1}\,\phi'_{i,1}(\prod_{m=2}^{j-1}\delta'_{i,m})p'_{i,r}$
$\sum_{s=2}^{j-1} R_{1,s}\phi'_{i,s}(\prod_{m=s+1}^{j-1}\delta_{i,m})p_{ij}$	$\sum_{s=2}^{j-1}\sum_{r=j+1}^{n} R_{1,s}\phi_{i,s}(\prod_{m=s+1}^{j-1}\delta_{i,m})p_{i,r}$

with $\delta_{i,m}=(1-p_{i,m})\phi_{i,m}$ and $\delta'_{i,m}=(1-p'_{i,m})\phi_{i,m}$

with $p'_{ij}=p_{ij}$ for $j-i\geq f$ and with $\prod_{m=s+1}^{j-1}\delta_{i,m}=1$ for $j-1<s+1$.

For a single batch of individuals, two major components of the general goodness of fit test are provided by RELEASE. The first one, called Test 2, investigates differences in ϕ and p between cohorts. The second one, called Test 3, investigates differences in ϕ and p between sub-cohorts (inside one cohort, those which have a different past history). According to our model structure, ϕ'_{i1} and p'_{ij} only show up in cohort 1 (Table 2, example with k=1). Testing the hypothesis $p'_{ij}=p_{ij}$ will require a comparison between cohorts and thus a particular partition of Test 2, which we called Test 2I. In the example shown in Table 2, $p'_{13}=p_{13}$ is tested by comparing the proportion 135/47 (test 2I.2, Table 2) with the proportion (91+33+3)/(15+5+3). The sum of these X^2_1 statistics (Table 2) over occasions and batches will test an overall difference between the p'_{ij} and the p_{ij}.

A related partition of these contingency tables provides a test of trap dependence (Pradel, this volume).

Finally, the effect of the age at first breeding on subsequent p and ϕ can be examined by grouping accordingly the sub-cohorts in each element of Test 3.

5. Importance of heterogeneity

We investigated here the impact of heterogeneity of capture rate on the choice of f. 100 simulations were done on two batches of 500 individuals with capture probabilities respectively equal to 0.6 and 0.4. Adult survival rate was kept constant (ϕ=0.8) and was different from the local immature survival rate (ϕ'=0.4). The full breeding age f was equal to 1 (no delay in reproduction: all p'_{1j}=p) and we set the number of occasions to 6. For each simulation, we calculated the different elements of Test 2 and Test 3 through RELEASE. We also fitted an array of models with varying f through SURGE4.1.

As expected, with a time dependent p and ϕ, likelihood ratio tests between models with different f (f=3,6) were equivalent to Test 2I components. The hypothesis that at least one p'_{1j} (j=3,5) differ from the corresponding p_{ij} was rejected in 56 cases out of 100. Heterogeneity was found to have a potentially strong influence on model selection involving f. Since heterogeneity causes extra binomial variation, Burnham et al. (1987) and Lebreton et al. (1992) advocated correcting the likelihood ratio test between two models by a variance inflation factor (c) calculated from the goodness of fit test (Test 3 + Test 2). In our case, part 2I of Test 2 had to be removed since it was used for the comparison between p'_{ij} and p. The correction seemed extremely efficient when c was calculated using only Test 3 (at least 1 p'_{1j} differs from the corresponding p_{ij}, 11 cases of rejection for 15 expected), compared to when it was calculated using Test 3 + the remaining part of Test 2 (18 cases of rejection). An important part of the heterogeneity showing up in Test 2 was probably concentrated into its 2I components. When heterogeneity can be attributed to some known factors, it is always preferable to split the data according to the categories of these factors (Lebreton et al., 1992).

6. Example: the great tit

Since 1947, data have been collected on great tits in Wytham Wood (Oxfordshire), using nest-boxes (Perrins, 1965). Young are ringed shortly before fledging time and, since 1963, parents have been captured on a regular basis when feeding the young. After several years of intense nest predation by the weasel *Mustela nivalis* (Dunn, 1977), nest-boxes were made predator-proof around 1975. We will concentrate our analysis on the 63-75 period, the nest predation period, where a total of 8200 young were produced. For this example, we used only males as they were locally recruited in higher numbers than females (McCleery and Clobert, 1990). Unfortunately, there is no easy way to sex nestlings in this species. However, the sex ratio at birth is 0.5 and does not fluctuate from one year to the next (Perrins and McCleery, pers. com.) so that we used the number of offspring divided by two as the number of ringed young males.

6.1. Goodness of fit

We first analyzed the overall data, batch by batch, using RELEASE and calculated the 2I elements of Test 2 (Table 3). Their sum gives a X^2 value of 32.540 for 20 degree of freedom (p=.043), and indicates that some p'_{ij} differ from p_{ij} after age 1.

Table 3. Testing equality between p'_{ij} and p_{ij} by looking to components of Test 2J, batch by batch using all the tit data. Test 2 and 3 were calculated by RELEASE also on each batch separately. These tests have to be interpreted cautiously as a fair number of them were calculated on a small amount of data. In parentheses the number of degrees of freedom for the X^2.

f Batch	2I.2 $p'_{i,i+2}=p_{i,i+2}$	2I.3 $p'_{i,i+3}=p_{i,i+3}$	2I.4 $p'_{i,i+4}=p_{i,i+4}$	TEST 2	TEST 3
1	0.885 (1)	0.715 (1)	0.335 (1)	3.291 (4)	12.056 (6)
2	2.284 (1)	3.855 (1)	-	12.869 (3)	1.923 (3)
3	0.599 (1)	3.819 (1)	-	1.646 (2)	0.553 (1)
4	0.779 (1)	1.024 (1)	-	1.230 (2)	9.561 (2)
5	2.724 (1)	0.760 (1)	-	3.990 (3)	2.370 (4)
6	4.804 (1)	0.052 (1)	1.159 (1)	4.888 (3)	9.632 (5)
7	2.332 (1)	1.185 (1)	-	3.516 (2)	1.917 (2)
8	1.504 (1)	0.818 (1)	-	6.682 (2)	3.604 (3)
9	0.135 (1)	2.267 (1)	-	0.756 (3)	4.982 (3)
Total	16.045 (9) p=0.066	14.495 (9) p=0.107	1.494 (2) p=0.474	38.868 (24) p=0.022	46.609 (29) p=0.020

2I.2 + 2I.3: 30.540 (18) p=0.033

2I.2 + 2I.3 + 2I.4: 32.034 (20) p=0.043

The value of Test 3 calculated by RELEASE (X^2_{29}=46.61 p=.020) indicates that some heterogeneity is present also at the sub-cohort level, and may well affect component 2I of Test 2. In this case, heterogeneity may be explained by a different behaviour of individuals which start to breed successfully at age 1 (B1) and those which start to breed successfully at latter ages (B2), as suggested by the work of Gustafsson and Pärt (1990). Applying RELEASE on the two groups separately (splitting the data) leads to a decrease in the among-cohort and sub-cohort heterogeneity in survival and capture probabilities (RELEASE applied on all batches; TEST 2: B1 X^2_8=10.74 p=.217, B2 X^2_5=6.58 p=.254; Test 3: B1 X^2_{18}=23.64 p=0.177, B2: X^2_{17}=18.90 p=.334).

6.2. Group B1 and B2 survival and capture probabilities

Although RELEASE applied on all batches of B1 did not detect heterogeneity across cohorts or sub-cohorts, we nevertheless look to age variation in both survival and recapture probabilities because: 1) Test 2 and some elements of Test 3 were based on insufficient data, 2) previous analyses on this species showed age variation (Perrins and McCleery, 1985; Clobert et al., 1988; Dhondt, 1985). We defined 5 age classes with respect to survival probabilities (from age 1 to age 2, 2 to 3, 3 to 4, 4 to 5, 5 and after) and 4 with respect to recapture probabilities (age 2, 3, 4, 5 and after), and we let them be time-dependent. This was our baseline model from which we look to age and time variation on ϕ and p by fitting an array of models with a reduced number of parameters.

Table 4. Akaike information criteria (AIC) for an array of models fitted on birds first observed as breeders at age 1 (group 1). T: time effect included into the model. *: interaction with time included, +: additive effect of time. A(5): models with 5 age classes (5ac) where age1 ≠ age2...≠ age5 = age6... A(4): 4ac, age1 ≠ age2... ≠ age4 = age5 =... A(3): 3ac, age1 ≠ age2 ≠ age3 =age4 =... A(2): 2ac, age1 ≠ age2 = age3 = age4... A(4'): 4ac, age1 ≠ age2 ≠ age3 = age4 ≠ age5 = age6... A(3'): 3ac, age1 ≠ age2 = age3 = age4 ≠ age5 = age6... A(2'): 2ac, age1 =...= age4 ≠ age5 = age6... ∞: AIC. ▲ number of estimated parameters. Model with the smallest AIC in bold.

p / ϕ	A(4)*T	A(3)*T	A(2)*T	T	constant
A(5)*T	2259.73∞	2234.73	2220.87	2209.45	2207.33
	83▲	75	66	56	46
A(4)*T	2245.08	2232.49	2217.91	2205.94	2203.49
	76	68	59	49	39
A(4')*T	2237.51	2222.10	2207.90	2195.98	2194.07
	75	67	58	48	38
A(3')*T	2220.37	2208.40	no fit	2185.50	2183.00
	66	58	49	39	29
A(2')*T	2209.44	2198.90	2184.20	2174.50	**2170.24**
	56	48	39	29	**19**
A(2')+T	2209.55	2200.6	2187.40	2178.61	2171.75
	50	42	33	23	13
T	2210.48	no fit	2188.68	2180.41	2172.12

The smallest AIC (Akaike Information Criterion, see Lebreton et al., 1992, for further details) was always found for models involving no age and time variation in p (Table 4). Age variation in ϕ can be reduced to two age classes, lumping age 1 to age 4 into one age class. Birds aged 5 and older (a5+) have different survival probabilities than younger ones (a1=a2=a3=a4, a=age, model A[2'] in Table 4). Survival probabilities were time-dependent in the two age classes, but not in a parallel way.

A similar model could not account for "age" (here given as time elapsed since first recapture) variation for group B2. The model with the smallest AIC for B1 (two age classes: a2=a3=a4, a5+, with time dependent ϕ, but constant p) gave an AIC equal to 587.00 when applied on group B2 while the model with the smallest AIC value (562.10) had three "age" classes (a2$\neq$a3$\neq$a4+) with no time dependent ϕ and p. Groups B1 and B2 differed in their survival probabilities. This confirmed previous results on the goodness of fit test. However, they had equal capture probabilities (B1 p=.59$\pm$.05, B2 p=.63$\pm$.08). We therefore pursued our analysis using only B1, the most numerous group.

6.3. Modelling recapture probabilities of birds not yet observed as successfull breeders

The analysis was carried on from a model with two age classes for survival probabilities (a1=a2=a3=a4$\neq$a5+). These survival probabilities were time-dependent while the capture probabilities were constant. Group J was now part of the analysis. Results from the goodness of fit test indicate that the full age at successful breeding (f) may be as late as 4 (Table 3). We therefore started with f=4 and estimated p'_{ij} for j=i+1 to i+f-1 in each batch i. We will refer to them as p'_{1j}, p'_{2j} ... and $p'_{f-1,j}$ respectively for age 1, age 2 ... and age f-1 with j being the time index. We then fit models with progressively decreasing values of f. We also investigated intermediate models by dropping the time effect on one or several of the p'_{ij}s or by assuming that p'_{1j}, p'_{2j},...$p'_{f-1,j}$ were all equal (Table 5). The two

Table 5. Modelling capture probabilities of individuals not yet observed as breeders (p'). f: full breeding age. $p'_1 \neq p'_2 \neq p'_3$: the rate of accession to a successfull reproduction is not constant with age, $p'_1 = p'_2 = p'_3$: this rate is constant. * Akaike (AIC) information criteria. ** number of parameters. The models with the smallest AIC values are in bold.

	f=2	f=3		f=4	
	p'_1	$p'_1 \neq p'_2$	$p'_1 = p'_2$	$p'_1 \neq p'_2 \neq p'_3$	$p'_1 = p'_2 = p'_3$
with effect of time	12462.87*	12479.22	12459.65	12471.20	12463.40
	41**	60	41	51	41
without effect of time	12459.11	12455.44	**12451.91**	12453.84	**12451.84**
	31	33	**31**	32	**31**

models showing the smallest AIC values involve both and equality between the p' and no time effect. The equality between the p' indicates that, up to the age of full breeding, a constant proportion of birds were recruited into the successful part of the breeding population at each age ($[p'_1=p'_2=p'_3]/p=58\%$ or $[p'_1=p'_2]/p=64\%$, Table 7). The two models however differ by their full breeding age f, 3 and 4. A significant proportion of birds were not observed as successful breeders before age 3, and some may be started being successful as late as age 4. During the nest predation period, these proportions were apparently time independent.

6.4. Modelling local immature survival rate

We conducted the analysis of the local immature survival (ϕ'_{i1}) using two models, both without time dependence for the p': one with f=2 and $p'_{1.}=p'_{2.}$ and one with f=3 and $p'_{1.}=p'_{2.}=p'_{3.}$. Attempts to reduce time variation in local immature survival probabilities were unsuccessful (f=2 $X^2_{10}=61.50$ p=0.000, f=3 $X^2_{10}=65.61$ p=0.000, Table X). In addition, apparent survival probabilities of immature birds and adults did not fluctuate in parallel over time (age and time interaction on survival probabilities : f=3 $X^2_9=28.72$, p=0.001, f=4 $X^2_9=27.81$ p=0.001, Table 6). Time variation in the local immature survival probabilities cannot be reduced. Final estimates are given in Table 7.

Table 6. Modelling the local immature survival rate (ϕ'). The analyses were done using two different full breeding age f=3 and f=4. (*) Akaike criteria. (**) number of parameters estimated.

model for ϕ'	f=3	f=4
time dependence	12451.54(*) 31(**)	12389.91 31
parallellism with breeders survival probabilities (ϕ)	12462.57 22	12461.77 22
constant over time	12535.34 21	12539.57 21

7. Discussion

Recently, stochastic models were presented for the estimation of recruitment parameters (Lebreton et al., 1990; Clobert et al., 1990; Clobert et al., in press), with an emphasis on model selection and parameter estimation. We have explored ways of checking model hypotheses and consequences of their violation on model selection. To do this, two lines of analysis were shown to be helpful: first, a close examination or a rearrangement of the components of the goodness of fit test available in RELEASE (Test 2 and Test 3), and second, when heterogeneity could be attributed to some individual characteristic, a split of data into as many categories as needed. Most of the assumptions in our model are variants of two more general assumptions common to most capture-mark-recapture models, and are therefore presented in this more general framework.

Table 7. Estimates for the final model. Local immature survival probabilities (ϕ') are time-dependent. Breeders survival probabilities (ϕ) are described by two age classes, from age 1 to age 4 and older than 4. Both are time-dependent. Capture probability of breeders (p) are constant. Capture probability of birds not yet observed as successfull breeders (p') are constant over time. They diifer up to age 4, but they are equal between them. Cumulative age specific proportions of successfull breeders (α) are also given.(*) confidence intervals too large.

SURVIVAL PROBABILITIES

	ϕ'			ϕ age1=age2-...-age4			ϕ(*) age5+
year	estimate	95% c.	int.	estimate	95% c.	int.	estimate
63-64	0.078	0.050	0.120	0.682	0.232	0.938	
64-65	0.066	0.037	0.115	0.656	0.449	0.817	
65-66	0.105	0.058	0.182	0.697	0.369	0.901	
66-67	0.221	0.145	0.322	0.511	0.291	0.727	
67-68	0.148	0.092	0.228	0.481	0.295	0.672	0.000
68-69	0.309	0.208	0.432	0.494	0.299	0.691	1.000
69-70	0.153	0.095	0.236	0.404	0.259	0.568	1.000
70-71	0.211	0.137	0.310	0.370	0.220	0.551	0.370
71-72	0.138	0.088	0.209	0.468	0.288	0.656	0.000
72-73	0.215	0.138	0.320	0.412	0.249	0.596	1.000
73-74	0.066	0.035	0.122	0.465	0.285	0.654	0.000

CAPTURE PROBABILITIES

P			p'		
estimate	95% c.	int.	estimate	95% c.	int.
0.584	0.472	0.688	0.339	0.258	0.430

CUMULATIVE AGE-SPECIFIC PROPORTIONS BREEDERS

α_1	α_2	α_3
0.580(±0.091)	0.824(±0.081)	0.926(±0.048)

Individuals have to be captured with the same probability on any single capture occasion. Heterogeneity in capture rate has little effect on survival estimation (Nichols et al., 1982; Pollock and Raveling, 1982), and on model selection which involves survival probabilities (Burnham et al., 1987). The results of our simulations showed that heterogeneity has a big effect on model selection which involves the p'_{ij}, and, in particular, on the selection of an appropriate full breeding age. However, the correction by a variance inflation factor (Burnham et al., 1987) gives satisfactory results. Assessing the statistical validity of the chosen f is particularly important, since biologists are often persuaded that delaying reproduction is a common feature to most species. However, the time at which the recapture of the breeder is done influences greatly the biological meaning of the age specific proportion of breeders. In the tit example, recapture was done at the nestling stage, excluding individuals which layed clutches but failed before the nestling stage because of egg predation. In this case, all individuals may well have been breeders at the age of 1 year, and part of them successfull breeders for the first time only later on (as late as age 4). The fact that a constant recruitment into the class of successfull breeders is found is consistent with that idea.

Individuals are assumed to have the same probability of surviving a defined interval of time. In particular, survival is not dependent on their previous capture-recapture history. Two assumptions relate in part to this specific point in our model: 1) reproductive activity does not influence, either, positively or negatively, capture and survival probabilities in the following time step (no direct cost or 'benefit') 2) early reproduction does not influence negatively or positively the long term behaviour of survival and capture probabilities (no long term cost or 'benefit').

A direct cost ('benefit') of reproduction will result either in the decrease (increase) of survival during the next time step or in a decrease (increase) of the breeding probability in the following occasion, and thus of the capture probability. This is analogous to a trap response effect and will show up in component 2T of Test 2 (Pradel, this volume). This will be totally confounded with a true trap response effect unless this last possibility can be discarded from what is known from the experimental design and the species' biology. In our example, there was no sign of a direct influence of reproduction, at least for individuals having bred at least once. Unfortunately, a direct effect of first reproduction cannot be examined since, in most experimental designs, non breeding birds of the same age are not detectable. Appropriate data have yet to be collected.

Long term effects of early reproduction can be followed by analyzing Test 3 components where sub-cohorts have been rearranged according to age at first reproduction. In our example, there were some indications that individuals reproducing at age 1 behaved differently than those found as breeders at later ages. In this case, analyzing birds separately according to the age of first breeding proved useful for reducing heterogeneity in survival probabilities.

More generally, the relationship between reproduction and future survival or fecundity should be translated into a positive or negative relationship between survival and recapture probabilities. From studies on band recoveries, goodness of fit tests seem to detect with reasonably good power a negative relationship between ϕ and p (Pollock and Raveling, 1982; Nichols et al., 1982). Positive one are less easily detectable (Nichols et al., 1982). It remains to be seen if these results are applicable in the CMR context and to what extent the nature of the relationship between survival and capture probability will affect model selection.

In conclusion, not all the assumptions of this model can be fully investigated at present. The most important ones are: 1) all the individuals that started to breed will breed on each successive occasion, 2) there is no cost associated with a first reproduction, 3) the return probability to the natal ground is constant, regardless of the age at which individuals return to their natal area for the first time. In the future, it will be possible to test these assumptions either, through appropriate simulation studies (testing biases in estimates), through the development of suitable statistical methods (estimating the proportion of non breeding birds in any given year), or through the collection of appropriate data (checking the constancy of return probability to the natal area according to age, cost of first reproduction). Some of these weakness will also be removed when models developed to measure the rate of exchange among populations (Hetsbeck et al., 1991) incorporate estimates of age-specific proportions of breeders.

Acknowledgements: We thank K.P. Burnham, M. Hochberg, J.D. Lebreton and A. van Noordwijk for their constructive comments on an earlier draft of this manuscript. Financial supports were provided by the EEC (grant n° 88300402 JU1), the S.R.E.T.I.E and the P.E.R (C.N.R.S).

References

Begon, M.(1983).- Abuses of mathematical techniques in ecology: applications of Jolly's capture-recapture method. Oikos, 40: 155-158.

Brownie, C. and Robson, D.S.(1976).- Models allowing for age-dependent survival rates for band-return data. Biometrics, 32: 305-323.

Burnham, K.P., Anderson, D.R., White, G.C., Brownie, C., & Pollock, K.H. (1987).- Design and analysis methods for fish survival experiments based on release-recapture. American Fisheries Society Monograph n°5.

Clobert, J. and Lebreton, J.D. (1987).- Recent models for mark-recapture and mark resighting data. A response to C.Brownie. Biometrics, 43: 1019-1022.

Clobert, J., Lebreton, J.D. & Allainé, D. (1987).- A general approach to survival rate estimation by recaptures or resightings of marked birds. Ardea, 75: 133-142.

Clobert J., Lebreton, J.D. & Marzolin, G. (1990).- The estimation of local immature survival rates and of age-specific proportions of breeders in birds populations. pp 199-213 *in* Population biology of Passerines birds: an integrated approach J. Blondel, A. Gosler, J.D. Lebreton & R.H. McCleery (Eds). Springer-Verlag, Berlin.

Clobert, J., Lebreton, J.D., Allaine, D. & Gaillard, J.M. (in press).- The estimation of age-specific breeding probabilities from recaptures or resightings in vertebrate populations II.Longitudinal models. Biometrics.

Clobert, J., Perrins, C.M., McCleery, R.H. & Gosler, A.G. (1988).- Survival rate in the Great Tit *Parus major* in relation to sex, age, and immigration status. Journal of Animal Ecology, 57: 287-306.

Clutton-Brock, T.H. (Ed.)(1988).- Reproductive success. Studies of individual variation in contrasting breeding system. Chicago University Press, Chicago.

Cormack, R.M. (1964).- Estimates of survival from the sightings of marked animals. Biometrika, 51: 429-438.

Dhondt, A.A. (1985).- Do old Great Tits forego breeding? Auk, 102: 870-872.

Dunn, E.K. (1977).- Predation by weasels (*Mustela nivalis*) on breeding Tits (*Parus* spp) in relation to the density of Tits and rodents. Journal of Animal Ecology, 46: 633-652.

Gustafsson, L. and Pärt, T. (1990).- Acceleration of senescence in the collared flycatcher *Ficedula albicollis* by reproductive cost. Nature, 347: 279-281.

Hestbeck, J.B., Nichols, J.D. & Malecki, R.A. (1991).- Estimates of movement and site fidelity using mark-resight data of wintering Canada Geese. Ecology, 72: 523-533.

Jolly, G.M. (1965).- Explicit estimates for capture-recapture rates with both death and immigration - stochastic model. Biometrika, 52: 225-247.

Johnson, D.H., Burnham, K.P. & Nichols, J.D. (1986).- The role of heterogeneity in animal population dynamics. Proceedings of the thirteenth International Biometrics Conference. Biometric Society, University of Washington, Seattle.

Jolly, G.M. (1982).- Mark-recapture models with parameters constant in time. Biometrics, 38: 301-321.

Lebreton, J.D. (1987).- Régulation par le recrutement chez la Mouette rieuse Larus ridibundus. Revue d'Ecologie (Terre et Vie), suppl.4: 173-182.

Lebreton, J.D., Burnham, K.P., Clobert, J. & Anderson, D.R. (1992).- Modeling survival and testing biological hypotheses using marked animals : case studies and recent advances. Ecological Monographs, 62: 67-118.

Lebreton, J.D., Hemery, G., Clobert, J. & Coquillart, H. (1990).- The estimation of age-specific breeding probabilities from recaptures or resightings in vertebrate populations I.Transversal models. Biometrics, 46: 609-622.

Massot, M., Clobert, J., Pilorge, T., Lecomte, J. & Barbault, R. (1992).- Density dependence in the common lizard: demographic consequences of a density manipulation. Ecology (in press).

McCleery, R.H. and Clobert, J. (1990).- Differences in recruitment of young by immigrant and resident Great Tits in Whytham Wood. pp 423-440 *in* Population biology of Passerines birds: an integrated approach . J. Blondel, A. Gosler, J.D. Lebreton & R.H. McCleery (Eds). Springer-Verlag, Berlin.

McCleery, R.H. and Perrins, C.M. (1985).- The effect of age and pair bond on the breeding success of the Great Tit *Parus major*. Ibis, 127: 306-315.

Newton, I. (Ed)(1989).- Lifetime reproduction in birds. Academic Press, London.

Nichols, J.D., Stokes, S.L., Hines, J.E. & Conroy, M.J. (1982).- Additional comments on the assumption of homogeneous survival rates in modern bird banding estimation models. Journal of Wildlife Management, 46: 953-962.

Perrins, C.M. (1965).- Population fluctuations and clutch-size in the Great Tit, Parus major. Journal of Animal Ecology, 34: 601-647.

Pollock, K.H., Hines, J.E. & Nichols, J.D. (1985).- Goodness-of-fit tests for open capture-recapture models. Biometrics, 41: 399-410.

Pollock, K.H. and Raveling, D.G. (1982).- Assumptions of modern band-recovery models, with emphasis on heterogeneous survival rates. Journal of Wildlife Management, 46: 88-98.

Porter, J.M. and Coulson, J.C. (1987).- Long-term changes in recruitment to the breeding group and the quality of recruits at a kittiwake *Rissa tridactyla* colony. Journal of Animal Ecology, 56: 675-689.

Pradel R., Clobert, J. & Lebreton, J.D. (1990).- Recent developments for the analysis of multiple capture-recapture data sets. An example concerning two Blue Tits population. The Ring, 13: 193-204.

Seber, G.A.F. (1965).- A note on the multiple-recapture census. Biometrika, 52: 249-259.

Seber, G.A.F (1982).- The estimation of animal abundance and related parameters. Griffin, London.

White, G.C. (1983).- Numerical estimation of survival rates from band-recovery and biotelemetry data. Journal of Wildlife Management, 47: 716-728.

Marked Individuals in the
Study of Bird Population
J.-D. Lebreton and P. M. North (eds.)

Challenges in the Analysis of Recruitment and Spatial Organization of Populations

Fred Cooke[1] and Charles M. Francis[2]

[1]*Biology Department, Queen's University, Kingston, Ontario, Canada K7L 3N6*
[2]*Center for Tropical Conservation, Duke University, Durham, NC 27705, USA*

Summary. Understanding and measuring recruitment processes and the spatial organization of populations is an integral component of the study of population dynamics. The factors controlling variation in recruitment processes can be analysed by considering the individual components of recruitment, such as production of young, survival of young to reproductive age, and age at first breeding. Several of these components can be examined using methods similar to those for mark-recapture survival analysis. We demonstrate this approach using data from a long term study on Lesser Snow Geese. We show that most variation in annual recruitment rates was due to a three-fold variation in post-fledging survival, which in turn is strongly related to feeding conditions in the breeding colony. The spatial organization of populations influences their dynamics through the processes of immigration and emigration. Spatial organization is usually modelled in terms of movements in and out of a single population, or movements among discrete populations. However, movement within large continuous populations, or those with strong spatial heterogeneity, also influences population dynamics, but cannot readily be analysed using most existing models.

1. Introduction

Recent statistical developments in the analysis of population dynamics have led to a proliferation of powerful new methods for the analysis of survival rates, many of which are discussed in this volume. In contrast, methods for the analysis of recruitment and spatial organization of populations, which also influence population dynamics, have received less attention. Apart from their importance in measuring population dynamics, such data are necessary for many other biological investigations. For example, gene flow among and within populations can only be measured by determining the interactions between spatial distribution and recruitment rates. Also, studies of evolutionary processes require knowledge of the factors influencing variation among individuals in recruitment probabilities. On a more pragmatic level, an understanding of spatial distribution and movement among and within populations is necessary for many analysis techniques to differentiate true mortality from emigration.

In this paper, we present what we consider to be some of the most important challenges in the analysis of recruitment and spatial distributions of populations. We have framed much of our discussion around the long-term study of the Lesser Snow Goose, *Anser c. caerulescens*, at La Pérouse Bay, near Churchill, Manitoba to demonstrate how the various components of the problem interact in a study of the population dynamics of a single species (for details of that study see Rockwell et al., 1983; Cooke et al., 1984; Cooke and Rockwell, 1988). However, we have also drawn examples from other long-term studies of bird populations, to illustrate both variation among species, and the generality of the phenomena.

2. Recruitment

Recruitment can be defined as the entry of new individuals into a population. In the present paper, we shall be concerned primarily with entry into the breeding population, as this is the most relevant unit to the evolutionary biologist. (It would be equally valid to consider recruitment as starting from the moment of birth, in which case factors such as pre-breeding mortality and dispersal would be considered components of mortality.) From the perspective of biologists, studies of recruitment must be aimed not only at measuring total recruitment, but at assessing the variability in recruitment, and the factors that influence that variation. In this section we outline first some of the aspects of measuring total recruitment, followed by a more detailed discussion of the individual components of recruitment and their analysis.

2.1 Measures of overall recruitment

The total recruitment rate, including both immigration and local production, is difficult to measure with precision. In the mark-recapture models of Jolly (1965) and Seber (1965), recruitment is estimated simultaneously with other parameters such as population size and mortality rates. However, these estimates, like those of population size, are very sensitive to violations of model assumptions, such as unequal capture probabilities. While considerable progress has been made in refining these models to estimate survival rates with greater precision and robustness (e.g. Pollock, et al. 1990; Lebreton et al., 1992), much less attention has been paid to modelling other population parameters.

In some cases, recruitment rates can be calculated from the mortality and emigration rates, if independent estimates are available of changes in population size. Such estimates could come from any of a variety of survey or census techniques, such as those discussed by Seber (1982). Minta and Mangel (1989) present a flexible approach for modelling population size based on capture-resighting data. Unfortunately, recruitment estimates calculated this way carry the combined uncertainty of both the survival estimates and the population estimates, and thus may be very imprecise. Difficulties may also arise if the population is not well defined (e.g. not geographically closed), so that its size cannot be measured.

An alternative approach to estimating recruitment has been proposed by Pradel (unpublished manuscript), who suggested that capture history data can be used to estimate the probability that an individual will enter the marked population in the same way that they can be used to estimate departure probabilities (i.e. mortality and emigration). Although this requires the assumption that marked and unmarked individuals have the same capture probabilities (which is relaxed for survival analyses), the approach may provide a flexible framework for further progress in this field.

Within the context of the Lesser Snow Geese, total recruitment rates are very difficult to measure. Every year, late in the breeding season, a sample of breeding adults and their offspring has been captured and ringed, and records kept of all previously ringed birds. However, goodness of fit tests indicate substantial heterogeneity in capture or survival probabilities (Francis and Cooke, this volume), thus invalidating the standard Jolly-Seber models. Reliable estimates of annual survival rates are available from recovery data (Francis et al., 1992a), so it should be possible to estimate recruitment if the population size could be measured. The total number of nests in the colony can potentially be estimated by transect sampling, but this technique is complicated by heterogeneity of nest densities, and difficulties in defining the boundaries of the population. Better estimates can be obtained from aerial photography, but unfortunately such data are not available for most years. Furthermore, adjustments would need to be made for the number of pairs in the population that chose not to breed or failed prior to the start of the census (see below). As a result, we do not have a good measure of total annual recruitment into the population.

2.2 Components of recruitment

Although total recruitment must ultimately be measured to understand population dynamics, the processes that influence recruitment can only be understood by examining its individual components. These are most easily measured with respect to local recruitment, which can be defined as recruitment from offspring of adults within the study population. This bias is primarily pragmatic, because most studies only deal with processes within a single population. However, the same principles and processes are relevant to measuring and understanding recruitment from outside the population.

The reproductive cycle of an organism can be divided into a series of fitness components based upon parameters that can readily be measured by the biologist. Here we discuss those components relevant to recruitment, derived from the diagram of fitness components for the Lesser Snow Goose presented by Cooke and Rockwell (1988). For convenience, we divide these components into three broad categories: production of young, survival of the young, and age of maturity.

2.2.1 Production of young

For geese, the number of young produced is determined by the proportion of breeding adults that nest in a particular year, the number of eggs laid in each nest, the failure rate of those nests, and the hatching success of the eggs. For many other species of birds, an important additional component is the number of nesting attempts within a year.

Estimation of the first of these components, the proportion of adults that are actually breeding, is challenging both biologically and statistically. For most species of birds, non-breeding individuals are difficult to identify or observe, because they generally have less conspicuous behaviour, are less tied to particular areas or, in the case of Snow Geese, often migrate away from the main study area. Such birds could represent adults that opt not to breed in a particular year or younger birds that have not yet started to breed. Some approaches to estimating age-specific breeding probabilities based on relative capture/resighting probabilities (discussed below under "Age at first breeding") could be adapted to estimate numbers of non-breeding adults. For example, Blondel et al. (1992) were able to estimate the proportion of adult Blue Tits, *Parus caeruleus*, not nesting in boxes, based on differences between calculated capture probabilities from mark-recapture models, and the known proportion of birds captured in boxes. However, they were not able to differentiate between temporary emigration, nesting in natural cavities within the study area, and failing to breed at all.

For some purposes, such as estimation of total recruitment rates, it may often be sufficient to ignore non-breeders in the calculations. However, in studies such as that of the Blue Tit just cited, this would tend to underestimate local production because of nesting in natural cavities. Also, when comparing the relative fitness of different phenotypes, it is important to consider possible differences in breeding proclivity.

Some of the other components of production are relatively easily measured, such as the mean clutch size within a nest, and the hatching success of the eggs, although undetected egg loss during laying could lead to slight underestimates of clutch size and overestimates of hatching success.

Total nest failure rates can be accurately measured if all nests are monitored from the start of egg-laying, as is done for a large sample of nests every year in the Snow Goose study. However, in many studies, nests are found at various stages during laying or incubation, and are not necessarily monitored until hatching. In such cases, nest success must be estimated using survival models to calculate daily nest survival rates. In the past, simple models have often been used that assume constant daily failure rates (e.g. Mayfield, 1961). However, such an assumption may often be invalid, and should at least be tested. Ideally, the principle of parsimony should be applied, whereby a series of models are compared, and the simplest model that adequately fits the data used for estimation, following the example of mark-recapture analyses (Clobert et al., 1987; Lebreton et al., 1992). Because capture probabilities are not relevant to nest studies, many of the standard survival models from the medical literature (e.g. Cox and Oakes, 1984), could be adapted, after allowing for truncated data (nests that are not followed until failure or hatch), and uncertainty in the precise date of failure if nests were not checked on a daily basis.

Measuring the number of breeding attempts within a season again provides a challenge. In some cases, measurements may be fairly straightforward, if all nests within a study area can be located, or if individuals consistently re-use the same nest throughout a season. However, in other cases estimates must be based on locating as many nests as possible for marked individuals. Naive estimates based upon individuals that are successfully followed throughout the season may be biased if, for example, individuals with few renesting attempts are difficult to follow. Proper analysis models must allow for variable resighting probabilities of individuals. These models cannot easily be adapted from capture-recapture models because nesting attempts do not occur in discrete periods–intervals between attempts vary among individuals, such that, for example, the second nesting attempt for one individual may overlap the first attempt of others. Also, both the timing and the probability of renesting are usually conditional on the success of the first attempt.

Possibly the most viable approach would be to use radio-tags, thus ensuring that all individuals will be followed. Unfortunately, this is currently only feasible for some of the larger species of birds, and care must be taken to ensure that the radios do not modify nesting success or behaviour. Such an approach is currently being tested for the Lesser Snow Geese to determine whether some individuals renest (or continue laying), if their nest is depredated during the early laying period.

2.2.2 Estimating immature survival rates

Measures of immature survival can again be divided for most species into two stages: survival from hatch until fledging, and survival from fledging until breeding age. For altricial species, survival prior to fledging is equivalent to measuring post-hatch nesting success, and suitable analysis techniques have already been discussed above. For precocial species, the distinction between pre- and post-fledging survival is somewhat arbitrary, but at least for the Snow Goose study, is based on the pragmatic consideration that geese do not receive permanent marks until they are ringed at fledging.

Measurement of pre-fledging survival of the geese is based on two separate components: partial brood loss, and total brood loss. Every year, all goslings from a sample of a few thousand nests are marked with small numbered tags clipped to the webs of their feet. These tags allow us to match the goslings to both their natal nest and their parents, provided the latter were identified at the nest. Because broods usually stick together with their parents throughout the brood-rearing period, partial brood loss can be estimated from the difference between brood size at hatch and brood size at ringing. However, total brood loss is more difficult to estimate because parents that lose all of their goslings are likely to move away from the foraging areas and are thus less likely to get caught.

An alternative method that could potentially have been used would be based on mark-recapture analysis of the web tags using captures of web-tagged birds in subsequent years to estimate the proportion captured during the first season, and hence their tag-to-ringing survival rate. Unfortunately, however, web tags are not very durable and tend to become illegible or fall off within a few years. Because web tags were usually removed from the geese at ringing (to ensure that their numbers were correctly read), it has not been possible to estimate the rate of tag loss.

Measures of post-fledging survival rates can be based upon standard survival models for either recaptures (Lebreton et al., 1992) or recoveries (Brownie et al., 1985). For female Lesser Snow Geese, both techniques produced remarkably similar results (Francis and Cooke, this volume), suggesting little bias, as well as relatively little emigration. However, virtually all immature males emigrate from the colony so recaptures cannot be used for analysis, and even recovery data are biased by heterogeneity of recovery rates (Francis and Cooke, 1992a). Unfortunately, for many non-hunted birds, especially in North America, recovery rates are very low, and only recaptures are available for estimating survival rates. Developing appropriate techniques for estimating true immature survival rates in these cases provides a challenge for both biologists and statisticians.

2.2.3 Age at first breeding

Recent models for estimating age-specific breeding probabilities have been developed based upon transversal models (Lebreton et al., 1990) and longitudinal models (Clobert et al., in press). The former require only that the age of breeding birds can be identified, while the latter, more powerful models are based on known capture histories of individually identified birds.

The design of the Snow Goose study provides several opportunities for application of models such as these. Individual marked birds can be identified at their nests based on numbered plastic leg rings that can be read through binoculars. These data can be used to obtain individual capture histories for the longitudinal models. In addition, the rings are colour-coded by cohort, with birds of known age (first captured as goslings or yearlings) marked on the right leg, and birds of unknown age marked on the left. Because the colour of the ring can be determined from a greater distance than the numbers, a larger sample would be available for use of the transversal models, although this may not offset the reduced precision of those models. Birds captured during the annual ringing drives provide another opportunity to measure age-specific breeding propensities using longitudinal models, as these birds are mainly successfully breeding adults.

Although we have not yet applied these full models to the goose population, crude analyses of relative capture probabilities indicate that only about half the birds start breeding as two year olds, and they do not reach adult levels of breeding propensities until at least four years of age (Cooke et al., 1984; Cooke and Rockwell, 1988). More detailed analyses of Black-headed Gulls, *Larus ridibundus*, indicate that, at least in some years, gulls do not reach levels of adult breeding propensity until at least 6 years of age (Clobert et al., in press). In many long-lived species such as seabirds, individuals may not reach full breeding age until much later (Newton, 1989).

These models are not without their difficulties, particularly if the population is spatially heterogeneous. For example, the Lesser Snow Goose population at La Pérouse Bay is currently expanding, and younger birds appear to recruit more often into new areas of the colony, away from the concentrations of older birds. Thus estimates of breeding propensity based on birds observed at their nests may be biased if nests are not randomly selected. Apparent differences in the age structure of birds seen on the nest, and those captured at ringing suggest this is a problem at La Pérouse Bay.

Closely related to the problem of measuring age-specific breeding probabilities is the question of determining the age of first breeding, which could be considered the measure of actual recruitment. This cannot easily be measured without considering additional components such as possible trade-offs between breeding and survival rates (e.g. whether non-breeders have higher survival than breeders), whether breeding in one year affects the probability of breeding in the following year, and whether these measures are correlated with age. Some of these parameters are likely to be confounded in capture-recapture models based only on observations of breeding birds. Additional information, such as direct observations of non-breeding birds, may be required to estimate all these parameters. These data are required to test hypotheses such as whether individuals are constrained to delay breeding, or whether they refrain from breeding to increase later fitness (Newton, 1989).

2.3 Annual variation in recruitment

From an ecological viewpoint, measures of variation in recruitment, and the factors that control that variation may be more important than the total recruitment rates. Such variation influences population growth rates, and provides insight into the factors that may limit population size. One approach to such an analysis is to determine which components of the recruitment process have the greatest influence on total cohort specific recruitment (i.e. the total number of individuals from a particular cohort that eventually recruit into the breeding population), and then look for the factors that may be responsible for that variation.

For the Lesser Snow Goose, annual variation in productivity, as measured by clutch size, nest loss, hatching success and pre-fledging survival is relatively low. In contrast, survival for the first year after fledging fluctuates considerably from year to year, varying from under 20% to over 60% (Francis et al., 1992a). Survival rates at high arctic colonies show even greater annual variation (Francis et al., 1992b). The close correspondence between apparent survival estimates of female geese from recaptures and survival estimates from recoveries (Francis and Cooke, this volume) indicates that this variation does not result in differential emigration. This contrasts with the situation for another colonial bird, the Black-headed Gull, for which dispersal levels are very high in years of high productivity, resulting in a low proportion of the birds recruiting into their natal colony (Lebreton, 1987). Annual variation in age at first breeding also appears to be low, indicating that this tremendous fluctuation in first-year survival corresponds directly to fluctuations in cohort-specific recruitment rates.

As these fluctuations directly regulate population growth, it is important to determine which factors are causing this variation in first-year survival. Much mortality of Snow Geese in their first year of life is caused by hunting. However, two lines of evidence indicate that this has very little influence on annual variation in first-year survival. First, the proportion of geese shot each year has declined over the study period, while first-year mortality has increased (Francis et al., 1992a). Second, the ratio of goslings to adults shot each year is positively correlated with first-year survival rates, as measured independently from recapture data (Figure 1), indicating that more immatures were shot in years when the immature survival rate was high. This suggests that most post-fledging mortality occurs before the geese reach the hunting areas.

Most of this mortality probably occurs shortly after fledging, as is suspected for

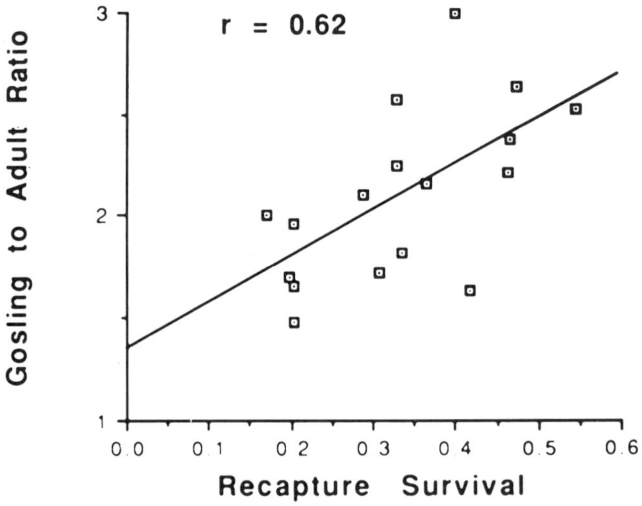

Figure 1. Relationship between the ratio of goslings to adults in recoveries of shot geese, and the apparent survival rates of immatures estimated from recaptures, 1970-1986.

many other bird species (some mortality may even occur prior to fledging, as we ring the geese just before they can fly). Annual variation in the lateness of the breeding season (determined largely by weather) affects plant growth, which in turn affects gosling growth rates (Cooch et al., 1991a). Also, there has also been a long term decline in gosling growth rate over the study, due to deterioration in the salt marsh vegetation from over-grazing by the geese (Cooch et al., 1991b). Annual variation in first-year survival correlates well with standardized mean gosling body mass, a measure of average growth rate (Figure 2). The long-term decline in average first-year survival also corresponds well with the similar decline in body size. These results suggest a causal relationship between food availability during the growth period, and post-fledging survival.

This example illustrates how measurements of the individual components of recruitment can be used to determine the major demographic factors that influence recruitment rate--in this case first-year survival. It further demonstrates how measuring the factors responsible for variation in these components can lead to an understanding of the environmental factors that influence recruitment rates.

Similar analyses can be used to compare the relative fitness of different phenotypes for evolutionary studies. For example, in the Snow Goose colony at La Pérouse Bay, there has been a gradual increase in the proportion of blue phase compared with white phase geese since the beginning of the study. However, detailed analysis of fitness components, including first-year survival rates, has failed to find any significant differences in any components between the two colour phases. This

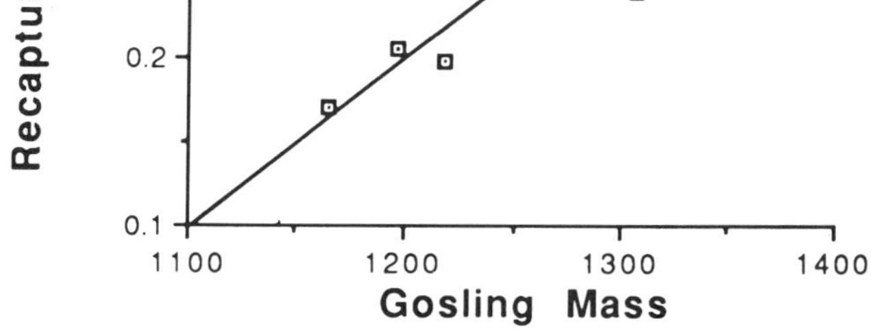

Figure 2. Relationship between standardized 35-day mean gosling mass and apparent first-year survival rates of goslings estimated from recaptures, 1981-1986.

suggests that differential immigration of blue phase geese from other colonies has been largely responsible for the change in phenotype frequencies, rather than differences in rates of local recruitment (Cooke et al., 1988).

2.4 External recruitment or immigration

Measurement of actual immigration rates is usually difficult, and can often only be calculated from differences between estimates of local recruitment and total recruitment rates. Unfortunately, such measures carry the combined uncertainty of both estimates which, as we have discussed above, may be considerable. In stable populations, it may often be possible to infer immigration rates from the difference between measures of productivity and annual survival rates. Only in small populations, such as those of tits nesting in boxes where nearly all locally produced young are ringed, can immigration rates be measured directly. Even then, caution is required to estimate correctly the proportion of locally produced young if, for example, some birds hatched in natural cavities within the area.

Estimation of the relative contribution of local recruitment and immigration is important in determining the stability and viability of populations. For example, analyses of some populations of passerines in North America suggest that productivity falls far short of that required to sustain the populations (Robinson, 1990). Without other populations producing an excess of young, some of these populations may well go extinct.

The distinction between local productivity and immigration is often a matter of scale. For many small passerines, return rates of young birds to their natal nest sites are extremely low, suggesting very little local recruitment if the study area is small. However, if they only disperse a short distance, then use of a larger area would result in a much higher estimate of local recruitment rates. This example highlights the importance of considering the spatial organization of populations to understand their dynamics.

3. Spatial organization of populations

As just suggested with respect to local recruitment, the spatial organization of populations, particularly as it relates to movements of individuals, affects measures of both recruitment and mortality/emigration. We shall consider here two aspects of spatial organization, particularly as they relate to demographic questions: philopatry, and movement among populations. These are differentiated mainly for pragmatic reasons: the first is based on analysis of data from a single population, while the latter requires data from several populations (or locations).

3.1 Philopatry

Natal philopatry can be defined as the tendency for animals to return to breed at the location where they were born. It has been measured as the proportion of marked individuals returning to the area in which they were raised, or as the average distance moved between birth location and site of breeding when the animal becomes an adult. Unfortunately, the first of these methods confounds return probabilities with survival, and the results are sensitive to the size of the study area. The latter must consider the availability of suitable habitats at varying distances from the original site, and will tend to underestimate long distance movements if it is based on recaptures by the observer. Nevertheless, such techniques have often been used to estimate relative philopatry among groups. For example, Greenwood (1980) and others have used these techniques to assess the relative natal philopatry between the sexes. However, their estimates may be biased if there were sexual differences in mortality rates.

Reliable methods of estimating natal philopatry would be valuable for assessing the evolution of philopatry. For example, are some individuals more philopatric than others, and if so does this have a genetic basis? In stable environments, philopatry might be favoured, while in unstable environments it would not (Cooch et al., 1992). In long term studies, one might expect different levels of natal philopatry under different environmental conditions.

Philopatry can often only be measured if additional data such as recoveries away from the study site or radio-tracking data are available for estimating survival independent of data from birds returning to the colony. For the Lesser Snow Geese, return probabilities of immature females as measured by recaptures, are extremely close to survival rates measured by recoveries (Francis and Cooke, this volume), indicating very high natal philopatry. There is no evidence of any decline in this philopatry despite the long-term deterioration of foraging conditions in parts of the colony (Cooch et al., 1991b). In contrast, virtually no young males ever return to the breeding colony, despite evidence that they have very similar survival rates to females (Francis and Cooke, 1992a). Since the sex ratio is 1:1, this means that almost all males breeding at the colony are immigrants from other colonies. Since breeding birds arrive at the colony already paired, this demonstrates that pair formation occurs away from the breeding grounds. This is an example of how assessment of emigration rates can be a valuable tool for demographic and evolutionary studies.

Ring recovery data also provide a potential means of assessing philopatry in terms of distance moved from the natal area. Moore and Dolbeer (1989) present an example estimating dispersal distances and gene flow for some North American blackbirds based upon recoveries of birds ringed during the breeding season. However, caution must be used in interpreting such data to consider the potential distribution of possible movements in relation to suitable habitats, as well as the distribution of people likely to find and report rings.

Adult philopatry, or site fidelity can be considered as the probability that an adult will return to the same nesting site in subsequent years, although it could also be considered in terms of wintering locations or migratory stop-over points. Measuring adult philopatry has the same logistic problems as measuring natal philopatry, but presents a number of biologically different questions. For example, adult philopatry may be influenced by experience, and many studies have suggested that individuals are less likely to return to a study area after breeding failure. However, as before, most of these studies were unable to determine whether this was due to lower survival, lower philopatry, or a combination of the two.

3.2 Movement among populations

Clearly, emigration from one population implies immigration into another population. However, direct measures of the rates of movement are only possible if several different populations are studied. Mark-recapture models have been developed for measuring rates of movement between discrete populations (Hestbeck et al., 1991; Nichols et al., this volume). When adequate capture-recapture data are not available, it is sometimes possible to infer movements based on indirect measures. For example, as already noted, most immature male Snow Geese emigrate from their natal colony. Although few recaptures are available from other colonies, the distribution of ring recoveries of these geese after they become adults could be used to infer the proportions moving to eastern and western Hudson Bay colonies based on differences in their migration routes (Francis and Cooke, 1992b). Similarly, measures of genetic differences among populations could be used to infer patterns of migration, although these may primarily reflect processes during evolutionary time rather than present conditions.

The greatest limitation of all these models is that they assume populations are discrete. There is a strong need for models of movement and exchange within geographically continuous populations. Such models must consider not only movement distances, but also spatial heterogeneity, and discontinuities. Within the Snow Goose population, for example, as already mentioned, traditional foraging areas have suffered severe degradation from over-grazing, and many geese are now using new areas where foraging conditions are better (Francis and Cooke, this volume). Although we only capture the geese in discrete locations, the geese are continuously, although non-uniformly, distributed over a large area with varying degrees of suitability for foraging. This distribution needs to be considered in the analysis of philopatry and movement in this colony. Similarly, nesting success of Black-legged Kittiwakes, *Rissa tridactyla*, varies from the centre to the edge of the colony (Coulson, 1968), implying that movement of birds within the colony will affect the population dynamics. Again, analysis must be based on models allowing continuous movements.

4. Conclusions

In this paper, we have outlined what we feel are some of the key biological questions related to the analysis of recruitment and spatial organization in populations. We have tried to indicate some of the older approaches that have been used in the past, highlighting their weaknesses, and pointing the direction for more appropriate analysis techniques. In some cases, further development of existing statistical methods may be sufficient to address the problems, while in other cases new and innovative approaches to study design and data collection may be required. However, in all cases, full development of approaches adequate to address these challenges will only be possible with a close working relationship between statisticians and biologists.

Acknowledgments. Snow Goose research at La Pérouse Bay has been supported in part by the Natural Sciences and Engineering Research Council of Canada (NSERC), the Canadian Wildlife Service (CWS), the Manitoba Department of Renewable Resources, the Wildlife Management Institute, the Mississippi and Central Flyway Councils, Ducks Unlimited Canada, and Ducks Unlimited Incorporated. FC was on sabbatical at the Centre d'Ecologie Fonctionnelle et Evolutive, C.N.R.S. while preparing this contribution, and gratefully acknowledges assistance from many of their personnel. CMF was supported by an NSERC post-doctoral fellowship, the CWS university research support fund, and the Chapman Fund of the American Museum (Natural History). J.-D. Lebreton, G. C. White, J. D. Nichols, and J. Hines kindly provided computer programs, while Queen's University computer science department, the Duke University Zoology Department Morphometrics Laboratory, and C.N.R.S. provided computing facilities.

References

Blondel, J., Pradel, R. & Lebreton, J.-D. (1992).- Low fecundity insular blue tits do not survive better as adults than high fecundity mainland ones. Journal of Animal Ecology, 61: 205-213.

Brownie, C., Anderson, D. R., Burnham, K. P. & Robson, D. S. (1985).- Statistical inference from band recovery data--a handbook. Second Edition. United States Fish and Wildlife Service Resource Publication 156.

Clobert, J., Lebreton, J.-D. & Allaine, D. (1987).- A general approach to survival rate estimation by recaptures or resightings of marked birds. Ardea, 75: 133-142.

Clobert, J., Lebreton, J.-D., Allaine, D. & Gaillard, J. M. (in press).- The estimation of age-specific breeding probabilities from recaptures or resightings in vertebrate populations. II Longitudinal models. Biometrics.

Cooch, E. G., Jefferies, R. L., Rockwell, R. F. & Cooke, F. (1992).- Environmental change and the cost of philopatry: an example in the lesser snow goose. Oecologia, in press.

Cooch, E. G., Lank, D. B., Dzubin, A., Rockwell, R. F. & Cooke, F. (1991a).- Body size variation in Lesser Snow Geese: environmental plasticity in gosling growth rates. Ecology, 72: 503-512.

Cooch, E. G., Lank, D. B., Rockwell, R. F. & Cooke, F. (1991b).- Long term decline in body size in a snow goose population: evidence of environmental degradation? Journal of Animal Ecology, 60: 483-496.

Cooke, F., Findlay, C. S. & Rockwell, R. F. (1984).- Recruitment and the timing of reproduction in lesser snow geese (*Chen caerulescens caerulescens*). Auk, 101: 451-458.

Cooke, F., Parkin, D. T. & Rockwell, R. F. (1988).- Evidence of former allopatry of the two color phases of Lesser Snow Geese (*Chen caerulescens caerulescens*). Auk, 105: 467-479.

Cooke, F. & Rockwell, R.F. (1988).- Reproductive success in a Lesser Snow Goose population. In Reproductive Success, T. H. Clutton-Brock (ed), 237-250. Chicago, University of Chicago Press.

Coulson, J. C. (1968).- Differences in the quality of birds nesting in the centre and on the edges of a colony. Nature, 217: 478-479.

Cox, D. R. & Oakes, D. (1984).- Analysis of survival data. London, Chapman and Hall.

Francis, C. M. & Cooke, F. (1992a).- Sexual differences in survival and recovery rates of lesser snow geese. Journal of Wildlife Management, 56: 287-296.

Francis, C. M. & Cooke, F. (1992b).- Migration routes and recovery rates of lesser snow geese from southwestern Hudson Bay. Journal of Wildlife Management, 56: 279-286.

Francis, C. M., Richards, M. H., Cooke, F. & Rockwell, R. F. (1992a).- Long-term changes in survival rates of Lesser Snow Geese. Ecology, 73: 1346-1362.

Francis, C. M., Richards, M. H., Cooke, F. & Rockwell, R. F. (1992b).- Changes in survival rates of Lesser Snow Geese with age and breeding status. Auk, in press.

Greenwood, P.J. (1980).- Mating systems, philopatry and dispersal in birds and mammals. Animal Behaviour 28: 1140-1162.

Hestbeck, J. B., Nichols, J. D., & Malecki, R. A. (1991). Estimates of movement and site fidelity using mark-resight data of wintering Canada Geese. Ecology, 72: 523-533.

Jolly, G. M. 1965. Explicit estimates from capture-recapture data with both death and immigration–stochastic model. Biometrika, 52: 225-247.

Lebreton, J.-D. (1987).- Regulation par le recrutement chez la Mouette Rieuse *Larus ridibundus*. Revue d'Ecologie (Terre et Vie) suppl. 4, 173-182.

Lebreton, J.-D., Hemery, G., Clobert, J. & Coquillart, H. (1990).- The estimation of age-specific breeding probabilities from recaptures or resightings in vertebrate populations. I. Transversal models. Biometrics 46: 609-622.

Lebreton, J.-D., Burnham, K. P., Clobert, J. & Anderson, D. R. (1992).- Modeling survival and testing biological hypotheses using marked animals: a unified approach with case studies. Ecological Monographs, 62: 67-118.

Mayfield, H. (1961).- Nesting success calculated from exposure. Wilson Bulletin, 73: 255-261.

Minta, S. & Mangel, M. (1989).- A simple population estimate based on simulation for capture-recapture and capture-resight data. Ecology, 70: 1738-1751.

Moore, W. S. & Dolbeer, R. A. (1989).- The use of banding recovery data to estimate dispersal rates and gene flow in avian species: case studies in the Red-winged Blackbird and Common Grackle. Condor, 91: 242-253.

Newton, I. (1989).- Synthesis. In Lifetime reproduction in birds, I. Newton (ed.), 441-469. London, Academic Press.

Pollock, K. H., Nichols, J. D., Brownie, C. & Hines, J. E. (1990).- Statistical inference for capture-recapture experiments. Wildlife Monographs 107.

Robinson, S. K. (1990).- Effects of forest fragmentation on nesting songbirds. Illinois Natural History Reports, No. 296.

Rockwell, R. F., Findlay, C. S. & Cooke, F. (1983).- Life history studies of the lesser snow goose (*Anser caerulescens caerulescens*) I. The influence of age and time on fecundity. Oecologia, 56: 318-322.

Seber, G. A. F. (1965).- A note on the multiple recapture census. Biometrika, 52: 249-259.

Seber, G. A. F. (1982).- The estimation of animal abundance and related parameters. Second edition. Macmillan, New York, New York.

Marked Individuals in the
Study of Bird Population
J.-D. Lebreton and P.M. North (eds.)

A Model for Mark-Recapture Data Allowing for Animal Movement

Bryan F.J. Manly and Chhondira Chatterjee

Department of Mathematics and Statistics, University of Otago, P.O. Box 56, Dunedin, New Zealand

Summary. A generalized random walk model for mark-recapture data is described which accounts for the positions of releases and recaptures, the probability of survival per unit time, and the effects of environmental variables on recapture probabilities.

1. Introduction

A recent review by Pollock et al. (1990) shows that many models have been proposed for mark-recapture data obtained from populations of mobile animals, where these have mostly been directed towards the estimation of population sizes or survival probabilities. One aspect of modelling that has received relatively little attention relates to the movement of animals, where the work that has been done has had several strands: (1) experiments have been conducted to gain information about how far animals move (e.g., Powell and Dobzhansky, 1976); (2) methods have been developed to adjust survival estimates for migration (e.g., Sheppard et al., 1969); (3) methods have been developed to estimate migration rates between different parts of a study area (e.g, Dorazio and Rago, 1991); and (4) traps have been set out in a fixed configuration such as a rectangular grid or a web, and captures and recaptures have been used to estimate the effective area covered by a population (e.g., Parmenter et al., 1989).

In this paper we concentrate on situations of type (1), although situations of types (2) to (4) can also be handled providing that only the first recaptures of animals after they are released are considered. Essentially our models give the probability of the first recapture of an animal as a function of the distance of the recapture point from the release point, a probability of recapture, and a survival probability.

Suppose that at time 0 an animal is released at the point (x_0,y_0) in a plane, and then performs a random walk. The distribution of the position (X,Y) of the animal at time

$t > 0$ is then bivariate normal with probability density function (pdf) $f(x,y;t) = (2\pi\sigma^2 t)^{-1}\exp[-\frac{1}{2}\{(x-x_0)^2+(y-y_0)^2\}/(\sigma^2 t)]$, and the distribution of the distance moved from the origin by time t has pdf $h(r;t) = r(\sigma^2 t)^{-1}\exp\{-\frac{1}{2}r^2/(\sigma^2 t)\}$, $r \geq 0$ (Skellam, 1973). This model has been used several times for the analysis of mark-recapture data (e.g., Dobzhansky and Wright, 1943; Kareiva, 1983).

One problem is that if animals are released at several locations then the probability of a recovery at a particular distance from the release point may depend on where the animal was released. Barrowclough (1978) proposed a method for allowing for this that is exact for circular study areas and approximate otherwise. See also the computer intensive solution to the problem that was used by Smith et al. (1990) to analyze data on recoveries of tagged fish.

Another problem is that the distribution of distance moved by animals is often not what is expected for a random walk. This led Manly (1977) to propose a model which assumes that the observed distance from the origin moved by an animal is given by $Z = R^{1/\alpha}$, where R is a random walk distance. In this case the pdf of Z is

$$S(z;t) = 2\{\alpha/\tau(t)\}\{z/\tau(t)\}^{2\alpha-1}\exp[-\{z/\tau(t)\}^{2\alpha}], \qquad (1)$$

where $\tau(t)^{2\alpha} = 2\sigma^2 t$. Thus Z has a Weibull distribution with shape parameter 2α and scale parameter $\tau(t)$ and the probability that Z exceeds z is $F(z;t) = \exp[-\{z/\tau(t)\}^{2\alpha}]$.

With this formulation, α is a parameter that measures the extent to which movement follows a random walk. If $\alpha < 1$, then there is a tendency for animals to move away from the origin more quickly than for a random walk. If $\alpha = 1$ then movement corresponds to a random walk. If $\alpha > 1$ then animals tend to stay closer to the origin than they would for a random walk. Hence a range of movement patterns are modelled by equation (1).

Manly (1977) considered non-linear least squares methods for estimating the parameters of this distribution from recapture data. Here we extend this work by setting up multinomial models from which maximum likelihood (ML) estimates of parameters can be calculated. The problem noted by Barrowclough (1978) in not addressed.

2. The Case Where Recaptures Occur Anywhere

Taking banded birds as an example, assume that: (a) all ringed birds have a survival probability of ϕ per year; (b) all directions of movement are equally likely, and the distance moved from the point of ringing has the distribution of equation (1); (c) birds behave independently of each other; and (d) the probability of a bird dying and being recovered in any year is λ, wherever that bird is. Also, let recoveries be classified into K distance classes such that class k consists of distances from z_{k-1} to z_k from the point of ringing, so that from equation (1) it follows that the probability of a bird being recovered in the ith year after ringing in distance class k is

$$\omega_{ik} = \phi^{i-1}\lambda[\exp\{-[z_{k-1}/\tau(t)]^{2\alpha}\} - \exp\{-[z_k/\tau(t)]^{2\alpha}\}], \tag{2}$$

where $\tau(t)$ is defined as for equation (1). In addition, let A be the total number of birds ringed, a_{ik} be the number of these that are recovered in year i in distance class k, a_0 be the number that are never recovered, and I be the last year after ringing for which recovery is possible. Then the probability of observing the set of recovery numbers $\{a_{ik}\}$ and a_0 is given by the multinomial distribution

$$P(\{a_{ik}\},a_0) = A!/\{a_{11}!a_{12}!...a_{IK}!a_0!\}\omega_{11}{}^{a_{11}}\omega_{12}{}^{a_{12}}...\omega_{IK}{}^{a_{IK}}\omega_0{}^{a_0}, \tag{3}$$

where $\omega_0 = 1-\omega_{11}-\omega_{12}-...-\omega_{IK}$ is the probability that an animal is not recovered. It follows that ML estimates of the parameters ϕ, λ, α and σ^2, with their approximate variances and covariances, can be obtained for a set of recapture data using a suitable computer program.

If the total number of birds banded is not known, then it is necessary to estimate parameters using the distribution of numbers recovered at different times and distances, conditional on a recovery occurring. To that end, note that the probability of a recovery taking place in year i at a distance from z_{k-1} to z_k from the point of ringing, conditional on a recovery, is $\omega^*_{ik} = \omega_{ik}/(1-\omega_0)$. Hence, the conditional multinomial distribution for the numbers recaptured $\{a_{ik}\}$ is

$$P(\{a_{ik}\}) = A^*!/\{a_{11}!a_{12}!...a_{IK}!\}\omega^*_{11}{}^{a_{11}}\omega^*_{12}{}^{a_{12}}...\omega^*_{IK}{}^{a_{IK}}, \tag{4}$$

where A^* is the total number recovered. Since the parameter λ cancels out in the probability ω^*_{ij}, the parameters that can be estimated by maximizing the likelihood are in this case ϕ, α and σ^2.

If recoveries are of live birds then the probability of a bird being recovered for the first time in year i at a distance between z_{k-1} and z_k from the point of ringing is still given by equation (2) but with ϕ defined to be the probability of surviving and not being recovered in any year, and λ defined to be the probability that a bird is recovered in a year given that it survives to the start of a year. Thus equation (3) still gives the multinomial distribution for the numbers recovered at different times and places, and the number not recovered, while equation (4) still gives the distribution conditional on recovery.

In practice more complicated models may be needed. For example, the probability of recovery at time t_j might be assumed be

$$\lambda_j = \exp(\gamma_0 + \gamma_1 x_{j1} + \gamma_2 x_{j2} + \ldots + \gamma_p x_{jp}), \tag{5}$$

where x_{jk} is the value of an environmental variable X_k at time t_j. This would be reasonable, for example, if there is reason to believe that recovery rates are related to weather variables. Also, relationships such as $\alpha = \alpha_0 + \alpha_1 t$ and $\tau(t)^{2\alpha} = 2\sigma^2 t^\beta$ might be considered in cases where movement parameters are thought to be time-dependent. For example, there is evidence from experiments involving the release of large batches of *Drosophila* that the flies initially travel further from the origin than is expected if they are performing random walks. However, after several days the positions approach what is expected from a random walk. This can then be modelled by allowing the value of α to increase with the time from release (Manly, 1977).

3. Recaptures in Traps from a Single Release

Suppose that A animals are released at time zero at the position (x_0,y_0), and some of these animals are then recaptured alive at later times $t_1, t_2, \ldots, t_S$ in traps set up at or away from the release point. In order to determine the distribution of the numbers of recaptures in different traps at different times we will still make the assumptions (a) to (c) listed in the previous section, but add the following ones as well: (e) recapture traps, set up at positions $(x_1,y_1), (x_2,y_2), \ldots, (x_R,y_R)$, make recaptures in such a way that if an animal is within a distance δ_i of the position (x_i,y_i) at time t_j then the animal is available for capture in trap i at that time; (f) the traps are sufficiently far apart and the trap radii are sufficiently small to ensure that an animal can only be in the area of one trap at one time; and (g) the probability λ_j that an animal is captured at time t_j, given that it is within the area of a trap at that time, is dependent on the values for certain environmental variables $X_1, X_2, \ldots X_p$ according to equation (5); and (h) recaptured animals are not released again. The assumption that trap i has the radius δ_i is a simple way of modelling the capture process, but in practice it is likely that an animal will have a certain probability of being captured in trap i at time t_j that depends on how close the animal is to the centre of the trap at that time. Hence δ_i is best thought of as the 'effective radius' of the trap rather than the actual radius.

It has been shown by Manly (1977) that a trap centred at a distance z from the release point with radius δ has an area that is a fraction $\delta/(4z)$ of the total area from a distance $z-\delta$ to a distance $z+\delta$ from the release point, and that the probability of an animal being within this range of distances from the release point is approximately $2s(z;t)\delta$ at time t, with s(z;t) given by equation (1). This implies that if the centre of trap i is at a distance z_i from the release point, then the probability of an animal being in the area of that trap at a time t_j after its release is approximately given by $\{\delta_i/(4z_i)\}2s(z_i;t_j)\delta_i = (\delta_i^2/2)s(z_i;t_j)/z_i$, providing that the animal is still alive and has not been recaptured at an earlier time. Hence, the probability of the animal being recaptured in trap i at time t_j, conditional on its survival to that time and not having been recaptured previously, is

$$p_{ij} = [\delta_i^2\alpha/\{z_i\tau(t_j)\}\{z_i/\tau(t_j)\}^{2\alpha-1}\exp\{-\{z_i/\tau(t_j)\}^{2\alpha}\}]\exp(\gamma_0+\gamma_1x_{j1}+...+\gamma_px_{jp}). \quad (6)$$

Also, using equation (1), the probability of an animal being recaptured at time t_j in a trap with radius δ_i at the release point, given that it is still alive and not been captured previously is

$$p_{ij} = \{1-\exp[-\{\delta_i/\tau(t_j)\}^{2\alpha}]\}[\exp(\gamma_0+\gamma_1x_{j1}+...+\gamma_px_{jp})]. \quad (7)$$

The unconditional probability Ω_{ij} that an animal will be recaptured in trap i at time t_j is p_{ij}, multiplied by the probability that the animal is still alive at time t_j after release, multiplied by the probability that the animal has not already been recaptured, which is given by

$$\Omega_{ij} = \begin{cases} p_{i1}\phi^{t_1}, & j = 1, \\ p_{ij}\phi^{t_j}(1 - \sum_{i=1}^{R} p_{i1})(1 - \sum_{i=1}^{R} p_{i2})...(1 - \sum_{i=1}^{R} p_{ij-1}), & j > 1. \end{cases} \quad (8)$$

It is now possible to write down the multinomial distribution for the numbers recaptured and the number not recaptured from the release of A animals. To this end, let a_{ij} be the number recaptured in trap i at time t_j, and a_0 be the number that are never recaptured, for i = 1 to R and j = 1 to S. Then

$$P(\{a_{ij}\},a_0) = A!/\{a_{11}!a_{12}!...a_{RS}!a_0!\}\Omega_{11}^{a_{11}}\Omega_{12}^{a_{12}}...\Omega_{RS}^{a_{RS}}\Omega_0^{a_0}, \quad (9)$$

where $\Omega_0 = 1-\Omega_{11}-\Omega_{12}-...-\Omega_{RS}$ is the probability that an animal will not be recaptured. This equation can then be the basis for ML estimation of the parameters α, σ^2, δ_1,

$\delta_2, \ldots, \delta_R, \gamma_0, \gamma_1, \ldots, \gamma_p$ and ϕ. However, as the model stands, γ_0 is confounded with the δ parameters since an increase in γ_0 can be compensated for exactly by a decrease in the δ values. This can be overcome by setting $\gamma_0 = 0$, and recognizing that δ_i then incorporates a component of the probability of capture as well as the effective radius of trap i.

Some other modifications may also be considered reasonable. Thus, for the reasons discussed at the end of the last section, the parameter α may be allowed to change with time, by making $\alpha(t) = \alpha_0+\alpha_1 t$, where t is the time since release. In a similar way, it may be desirable to set $\tau^{2\alpha} = 2\sigma^2 t^\beta$ instead of $\tau^{2\alpha} = 2\sigma^2 t$. Furthermore, to avoid an excessive number of parameters it will often be reasonable to set $\delta_1 = \delta_2 = \ldots = \delta_R$. Alternatively, if there are several distinct types of trap then the δ parameters may be set equal for all traps of the same type.

4. Experiments with Several Releases

Equations (6) to (9) give the distribution for the numbers recaptured from a single release of A animals at time zero, with recaptures being made at times $t_1, t_2, \ldots t_S$. However, there will often be a series of K releases, not necessarily from the same location, with estimates being required from the data for the whole experiment. In that case, each release will result in a multinomial distribution, providing that animals are marked in such a way that recaptured animals can be assigned to the release that they came from. Suppose that the assumptions made in the last section apply for each release, and that the kth release, of A_k animals, is made at time T_k from the position (x_{k0},y_{k0}). Then equation (6) can be modified in obvious ways to give the probability p_{ijk} that one of the A_k animals is in the area of trap i at time t_j as a function of z_{ik}, the distance from the centre of trap i to release point k, t_j-T_k, the time interval between the kth release and the jth recapture sample, and any environmental variables that are assumed to have the same effect for animals released at any time before t_j. In a similar way, a modified equation (7) can be used to find probabilities of capture for traps at release points, and the probability Ω_{ijk} of an animal in the kth release being recaptured in trap i at time t_j can be found by modifying equation (8). See Chatterjee and Manly (1992) for more details.

Let a_{ijk} denote the number of recaptures from the kth release in the ith trap at time t_j, for i = 1 to R and j = f_k to S, where f_k is the first recapture sample after the kth release, and let a_{0k} denote the number not recaptured from the kth release. Then the probability of the observations is

$$P_k(\{a_{ijk}\},a_{0k}) = [A_k!/\{\prod_{i=1}^{R}\prod_{j=f_k}^{S} a_{ijk}!a_0!\}]\ [\prod_{i=1}^{R}\prod_{j=f_k}^{S}\Omega_{ijk}{}^{a_{ijk}}]\Omega_{0k}{}^{a_{0k}}, \tag{10}$$

where $\Omega_{0k} = 1-\Omega_{1f_kk}-\Omega_{1f_{k+1}k}-...-\Omega_{RSk}$ is the probability that an animal in the kth release is not recaptured. The joint probability function for all K releases is then the product of each of these multinomial terms, i.e.,

$$P(\{a_{ij1}\},a_{01},...,\{a_{ijK}\},a_{0K}) = P_1(\{a_{ij1}\},a_{01})...P_K(\{a_{ijK}\},a_{0K}) \tag{11}$$

The parameters are α, σ^2, δ_1, δ_2, ..., δ_R, γ_0, γ_1, ..., γ_p and ϕ, exactly the same as for the single release experiment.

5. Estimation of Extra Movement Parameters

An important parameter that can be derived from equation (1) is the mean distance moved from the position of release by time t, $\mu(t) = \tau(t)\Gamma\{1+1/(2\alpha)\}$, where $\Gamma(x)$ is the gamma function. This can be estimated by $\hat{\mu}(t) = \hat{\tau}(t)\Gamma\{1 + 1/(2\hat{\alpha})\}$, for which a Taylor series approximation for the variance is

$$\begin{aligned}\mathrm{Var}\{\hat{\mu}(t)\} \approx& \{\mu(t)/\tau(t)\}^2\mathrm{Var}(\hat{\tau}) - \{\mu(t)/\alpha^2\}\Psi\{1+1/(2\alpha)\}\mathrm{Cov}\{\hat{\tau}(t),\hat{\alpha}\}\\ &+ \tfrac{1}{4}\{\tau(t)^2/\alpha^4\}[\Psi\{1 + 1/(2\alpha)\}]^2\mathrm{Var}(\hat{\alpha}),\end{aligned}$$

where $\Psi(x)$ is the digamma function. Also, the probability that the distance moved exceeds some value z can be estimated by $\hat{F}(z;t) = \exp[-\{z/\hat{\tau}(t)\}^{2\hat{\alpha}}]$, with the approximate variance

$$\begin{aligned}\mathrm{Var}\{\hat{F}(z;t)\} \approx& 4[F(z;t)\ln\{F(z;t)]^2[\{\alpha/\tau(t)\}^2\mathrm{Var}\{\hat{\tau}(t)\}\\ &+ 2\{\alpha/\tau(t)\}\ln\{z/\tau(t)\}\mathrm{Cov}\{\hat{\alpha},\hat{\tau}(t)\} + [\ln\{z/\tau(t)\}]^2\mathrm{Var}(\hat{\alpha})].\end{aligned}$$

There are some complications involved in using these variance equations since $\tau(t)$ will be estimated indirectly either by $\hat{\tau}(t) = (2\hat{\sigma}^2t)^{1/(2\hat{\alpha})}$, or by $\hat{\tau}(t) = (2\hat{\sigma}^2t^{\hat{\beta}})^{1/(2\hat{\alpha})}$. Hence $\mathrm{Var}\{\hat{\tau}(t)\}$ and $\mathrm{Cov}\{\hat{\alpha},\hat{\tau}(t)\}$ will not be obtained directly from the ML estimation procedure. Also, if α is estimated by $\hat{\alpha} = \hat{\alpha}_0 + \hat{\alpha}_1t$, then $\mathrm{Var}(\hat{\alpha})$ will not be available from the ML procedure. However, in all cases the Taylor series method gives approximate equations for $\mathrm{Var}\{\hat{\tau}(t)\}$, $\mathrm{Cov}\{\hat{\tau}(t),\hat{\alpha}\}$ and $\mathrm{Var}(\hat{\alpha})$ in terms of the variances and covariances of the parameters that are estimated. See Chatterjee and Manly (1992) for the relevant equations.

6. Goodness of Fit

The goodness of fit of the models described above can be measured in the usual way by the log-likelihood chi-squared statistic $X^2 = 2\Sigma O_i \ln(O_i/E_i)$, where O_i is the observed frequency and E_i is the expected frequency in the ith recapture class ($i = 1, 2, ..., M$). Then, providing that the model is correct, and the expected frequencies are large enough, X^2 will have an approximately chi-squared distribution with degrees of freedom (df) $\nu = M\text{-}p\text{-}1$, where p is the number of parameters estimated in fitting the model. Also, if two models are being considered, where the first model is a special case of the second, then the reduction in chi-squared due to fitting the more general model can be tested against the chi-squared distribution (with df equal to the number of parameters in model 2 that are additional to those in model 1) to see whether the model 2 is a significantly better fit than model 1.

Table 1. Recoveries of dead blackbirds of different ages at different ranges of distance from the point of ringing.

	Males Distance point of ringing (km)							Females Distance point of ringing (km)						
Age	0-3	-7	-20	-55	-148	-403	-1097	0-3	-7	-20	-55	-148	-403	-1097
1	58	11	2	2	0	1	0	42	4	0	3	0	1	0
2	35	5	0	0	1	0	0	16	2	2	1	0	0	1
3	34	1	0	0	0	0	0	15	0	0	0	0	0	0
4	23	0	0	0	0	0	0	14	1	2	0	0	0	0
5	12	1	0	0	0	0	0	9	1	1	0	0	0	0
6	6	0	0	0	0	0	0	3	0	0	0	0	0	0
7	4	2	0	0	0	0	0	0	0	0	0	0	0	0
8	4	0	0	0	0	0	0	1	0	0	0	0	0	0
9	1	0	0	0	0	0	0	0	0	0	0	0	0	0
10	0	0	0	0	0	0	0	0	0	0	0	0	0	0
11	1	0	0	0	0	0	0	0	0	0	0	0	0	0

7. Examples

As a first example of the use of the proposed model, consider some data on the recoveries of blackbirds (*Turdus merula*) ringed as juveniles that were extracted by Greenwood and Harvey (1976) from records of the British Trust for Ornithology (Table 1). Greenwood and Harvey found that the data are consistent with the hypothesis that all directions of movement were equally likely to occur. Because it is not known how many birds were ringed it is appropriate to use equation (4) as the distribution for the numbers recovered at different distances from the origin at different times, conditional on birds being recovered.

Consider the male data. Assuming that α is constant, ML estimates of this and the other parameters are $\hat{\alpha} = 0.0148$, $\hat{\sigma}^2 = 0.615$ and $\hat{\phi} = 0.821$; goodness of fit statistic,

X^2 = 104.46 with 73 df. In fact, inspection of the data indicates that mean of the distribution of dispersion distances was not increasing steadily with time as is implied by a constant α value. Therefore, a second model was fitted with $\alpha = \alpha_0+\alpha_1 t$, where t is the time in years since a bird was ringed. Here estimated parameters are $\hat{\alpha}_0$ = -0.0671, $\hat{\alpha}_1$ = 0.183, $\hat{\sigma}^2$ = 0.292 and $\hat{\phi}$ = 0.638, and X^2 = 62.13, with 72 df. This represents a highly significant reduction in the goodness of fit statistic of 42.33 with 1 df due to allowing α to vary with time. Finally, the model with $\alpha = \alpha_0 + \alpha_1 t$ and $\tau = 2\sigma^2 t^\beta$ was fitted. This produced the estimated parameters $\hat{\alpha}_0$ = 0.151 (0.054), $\hat{\alpha}_1$ = 0.0160 (0.0294), $\hat{\sigma}^2$ = 0.432 (0.079), $\hat{\beta}$ = -0.284 (0.247) and $\hat{\phi}$ = 0.638 (0.0223), with standard errors shown in parenthesis, and X^2 = 40.04 with 71 df. There is a highly significant reduction in the goodness of fit statistic (22.09, with 1 df) obtained by allowing β to be different from 1.

Similar results were obtained when the three models that have been considered for the male data were fitted to the female data; it appears that α should be allowed to vary in the model and β should not be set equal to 1. Estimates were also obtained for the male and female data lumped on the assumption that both sexes have the same parameter values. In this case X^2 = 90.84 with 147 df for the model with α allowed to vary with time and β estimated. This compares with the total X^2 = 81.39 with 142 df obtained by allowing different parameters for each sex. The difference of 9.45 with 5 df is not significantly large at the 5% level, although it is approaching significance; there is not much evidence of a sex difference.

It might be thought that the chi-squared statistics that have been quoted indicate that the models with $\alpha = \alpha_0+\alpha_1 t$ and $\tau = 2\sigma^2 t^\beta$ fit the data quite well. However, these statistics are misleadingly small because to some extent the df are inflated by many observed and expected frequencies that are close to zero. In fact, if some frequencies are combined, then X^2 statistics indicate that the model with parameters estimated separately for males and females is a somewhat unsatisfactory fit to the data, but is distinctly better than the fit of the model with common parameters. More details about these tests and other aspects of this example are provided by Chatterjee and Manly (1992).

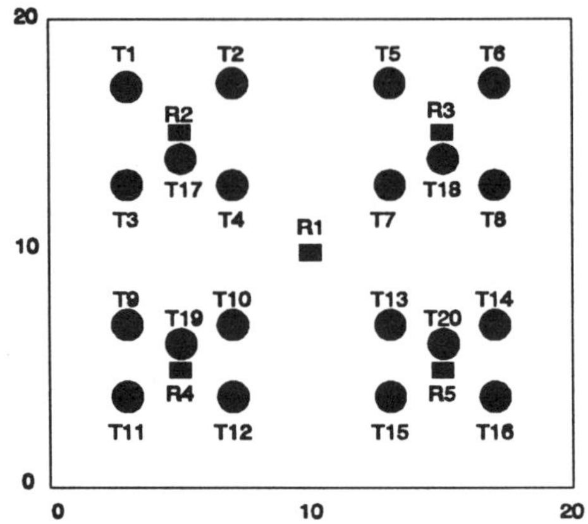

Fig. 1. The layout used for a computer simulation experiment involving 200 'animals' being released at each of the locations R1 to R5 at times 0, 1, ..., 4 and recaptures in traps T1 to T20 made at times 1, 2, ..., 10. Trap radii were 0.5 units so that the circles shown are drawn to scale.

We use simulated data to illustrate the use of our model in the situation where there are several releases and animals can only be recaptured in traps. These data were obtained using the experimental layout indicated in Fig. 1. Thus a square area of size 20 units by 20 units was considered, with five release points R1 to R5, and 20 recapture traps T1 to T20. At time 0, 200 'animals' were released at the position R1 in the figure. Of these, 15 were recaptured. Further releases of 200 'animals' were made at positions R2, R3, R4 and R5 at times 1, 2, 3 and 4, respectively, resulting in the recapture data shown in Table 2. All recapture traps shown had a radius of 0.5 units, and an 'animal' was considered to be recaptured in a trap at time i if it was in the circular trap area at that time. There were no 'environmental' variables affecting the probability of capture. However, to simulate the situation where an environmental variable with no effect is measured, the 'temperatures' 20, 21, 24, 23, 19, 21, 20, 21, 22 and 24°C were arbitrarily assigned to the recapture sample times 1 to 10. 'Animals' were made to move according to the model of equation (1) with $\alpha = 0.75$ and $\sigma^2 = 1$, and a survival probability of $\phi = 0.9$ per day was imposed.

We fitted a series of models to the data. Model 1 had α set equal to the random walk value of 1 so that the parameters estimated were σ^2 (for movement), ϕ (for survival), and δ (the trap radius). Estimates were obtained by maximizing the likelihood function of equation (11), and the resulting goodness of fit statistic was found to be X^2 = 295.62 with 797 df. Model 2 (X^2 = 261.46 with 796 df) was the same as the first except that α was estimated instead of being set equal to 1; the reduction in X^2 in moving from Model 1 to Model 2 is 34.16 with 1 df, which is very highly significant. For Model 3 (X^2 = 259.74) the parameter α was allowed to change with time t since the release of an 'animal' by setting $\alpha = \alpha_0 + \alpha_1 t$, but in other respects the model was the same as Model 2; the reduction in X^2 in moving from Model 2 to Model 3 is 1.72 with 1 df, which is not significantly large at the 5% level. Model 4 (X^2 = 258.46 with 794 df) was the same as Model 3 except that equation $\hat{\tau}(t) = (2\hat{\sigma}^2 t^{\hat{\beta}})^{1/(2\hat{\alpha})}$ was used instead of $\hat{\tau}(t) = (2\hat{\sigma}^2 t)^{1/(2\hat{\alpha})}$ to estimate $\tau(t)$; the reduction in X^2 in moving from Model 3 to Model 4 is 1.28 with 1 df, which is again a non-significant change. Model 5 (X^2 = 258.31 with 793 df) was the same as Model 4 except that the temperature was allowed to influence the number of recaptures. Thus the probability that an 'animal' was in a trap area at time j included a term of the form $\exp(\gamma_0 + \gamma_1 x_{j1})$ and γ_1 was estimated along with all the parameters in Model 4. The parameter γ_0 was set equal to zero rather than being estimated since, as noted earlier, this parameter is confounded with δ. The reduction in X^2 in moving from Model 4 to Model 5 is 0.15 with 1 df.

This analysis suggests that Model 2 is best, and indeed this was the model that we used to generate the data. Table 3 shows the true parameter values, their estimates and estimated standard errors, and goodness of fit statistics for each of the five models. Differences between the estimated and true values are mostly within the ranges that can be expected (true values ± two standard errors). The exceptions are that the estimates of σ^2 and α_0 seem to be too low for Model 2 and Model 3. More simulations are required to see whether this is accidental.

With real data we would of course want to know whether the model that seems best from chi-squared tests actually fits the data reasonably well. This raises the same problem as was encountered with the previous example: the df for goodness of fit statistics is inflated by the existence of large numbers of release time, trap number and recapture time combinations that have a vanishingly small probability of being observed. This problem is overcome most easily by grouping recaptures in appropriate ways. For example, the recaptures in traps at the same distance from the point of release could be combined, as could recaptures made at different times after release. However, we will not pursue these ideas here.

Table 2. Recapture data from a simulated experiment with five releases of 200 animals each, 20 recapture traps and 10 recapture samples at times 1, 2, ..., 10.

Release	Recapture Trap	1	2	3	4	5	6	7	8	9	10	Recapture total
(1) From position R1 at time 0	3										1	
	4	1		1					1			
	7		1		1				1			
	9				1							
	10			1								
	13		1			1						
	17						1	1				
	18									1		
	20										1	15
(2) From position R2 at time 1	1	-	3	2	1	1	2					
	2	-		1		1		2	1	1		
	3	-		1	2	2						
	4	-					1	2				
	7	-					1					
	9	-			1							
	17	-	12	7	4	3			1	2		54
(3) From position R3 at time 2	2	-	-					1				
	5	-	-	1	1	1			1			
	6	-	-	1	1	1						
	7	-	-		2	1		1		2		
	8	-	-	1		1		1		1		
	18	-	-	10	12	6	3		1	2	1	53
(4) From position R4 at time 3	3	-	-	-			1				1	
	4	-	-	-							1	
	9	-	-	-	1		2	1	2	1		
	10	-	-	-	2		2	1	1			
	11	-	-	-	1	2	1	1	1			
	12	-	-	-	1		1	1				
	13	-	-	-							1	
	19	-	-	-	7	4	1	3	1	1	1	44
(5) From position R5 at time 4	8	-	-	-	-				1			
	13	-	-	-	-	1	2	1	2		1	
	15	-	-	-	-	2	2	1				
	16	-	-	-	-	1	2	2		1	1	
	18	-	-	-	-				1			
	20	-	-	-	-	11	4	2	5	7	2	52

Table 3. Estimates of parameters for five models fitted to artificial mark-recapture data, with chi-squared goodness of fit statistics. Models 2 to 5 are all correct.

		Model 1		Model 2		Model 3		Model 4		Model 5	
Parameter	True	Est.	S.E.	Est.	S.E.	Est.	S.E.	Est.	S.E.	Est.	S.E.
ϕ	0.90	0.88	0.02	0.88	0.06	0.86	0.03	0.87	0.03	0.87	0.03
σ^2	1.00	1.69	0.14	0.72	0.13	0.66	0.12	0.83	0.22	0.82	0.22
δ	0.50	0.50	0.03	0.53	0.03	0.56	0.04	0.54	0.04	0.46	0.19
α_0	0.75	-	-	0.64	0.06	0.54	0.09	0.62	0.12	0.62	0.12
α_1	0.00	-	-	-	-	0.02	0.02	0.00	0.03	0.00	0.03
β_1	1.00	-	-	-	-	-	-	0.74	0.22	0.74	0.22
γ_1	0.00	-	-	-	-	-	-	-	-	0.01	0.04

8. Discussion

One major limitation of our model is the assumption that all directions of movement are equally likely since presumably if this is not true then estimates may be biased. We suspect that the bias may be quite small if traps are placed symmetrically round release points but we have not yet investigated that. If the movement away from release points tends to be in one direction then the biased random walk model of Vogt and Morton (1991) may be more useful than our model. See also Kareiva and Shigesada (1983), Bovet and Benhamou (1988), Marsh and Jones (1988) and Othmer et al. (1988) for other approaches.

Another limitation of the model is the restriction that only the first recaptures of animals can be used, which can only be overcome by setting up and maximizing a likelihood function that accounts for the positions of multiple recaptures. Of course, if the movement distribution and survival of recaptured and released animals is the same as for animals when they are first released then animals released for the second time can just be treated as new releases. Unfortunately, as far as movement is concerned this is not really consistent with our model unless either $\alpha = 1$ or animals lose their memory of earlier releases when the are released for a second or subsequent time.

A third limitation is that the model does not easily allow for natural barriers in the study area. For example, if there is an absorbing barrier such that all animals passing this are captured, then this is not easily allowed for, except perhaps by approximating the absorbing area by a grid of traps.

Finally, we note that one of our assumptions is that marked animals move independently of each other. Often the true situation is that there are groups of animals that move more or less independently of each other. It seems likely that the main effect of grouping of animals will be that the standard errors calculated from our model will be too small, but that biases in estimators will not necessarily occur unless movement behaviour is related to the size of groups.

Acknowledgements. We thank John Rayner, Richard Morton, and anonymous referees for useful comments on earlier drafts of the manuscript.

References

Barrowclough, G.F. (1978). Sampling bias in dispersal studies based on finite area. Bird Band. 49: 333-341.

Bovet, P. and Benhamou, S. (1988). Spatial analysis of animals' movements using a correlated random walk model. J. Theor. Biol. 131: 419-433.

Chatterjee, C. and Manly, B.F.J. (1992). A model for mark-recapture data that allows for animal movement. Res. Rep. 1992: 1, Dept. Math. Statist., University of Otago.

Dorazio, R.M. and Rago, P.J. (1991). Evaluation of a mark-recapture method for estimating mortality and migration rates of stratified populations. Can. J. Fish. Aquat. Sci. 48: 254-260.

Dobzhansky, T. and Wright, S. (1943). Genetics of natural populations. X. Dispersion rates in *Drosophila pseudoobscura*. Genetics 28: 304-340.

Greenwood, P.J. and Harvey, P.H. (1976). The adaptive significance of variation in breeding area fidelity of the blackbird (*Turdus merula* L.). J. Anim. Ecol. 45: 887-898.

Kareiva, P.M. (1983). Local movement in herbivorous insects: applying a passive diffusion model to mark-recapture field experiments. Oecologia 57: 322-327.

Kareiva, P.M. and Shigesada, N. (1983). Analyzing insect movement as a correlated random walk. Oecologia 56: 234-238.

Manly, B.F.J. (1977). A model for dispersion experiments. Oecologia 31: 119-130.

Marsh, L.M. and Jones, R.E. (1988). The form and consequences of random walk models. J. Theor. Biol. 133: 113-131.

Othmer, H.G., Dunbar, S.R. and Alt, W. (1988). Models of dispersal in biological systems. J. Math. Biol. 26: 263-298.

Parmenter, R.R., MacMahon, J.A. and Anderson, D.R. (1989). Animal density estimation using a trapping web design: field validation experiments. Ecology 70: 169-179.

Pollock, K.H., Nichols, J.D., Brownie, C. and Hines, J.E. (1990). Statistical inference from capture-recapture experiments. Wildl. Monog. 107.

Powell, J.R. and Dobzhansky, T. (1976). How far do flies fly? Amer. Sci. 64: 179-185.

Sheppard, P.M., MacDonald, W.W., Tonn, R.J. and Grab, B. (1969). The dynamics of an adult population of *Aedes aegypti* in relation to dengue haemorrhagic fever in Bangkok. J. Anim. Ecol. 38: 661-702.

Skellam, J.G. (1973). The formulation and interpretation of mathematical models of diffusionary processes in population biology. In The Mathematical Theory of the Dynamics of Biological Populations (eds. M.S. Bartlett and R.W. Hiorns), pp. 63-85. Academic Press, London.

Smith, B.D., McFarlane, G.A. and Cass, A.J. (1990). Movement and mortality of tagged male and female Lingcod in the Strait of Georgia, British Columbia. Trans. Amer. Fish. Soc. 119: 813-824.

Vogt, W.G. and Morton, R. (1991). Estimation of population size and survival of sheep blowfly, *Lucilia cuprina*, in the field from serial recoveries of marked flies affected by weather, dispersal and age-dependent trappability. Res. Popul. Ecol. 33: 141-163

Marked Individuals in the
Study of Bird Population
J.-D. Lebreton and P. M. North (eds.)

On the role of ringing schemes in the measurement of dispersal.

Arie J. van Noordwijk

Netherlands Institute of Ecology, P.O. box 40, NL 6666 ZG Heteren, The Netherlands

Summary. Bird-rings are marks to link two or more observations of the same individual. Such observations are interesting when considerable time and or distance lies between them. There are many biological questions that can be answered by a single group of amateur or professional observers using marked individuals. In addition, there are three major biological problems for which the coordinating activities of ringing schemes, together with the effort of many ringers, are essential, to satisfy the need for large quantities of data or to enlarge the space covered beyond the scope of any single group. These three problems are: bird migration, geographic patterns in survival rates and dispersal.

I discuss the role that ringing centres could play in organizing the production of data needed to analyse geographic patterns in survival estimates and to analyse dispersal from site of birth to site of reproduction. The most important new element is that ringers should be directed to increase the recaptures of living birds. Designs of ringing projects should be judged on the expected frequency of recaptures. It is recommended to exploit the site fidelity of breeding birds during the breeding season by concentrating ringing activities on nestlings and breeding adults. Studies along these lines should provide the data necessary for all three major problems.

1. Introduction

Most of the chapters in this volume deal with methods to calculate survival rates. These rates are convenient summary statistics which can be used in the calculation of population growth rates or sustainable harvest rate, provided that some additional information is available on reproductive rates. However, for many or even most biological questions, these rates are most interesting when the variation in them can be related to other variables. In principle, such comparisons could be made at different levels of biological complexity. However, since the survival rate is a property of a group of individuals, which must be assumed to be identical, the population level within species is the lowest level of taxonomic hierarchy for which such comparisons can be made. At the same time, the level of populations within species knows the fewest complicating other differences among groups, and thus the interpretation of differences in other traits associated with differences in survival rate is more straightforward. It would thus be interesting to have sufficient data on a single species to perform analyses on separate populations within it. In principle, a ringing scheme could organise several groups of ringers to collect data in a standardized way, in order to collect sufficiently large datasets.

In the past, ringing schemes have concentrated their efforts on bird migration. One aspect that has been relatively neglected is the organisation of ringing activities in a manner that the ringers can provide regular recaptures or resightings. Recently, some schemes have initiated programmes in which the recaptures form a key element (e.g. Baillie, 1990) and an increasing number of studies exploit the possibility of organising large groups of observers to provide resightings. Apart from the collection of data for differentiated survival estimates, the same data collection could also be used to provide the necessary data for studies of dispersal. Below, I will first discuss the importance of studying dispersal, I will then outline some of the problems in the analysis of dispersal data and finally I will discuss the role that ringing schemes could and should play.

2. Dispersal

In any population study, differences in dispersal between groups are a major source of potential bias in estimating survival rates or selection processes (e.g. Dhondt and Hublé, 1968). Further, dispersal is of major interest in evolutionary biology. Dispersal, here defined as the displacement between site of birth and site of reproduction (also called natal dispersal), is equivalent to gene flow. Gene flow between areas with different selection regimes is a powerful mechanism to prevent populations from becoming better adapted to the local conditions (e.g. Dobzhansky, 1937; Dhondt et al., 1990; Perrins, 1990). At the same time, however, gene flow is capable of maintaining significant amounts of genetic variation (e.g. Bulmer 1980) and thus the potential to respond in an evolutionary way to changes in environmental conditions (e.g. Wright, 1977).

Dispersal is a description of the behaviour of an individual. The net effect at the population level is often described by immigration and emigration (e.g. Rockwell and Barrowclough, 1987). Of the four basic components of changes in population size, much more is known about birth and death than about immigration and emigration. There is also a link between the importance of dispersal and the size of the population. It is to be expected that in smaller populations the migration component becomes quantitatively more important in population dynamics. Therefore, habitat fragmentation makes the study of dispersal all the more urgent. A further interaction between habitat fragmentation and the role of gene flow is that habitat fragmentation is likely to increase the number of different selection regimes compared to a more continuous habitat.

3. Problems in the study of dispersal

Heterogeneity in the chances of being observed (captured, recaptured, resighted, retrieved) is a key problem in determining the reliability of survival estimates (e.g. Lebreton et al., 1992). This problem is equally important, but more complex in the study of dispersal due to the added spatial complexity (North, 1988). An essential type of information lies in the negative observations: "Individual X would have been reported if it had been there at that time". It is virtually impossible to draw firm conclusions from the absence of observations on the absence of individuals. However, there are three types of observations from which approximate inferences that an individual was not at location **A** at time **T** can be made:

1) The focal individual was observed elsewhere at the same time.
2) If a different individual was reported from the place and time in question, the focal individual might also have been reported, with some probability to near certainty depending on the exact circumstances.
3) In some cases, one may assume that all individuals must be in a certain type of place at a certain time of the year. A failure to report the focal individual after an exhaustive search of all suitable places may provide positive evidence of its absence. Nestsites are probably the most important category of such places.

In each of these cases a description of the set of potential observations can be generated and then be used as a filter for the effects of observer distribution on the observations actually made. The principal idea is to learn about the dispersal behaviour of the birds by regarding the difference between the potential and the actual observations. Van Noordwijk (1984) has provided an example of the analysis of small scale dispersal with the help of explicitly generating the set of potential observations (see also van Tienderen and van Noordwijk, 1988). Attempts to apply the same idea to data from the EURING databank are still in an early stage.

4. The role of ringing centres.

Bird ringing is a technique that has many different applications. From an organisational point of view, there is a dichotomy between those applications in which the same (group of) people make the repeated observations on the same individual bird and those applications in which different (groups of) people observe the same individual. Whereas the range of scientific questions addressed with the help of ringing on the basis of 'within group data collection' is very broad, the involvement of ringing centres was and still is largely related to migration. Here the ringing centres provide the link between the repeated observations, most commonly a first ringing event and a later report on a corpse or a recapture.

The unique property of ringing centres is their ability to collate and coordinate observations made on single individual birds. There is no other group of organisms for which a comparable organisational structure exists. On the one hand there is less need for a coordinating organisation in less mobile creatures, on the other hand it provides unique possibilities for answering general questions e.g. about the origin of immigrants, which play an essential role in the question at what spatial scale local adaptation is possible. It is natural that ringing centres have concentrated on those types of data in which they are essential, namely observation of the same individuals in different places. A second opportunity, the coordination of data-collection on a scale that exceeds the capacity of any individual (group) has received less attention.

Whereas ringing centres have to some extent regulated and coordinated the ringing of birds, efforts to direct the collection of subsequent observations on ringed individuals are few and recent. In principle, a bird-ring only provides useful information if its carrier is observed at least twice. Thus, any effort to increase the number of subsequent observations on ringed individuals increases the efficiency of data collection. Moreover, an organised collection of resightings is an essential condition for any attempt to draw conclusions from the absence of resightings.

These considerations lead to a new agenda for bird-ringing that has three major questions:

1) DISPERSAL: what is the relation between site of birth and site of breeding.
2) SURVIVAL: estimates with sufficient data to allow analyses of the variation in survival rates per region, per habitat and per cohort.
3) MIGRATION: ringing is still a major technique for the documentation of any changes in migration patterns. Any documentation of quantitative changes can only result from further research, but 'random ringing' can certainly provide clues to be used in the direction of this research.

For each of these questions, ringing birds of known age and known origin and concentrating retrapping efforts in those places and during those periods in which a high proportion of the individuals present was previously ringed, will bring the largest amount of information per unit of effort. This implies that the ringing of pulli, together with a (re)trapping of the breeding birds is a form of data collection that is useful for each of the three questions and is thus to be promoted. The choice of ringing at the site of birth and organising recaptures at the site of reproduction is essential to the study of dispersal. The fact that many if not most birds have a high site fidelity once they start reproducing (Greenwood 1987), implies that regular recaptures are relatively easy at the breeding site. Especially in connection with inventories of breeding pairs in a given area, one might even obtain an independent estimate of the trapping efficiency. The study of migration has traditionally relied on the ringing of birds during their migration. The disadvantage is that relations with the breeding grounds can only be drawn from recoveries or recaptures there, or in a round about way based on differences in biometry among individuals from different populations.

There is one major practical problem. Whereas visits to the nest can be made without any risk of disturbance in some species, in other species it is impossible to visit nests of other species without damaging the vegetation around the nest and thereby increasing the predation risk on these nests. In precocial birds, the nest is left before the young are big enough to be ringed. It is therefore necessary to develop different strategies for different groups of bird species. In essence, however, the ringing of parents and offspring at or near the breeding site is possible for all species.

For the study of dispersal, the spatial structure of the study is important. The best compromise is probably a checker-board of study plots, in which both the within plot movements and the between plot movements can be studied on a continuous scale. If the plots are similar in size and roughly equidistant, a second coarser scale for analyses is in the form of between plot movements of individuals in a transition matrix form. There are two special problems in the optimal design of study areas for dispersal. The first is that the optimal design depends strongly on the dispersal behaviour of the species to be studied and the second problem is that the from the viewpoint of gene-flow the longest distances have the greatest effect on the net gene flow, but these are also the most difficult to observe. The latter is even made worse by the fact that discontinuities in the habitat are more likely and more likely to have great effects, the larger the spatial scale considered becomes.

5. International dimensions (EURING)

National boundaries are non-existent for birds. The cooperation between the various ringing centres in Europe is relatively old and intensive. In Europe, bird migration is automatically international. However, even for the collection of data on dispersal and variation in survival rates of resident species, international cooperation is extremely valuable because of the quantities of data that can be collected. This requires that data are collected and registred in compatible ways. The EURING organisation has already provided a standard code for the registration of ringing data. Where ringing centres are moving towards a further steering of ringing and retrapping activities, a coordination on an international scale could greatly enhance the value of the data collected. At EURING meetings many ringing centres have expressed a great interest in a coordinated effort in this direction. The challenge for the next decade is to redirect ringers away from ringing birds towards recapturing birds already ringed and to reorganise the ringing centres to make effective use of these data.

References

Baillie, S.R. (1990).- Integrated population monitoring of breeding birds in Britain and Ireland. Ibis 132: 151-166

Bulmer, M.G. (1980).- The mathematical theory of quantitative genetics. Clarendon Press, Oxford.

Dhondt, A.A., Adriaensen, F., Matthysen, E. and Kempenaers, B. (1990).- Nonadaptive clutch-sizes in tits: evidence for the gene flow hypothesis. Nature, London 348: 723-725.

Dhondt, A.A. and Hublé, J. (1968).- Fledging date and sex in relation to dispersal in young Great Tits. Bird study 15: 127-134

Dobzhansky, T. (1937).- Genetics and the origin of species. Columbia University Press, New York.

Greenwood, P.J. (1987).- Inbreeding, philopatry and optimal outbreeding in birds. pp 207-222 in: Avian Genetics. F. Cooke and P.A. Buckley (Eds). Academic Press, London.

Lebreton, J.-D., Burnham, K.P., Clobert, J. and Anderson, D.R. (1992).- Modelling survival and testing biological hypotheses using marked animals: a unified approach with case studies. Ecol. Monogr. 62: 67-118.

North, P.M. (1988).- A brief review of the (lack of) statistics of bird dispersal. Acta Ornithol. 24: 63-74.

Perrins, C.M. (1990).- Concluding remarks: dispersal and gene flow. pp 475-480 in Population Biology of Passerine birds. J. Blondel et al. (Eds) Springer Nato ASI series G: 24: 475-480.

Rockwell, R.F. and Barrowclough, G.F. (1987).- Gene flow and the genetic structure of populations pp 223- 255 in: Avian Genetics. F. Cooke and P.A. Buckley (Eds). Academic Press, London.

van Noordwijk, A.J. (1984).- Problems in the analysis of dispersal and a critique on its "heritability" in the Great Tit. J. Anim. Ecol. 53: 533-544.

van Tienderen, P.H. and van Noordwijk, A.J. (1988).- Dispersal, Kinship and Inbreeding in an island population of the Great Tit. J. Evol. Biol. 1: 117-137.

Wright, S. (1977).- Evolution and the genetics of Populations. Vol. 3 Experimental results and evolutionary deductions. Univ. of Chicago Press, Chicago.

Survival in relation with habitat

Marked Individuals in the
Study of Bird Population
J.-D. Lebreton and P.M. North (eds.)

Testing Hypotheses about the Relationship of Habitat to Animal Survivorship

Michael J. Conroy

U.S. Fish and Wildlife Service, Georgia Cooperative Fish and Wildlife Research Unit, School of Forest Resources, University of Georgia, Athens, GA 30602 USA

Summary: Inferences about the effects of habitat quality and fragmentation on populations require habitat-specific estimates of survivorship. I discuss several approaches using marked animals under a range in spatial and temporal resolutions, and assumptions about animal mobility. Single mark-encounter (e.g., ring-recovery) studies may be used to address relatively coarse questions. Pre-stratification models (e.g., Brownie et al., 1985), can test for presumed habitat effects during the period just after marking; post-stratification models (Schwarz et al., 1988) can be used if habitat effects occur closer to the time of recovery. Mark-resighting models (Hestbeck et al., 1991) enable estimation of time- and habitat-specific survival and movement rates. Continuous observations of animals marked with radio transmitters offer the best promise for fine-scaled inferences about habitat effects using proportional hazards (Cox, 1972) methods. I evaluate power to detect a specified alternative for the last 2 methods, and close with some recommendations for the design of studies to test for habitat-specific survivorship.

1. Introduction

Few studies of habitat choice in animals have directly compared demographic parameters determining fitness, but rather have inferred benefit from the relative use or type of activity (e.g., time budgeted) in various habitats (Van Horne, 1983; Gavin, 1991; Murphy and Noon, 1991). Inference based on such data without corresponding demographic data may lead to spurious conclusions about habitat "quality" (Van Horne, 1983; Pulliam, 1988) and to faulty management regimes. Habitat-specific estimates of reproduction, survival, and movement rates among habitat patches are needed.

In this paper I address habitat-specific inference about survival rates using marked animals. I discuss how assumptions about animal mobility, the hypothesised relationship of habitat to survival, and the spatial and temporal resolution of data all determine the types of questions that can be addressed. I provide a hierarchical approach, leading from very simple data structures and unrealistic assumptions, to intensive sampling involving realistic modeling. I use simulations to provide estimates of power, under various scenarios, to detect a specified alternative, and end with recommendations for experimental designs aimed at testing hypotheses about habitat-specific survivorship.

2. Data structures and models considered

Inference about the relationship of habitat to survival is determined by both the spatio-temporal resolution of data available to estimate survival rates and define habitat, and the assumptions one is able (or willing) to make about such factors as animal mobility, and the specific role of habitat in relation to survival. The data structures and corresponding models considered below address these points, and suggest a hierarchy in which simple data structures can be used to answer certain rather coarse-grained questions about the relationship between habitat and survivorship, and more complex data structures can be used to answer better focussed questions.

2.1 Ring-recovery data

The data structure in ring-recovery models consists of 2 broad cases: either the fates of all marked animals are known, and all marks are recovered (very rare), or some marks are not recovered, and therefore the fates of some animals unknown. I have not specified the time interval between marking and recovery, but ordinarily this would be 1 calendar year or less. I will consider the first case only because it is a logical foundation on which to build more complex and realistic structures.

2.1.1 Fates all known

In this cases, N animals are marked; R die and are recovered, and $N\text{-}R$ live (e.g. are recaptured or otherwise known to be alive). I consider 2 ways in which habitat might be related to survivorship. The first, appropriate for non-migratory animals or animals for which the critical mortality period occurs before migration, assumes that survivorship is related *to the habitat in which the animal was marked.* A straightforward method is then to classify the N marked animals by the habitat at marking and by fate, and perform a test of homogeneity of proportions of animals dead or alive.

The second situation assumes that survivorship is related *to the habitat in which the animal was recovered*, and is more appropriate for migratory animals in which the critical mortality period occurs after migration. Tests of binomial proportions will work here as well, but the N animals are classified by fate and habitat of recovery.

A different approach is to divide the N animals into 2 samples (R and $N\text{-}R$) based on fate, and perform a bivariate test of location (e.g., Mardia, 1967). The latter approach does not directly consider classes of habitat, but it might be possible to infer habitat effects from distributional differences.

Several assumptions are common to both situations, and to either the tests of proportions or distributional tests: (1) All marked animals have known fates over some time interval. (2) Any mortality effect relates to either the habitat in which the animal is marked, or that in which it is recovered. Other habitats visited between marking and recovery have no differential effect on survival (or if they do it is ignored). (3) There is no "switching" of habitats during the period over which the hypothesised habitat effect occurs. Thus, habitat is a fixed attribute of each animal, at least over the period of interest.

2.1.2 Fates of some marked animals unknown

More commonly, N animals are marked and R animals are recovered (e.g. found dead or reported by hunters). The rest (N-R) are never recovered and have unknown fates: some may have died, but some are still alive and simply were not recovered. An extension of this situation involves the release of N_t marked animals at each of t (t=1,...T) sampling occasions, from which R_{tl} are recovered in each subsequent recovery period l (l=1,...L) and N_t - R_t are never recovered, and corresponds to the ring-recovery models of Brownie et al. (1985). Again, there are 2 cases. In the first, the habitat in which the animal was marked has the hypothesised relationship to survivorship. These data can be modeled by the Brownie et al. approach, in which habitat is considered a stratum or attribute at the time of marking (pre-stratification). Tests can be performed either by contingency tables (Brownie et al., 1985, pp. 144-148) or by parameterisation of a habitat effect in program SURVIV (White, 1983). In the second case, it is the habitat in which the animal is recovered that is assumed to be related to survivorship. This analysis uses the post-stratification ring-recovery model (Schwarz et al., 1988), in which recoveries are grouped by the habitat of recovery, to which animals are assumed to remain faithful (migrate to each year).

Ring-recovery models allow a more realistic sampling situation, in that not all animal fates are required to be known. However, both pre- and post-stratification models require that habitat is an unchanging attribute of the animal, so that each animal is completely faithful to 1 and only 1 habitat (whether defined at marking or at recovery). More importantly, ring-recovery models also require assuming that all differential habitat effects are defined by the habitat in which the animal happens to be marked or recovered, ignoring all other periods and habitats. Often, long periods of time elapse between marking and recovery, over which animals are exposed to multiple risk factors; it is unrealistic to ignore other habitats that may effect survival over these periods. Finally, recovery data ordinarily are obtained fortuitously, e.g. by hunters who shoot ringed gamebirds, and are unlikely to provide usable data on other than very crude spatial scales, especially when stratification is after recovery; for example, waterfowl flyways were used by Schwarz et al. (1988).

2.2 Mark-resighting data

The above limitations on recovery data suggest data structures in which multiple observations are made over time on both animal fate, and animal location. Here I consider a data structure in which observations are made on marked animals over T

sampling periods. Ordinarily these will be resightings of marked animals, but recaptures of both marked and unmarked animals also could be used. During each sampling period animals that are still alive are also classified by the habitat in which they occur, providing a data structure corresponding to the mark-resighting and movement model of Hestbeck et al. (1991). The data are "resighting histories" of animals at discrete sampling periods and location categories ("habitats"). For example, "1023102", "0231111" represents 2 observation histories over 7 sampling periods (the columns) where "0" signifies "not observed", "1" is "observed in habitat 1", "2" is "observed in habitat 2", etc. Hestbeck et al. (1991) modeled these as multinomial likelihoods, the probability of each history being represented as the product of transition and sighting probabilities, for example:

$$\Pr\{112 \mid \text{Released in time 1, habitat 1}\} = \Phi_{111} p_{21} \Phi_{212} p_{32},$$

where

p_{tj}= the probability that an animal present in habitat j (j=1,..., J habitats) during period t (t=1,..., T sample periods) is observed in that period;

Φ_{tjk} = the probability that an animal alive and present in habitat j during period t survives and is present in habitat k during period t+1. Under certain assumptions (independence of S_{tj} and Ψ_{tjk}) the transition probabilities (Φ_{tjk}) can be written as products of survival and movement probabilities :

$$\Phi_{tjk} = S_{tj}\, \Psi_{tjk} \qquad \textbf{(1)}$$

where

S_{tj} is the probability that an animal alive in habitat j at period t survives to period t+1, and

Ψ_{tjk} is the probability that an animal in habitat j during time t moves to habitat k, given that it survives.

The observed resighting histories can be used together with these probabilities to estimate the parameters, using the numerical methods of White (1983). Hestbeck et al. (1991) apply this approach to Canada geese (*Branta canadensis*) neck-collared and resighted in 3 wintering areas over 3 years. The Hestbeck et al. (1991) design deals with broad geographic areas (e.g., flyways), but conceptually the approach works equally well for discrete habitats.

The assumptions required of the Hestbeck et al. (1991) approach are (1) that each animal is observed in only 1 habitat per sampling period, (2) time- and habitat-specific sighting and transition probabilities are equal for all animals in a particular habitat and sampling period, (3) the usual assumptions about independence, mark loss, and instantaneous sampling apply, and (4) that any loss to the population is permanent (death or emigration). With the exception of (1) these are standard Jolly-Seber assumptions (Seber, 1982, p. 196). Unlike the recovery models, animals are explicitly allowed to move among habitats *between* sampling periods; they are assumed to be in only 1 habitat *within* a sampling period. Movement among habitats, although not of primary interest in this paper, is an important biological parameter, and also can be estimated. Hypotheses about the survival differences related to habitat can be directly tested, via likelihood-ratio.

There are potential drawbacks to the mark-resighting approach for inference about habitat-specific survivorship. First, unless sampling periods are very short or habitats are very coarse-grained, it is unlikely that animals will remain in a single habitat. Second, the meaning of habitat-specific survival estimates is not clear if animals move among habitats *between* sampling periods. In the Hestbeck et al. (1991) example, birds were relatively sedentary with respect to the rather large geographic "regions"; region-specific inference about survival was probably reasonable. Also, single parameters are used to estimate constant survivorship probability between sampling periods. In many situations (e.g., wintering birds) mortality risk is *not* constant over time, but rather occurs in "bursts" over high-risk periods. Further, unless sampling occasions are numerous (e.g., every few days, or more frequently), meaningful data on either changing mortality risk (possibly related to habitat) or habitat use are unlikely to be obtained. The mark-resighting approach also requires estimation of many parameters in such a case. Because both mortality risk, and decisions about habitat use occur continuously through time, and not as discrete packages, it is logical to pursue data structures and models that treat both phenomena as functions of continuous time.

2.3 Continuous multiple encounter data

This approach treats mortality risk as an instantantaneous process, even though the data are usually discrete at some level of resolution (e.g., a day). For each individual in the sample, the data consist of a "time-to-failure", usually the number of elapsed days (or other time units) between release and the eventual death of the animal. In its simplest form, the approach assumes that all animals are followed until death, but various kinds of "censoring" are also allowed. Applications in animal ecology generally involve the use of radio telemetry to continuously monitor individual animals (Conroy, Costanzo, and Stotts, 1989; Pollock et al., 1989a,b). Censoring occurs because an individual survives to the end of the time interval of interest (e.g., over winter), or is lost to the study before that time (e.g., because of radio failure).

A variety of models have been developed to enable inference from failure time data. All share a fundamental connection between $S(t)$, the probability of survival from initial time t_0 to time (e.g., $t_0=0$), and $h(t)$, the conditional probability of death over the next very short time interval, given that the subject is alive at t:

$$S(t)=\exp\left[-\int_0^t h(u)\,du\right] \tag{2}$$

Parametric models assume an underlying mathematical form for the hazard rate. For example, if it is assumed that hazard is constant over time, then $h(t)= h$, and:

$$S(t) = \exp(-ht) \tag{3}$$

which is the exponential model for survival.

Nonparametric models such as the Kaplan-Meier (Kaplan and Meier, 1958) approach (e.g., Pollock et al., 1989 a,b) assume no underlying form for the hazard, but instead estimate a survival function from observed failure times. A "semi-parametric" model for hazard is the Cox proportional hazard model (Cox, 1972,1975), which assumes a baseline hazard function, $h^{(0)}(t)$, but does not specify its form. The observed hazard for each animal ($h_i(t)$) is proportionally related to the baseline hazard, dependent on predictors $\underline{x}'_i$, which may be either continuous or discrete (classification) factors:

$$h_i(t) = h_0(t)\exp[\underline{\beta}'\underline{x}_i] \tag{4}$$

where $i=1, ...I$ animals, and $t \in [0,\infty]$.

Note that $\underline{x}'_i$ is not time-dependent, so that each animal has a covariate vector that is fixed over the duration of the study. The hazard may vary with time, but always in the same (proportional) manner. As a special case, if x_i has values 0 or 1, then the hazard is:

$$h_i(t) = \begin{cases} h^{(0)}(t), & x_i=0 \\ h^{(0)}(t)\exp(\beta), & x_i=1 \end{cases} \tag{5}$$

Cox (1972, 1975) generalised this model to allow for time-dependent covariates, in which the hazard for each individual may change over time, resulting in time-dependent shifts in hazard in addition to the specified time-variation in the baseline hazard. Thus:

$$h_i(t) = h^{(0)}(t)\exp[\underline{\beta}'\underline{x}_i(t)] \tag{6}$$

In the present context, I propose $\underline{x}'_i(t)$ as a vector of classification variables, such that $x_{ij}(t) = 1$ if the ith animal is present in the jth habitat at time t, $x_{ij}(t) = 0$ otherwise, and the presence in habitats $j=1, ...J$ at time t is mutually exclusive. For the special case of 2 habitats, this gives:

$$h_i(t) = \begin{cases} h^{(0)}(t), & x_i(t)=0 \\ h^{(0)}(t)\exp(\beta) \equiv h^{(1)}(t), & x_i(t)=1 \end{cases} \tag{7}$$

where habitats are mutually exclusive.

This model contains 1 parameter, ß, which expresses the relationship of the observed hazard to the baseline hazard, through the habitat covariate ($x_i(t)$). When an animal is in habitat 0, the hazard is the baseline hazard ($h^{(0)}(t)$) for that time; when it is in habitat 1, the hazard ($h^{(1)}(t)$) is proportional to the baseline hazard:

$$h^{(1)}(t)/h^{(0)}(t) = \exp(\beta) = \gamma \qquad (8)$$

3. Ability to detect a habitat effect on survival

A key issue to consider, given that a true difference in mortality risk (hazard) because of habitat occurs, is whether one is likely to detect such an effect given observational data. In the context of the proportional hazards model described above for 2 habitat classes, this is equivalent to rejecting H_0:ß=0 or γ=1, versus a specified alternative ß or γ, where $h^{(1)}(t)$, $h^{(0)}(t)$, and γ are defined in expressions (7) and (8).

3.1 Methods

I examined the ability of proportional hazards testing to detect specified habitat effects on hazard, via Monte Carlo simulation. I chose a scenario in which N animals were continuously monitored over a 90-d period. I specified a baseline hazard ($h^{(0)}(t)$) which increased linearly for the first 30 d, was constant for days 30-60, then decreased at the same rate as the increase over the first 30 d. The shape of this hazard function is arbitrary, but resembles the hazard one might anticipate in a wintering animal (e.g. Conroy et al., 1989). Initially, I chose values for the slope and intercept of the hazard function to provide a baseline survival over 90d ($S^{(0)}(90)$) of 0.50 (Fig. 1) ; later I also generated baseline hazards providing $S^{(0)}(90)$=0.75. Hazard was obtained as function of habitat (0 or 1) at each day by expression (7), using a range of assumed proportional effects of habitat. Each animal's habitat use was time-dependent as described below. Each animal's fate was determined at each day by comparing a uniform random deviate to the hazard probability. If death did not occur by day 90, the animal was right-censored at day 90; no censoring occurred before day 90.

Habitat usage at each day was simulated as a Markov process, in which animals had an initial probability of selecting a habitat (π) and Markovian probabilities of either remaining (p_{11}, p_{00}) or switching ($p_{10}=1-p_{11}$, $p_{01}=1-p_{00}$) habitats. I chose values π=0.5 or π=0.75, and of $p_{10}=p_{01}$=0.5, 0.2, and 0 representing a range of daily transition probabilities from "random" (p_{10}=0.5) to perfect "fidelity" (p_{10}=0).

I considered a range of possible effects of habitat on survival, ranging from no effect (H_0) to an effect resulting in 90-day survival ($S^{(1)}(90)$) in habitat 1, the "better" habitat, of 0.75, compared to a baseline survival ($S^{(0)}(90)$) of 0.50 in the "poorer" habitat, corresponding to γ in the range 1 (no effect) to 0.42 (maximal effect). For the case in which $S^{(0)}(90)$=0.75, these same ranges in γ gave rise to $S^{(1)}(90)$ in the range 0.75 (no effect) to 0.89 (maximal effect).

Initially I examined a range of sample sizes (N) of marked animals of 25 to 100, which brackets some of the largest studies of radio-marked animals conducted; I later

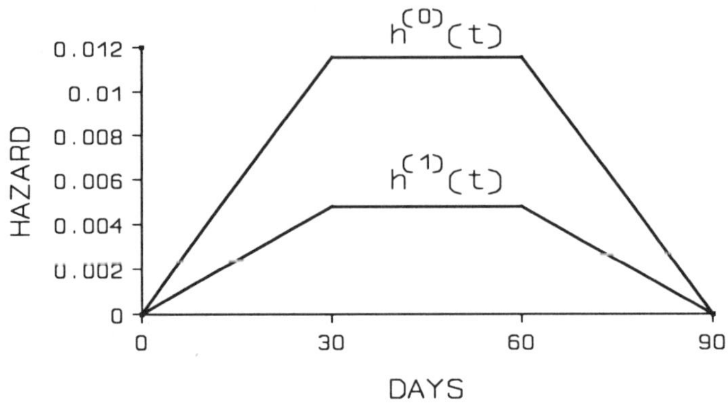

Figure 1. Example hazards $h(t)$ to simulate death times from time-specific habitat useage, maximal habitat effect for habitat 0 vs. habitat 1 (see text).

conducted simulations of up to N=300. For each trial (combinations of N, γ, $S^0(90)$, π, and p_{10}) I conducted 250 replications. At each replication, I tested H_0: $\beta=0$ ($\gamma=1$) using PROC PHREG in SAS. The time-dependent predictor was the habitat (0 or 1) in which each animal occurred, at each failure time (time at which a death occurred). Power was estimated as the proportion of simulations in which rejection of Wald's test (SAS Inst., 1991) occurred at various alpha levels. I also conducted tests using PROC LIFETEST and a log-rank statistic (SAS Inst., 1988) in which the predictor (time invariant) was the habitat in which each animal occurred at that animal's death time. All simulations were conducted using SAS 6.07 for a Unix operating system on a SunSparc workstation.

For comparison, I also conducted simulations in which the test of habitat effect was conducted using a discrete model similar to that of Hestbeck et al. (1991). I simulated death times as above, but divided the 90-d interval into 8 sub-intervals of approximately 11d each. Within each 11-d interval, the habitat occurrence (again, simulated as above) of each animal was the habitat in which the animal occurred on the approximate mid-point (day 5) of each interval. An 8-period resighting history was thus determined for each animal. I assumed that each animal's fate and habitat location was known at each of these sub-intervals. Thus, the sampling situation corresponded to a re-sighting experiment in which habitats are visited periodically (e.g. every 1-2 weeks) and each animal's location and fate (alive or dead) determined at each visit (resighting probabilities = 1). I used a cohort analysis and program SURVIV (White, 1983) to test H_O:$S_{t0}=S_{t1}$, where S_{t0}, S_{t1} are the probabilities of survival for animals alive in habitats j=0,1 at time t, over the interval [t, t+1] (expression 1). For these simulations only, PC-SAS Version 6.03 was used on a SWAN 386/33; each trial was replicated 100 times.

3.2 Results and discussion

Results presented here are for the α=0.10 level of rejection; power was also evaluated at critical levels of 0.05-0.25. Analyses of variance indicated that π, and p_{10} in the range 0.5-0.2 had little effect (P>0.10) on power of PHREG tests, whereas N, $S^{(0)}(90)$ and γ all influenced power. Plots of power versus N for different levels of γ at $S^{(0)}(90)$=0.50 (Fig. 2) suggest that power to detect all but very large hazard effects will not exceed 0.5 unless sample sizes are very large (N>200). For $S^{(0)}(90)$=0.75, power in excess of 0.5 was only achieved for large sample sizes and the greatest habitat effect considered (N>200, γ= 0.42). For a population with a 90-d baseline

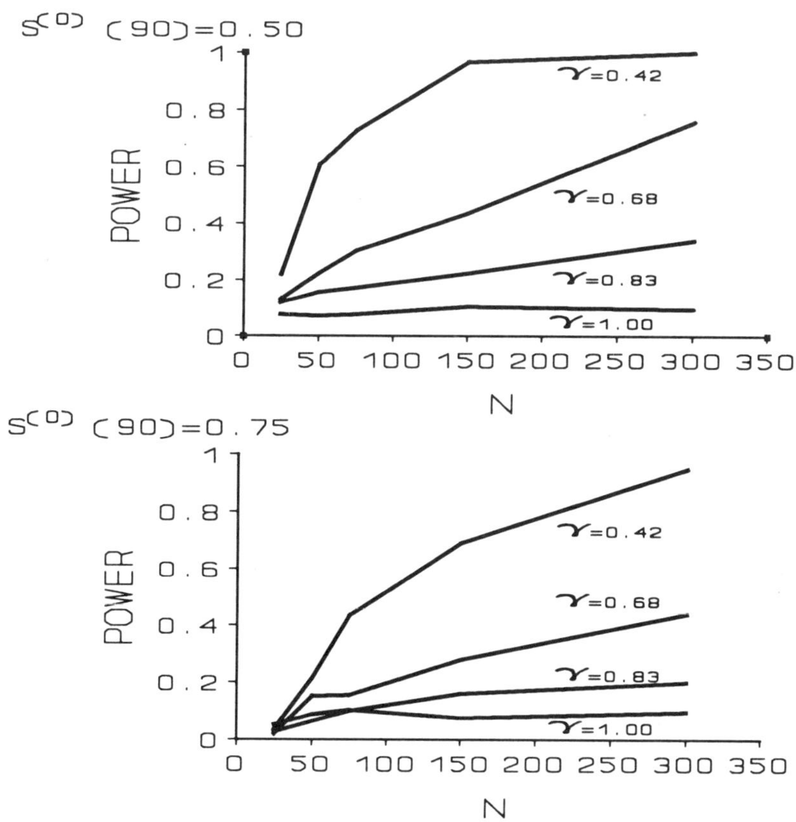

Figure 2. Estimated power to detect specified differences in hazard (γ) due to a time-specific habitat covariate, 0.10 critical level (see text).

survival of 0.5 (probably unrealistically low for many animal populations), a habitat effect enhancing survival to 0.625 (γ=0.68) or higher is detectable with powers of 0.3-0.5 for sample sizes likely to be achievable in telemetry studies (N=80-140). For populations with 90-d baseline survival of 0.75 (probably more realistic for many populations) similar habitat effects would require enormous investment of resources. In such cases, investigators might be willing to trade a higher risk of Type I error in exchange for greater power: power of 0.3-to 0.5 was achieved for the simulated $S^{(0)}(90)$=0.75 population for modest habitat effects (γ=0.83-0.67), at the α=0.25 level, for sample sizes of around 100 marked animals. Investigators making this choice would do so advised that Type I error probabilities of this magnitude ordinarily are not acceptable.

As noted, movement probabilities of 0.2-0.5 had little effect on estimated power for either PHREG or LIFETEST. However, I also examined an extreme situation in which movement probabilities were zero; that is, once having selected a habitat, an animal stayed in it for the 90-d period. Under this scenario, one would expect that PHREG and LIFETEST would perform similarly, the time-dependent nature of the covariate no longer being at issue. In general, this was borne out by the simulation results, with PHREG performing better than LIFETEST when movement probabilities were 0.5 (Fig. 3) and performance essentially identical when movement probabilities were 0. In contrast, similar performance of the SURVIV tests required N=300 or higher, probably because the time-varying hazard required a model in which 7 survival rates were estimated, even under H_o, compared to 1 hazard coefficient. However, this is not a comparison of the SURVIV estimation procedure versus PHREG/LIFETEST, but rather a reflection of the assumed coarse-grained sampling procedures.

4. Summary and recommendations

Hopefully, this presentation will have made clear the need for biologists to clearly define the temporal and spatial scales of the habitat-specific inferences they seek to address. This should lead one to then carefully consider the appropriate sampling and inferential methods needed, versus those possible or available. For very coarse spatio-temporal scales, and fairly general questions, ring-recovery data may be useful; certainly, this approach should not be discarded out of hand. For finer-scaled inference, more intensive monitoring schemes involving mark-resighting and telemetry are probably needed. These, particularly the latter, can be very demanding of resources. No matter the approach, investigators are encouraged to consider the probable outcome in terms of power, via simulation or other methods, before investing resources to address these important questions.

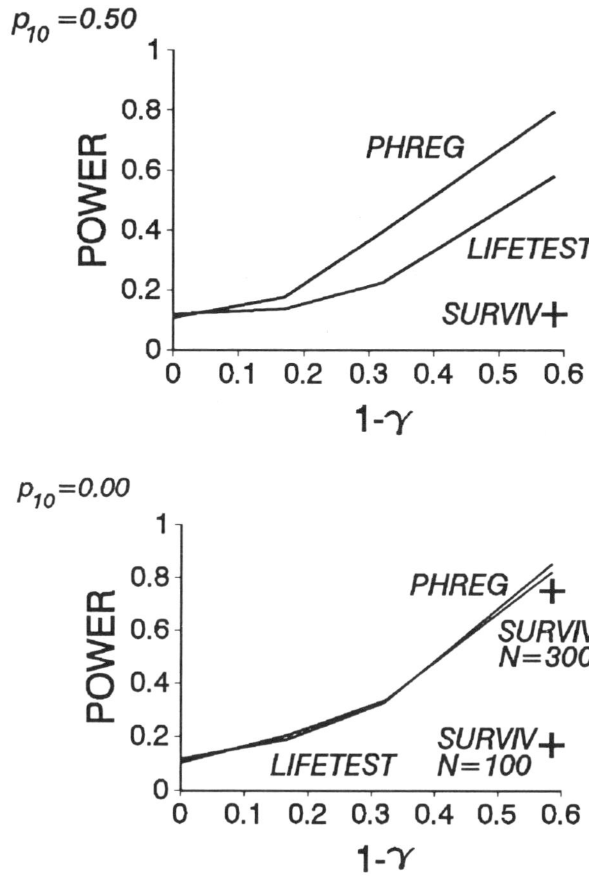

Figure 3. Estimated power to detect specified differences in hazard (γ) due to a time-specific habitat covariate. N =100, $S^{(0)}(90)$=0.50 (see text).

Acknowledgements. I thank Gary White for earlier hallway discussions that encouraged my further thoughts about proportional hazards modeling of habitat effects. I am indebted to Jon Anderson for use of the work stations in the Statistics Department of UGA; Phil Dixon and Jon Anderson for assisting me in setting up the simulations; Jim Hines for several clever ideas about programming; and Cathy Thomas, Sue Anthony, and Dave Smith for manuscript preparation. Jon Anderson, Jim Nichols, and Dave Smith provided helpful comments on earlier drafts. The Georgia Cooperative Fish and Wildlife Research Unit is jointly sponsored by the U.S. Fish and Wildlife Service, the University of Georgia, the Georgia Department of Natural Resources, and the Wildlife Management Institute.

References

Brownie, C., Anderson, D.R., Burnham, K.P., and Robson, D.S. (1985).- Statistical inference from band recovery data - a handbook. US Fish and Wildlife Service, Research n°156, 2nd edition, Washington, DC. 305pp.

Conroy, M.J., Costanzo, G.R., and Stotts, D.B. (1989).- Winter survival of female American black ducks on the Atlantic Coast. Journal of Wildlife Management, 53: 99-109.

Cox, D.R. (1972).- Regression models and life tables (with discussion). Journal of the Royal Statistical Society, B34: 187-220.

Cox, D.R. (1975).- Partial likelihood. Biometrika, 62: 209-276.

Gavin, T. (1991).- Why ask "why": the importance of evolutionary biology in wildlife science. Journal of Wildlife Management, 55: 760-768.

Hestbeck, J.B., Nichols, J.D., and Malecki, R.A. (1991).- Estimates of movement and site fidelity using mark-resight data of wintering Canada geese. Ecology, 72: 523-533.

Kaplan, E.L. and Meier, P. (1958).- Nonparametric estimation from incomplete observations. Journal of the American Statistical Association, 53: 457-481.

Mardia K.V. (1967).- A non-parametric test for the bivariate two-sample location problem in the normal case. Journal of the Royal Statistical Society, B30: 83-92.

Murphy, D.D. and Noon, B.R. (1991).- Coping with uncertainty in wildlife biology. Journal of Wildlife Management, 55: 773-782.

Pollock, K.H., Winterstein, S.R., Bunck, C.M., and Curtis, P.D. (1989a).- Survival analysis in telemetry studies: the staggered entry design. Journal of Wildlife Management, 53: 7-15.

Pollock, K.H., Winterstein, S.R., and Conroy, M.J. (1989b).- Estimation and analysis of survival distributions for radio-tagged animals. Biometrics, 45: 99-109.

Pulliam, H.R. (1988).- Sources, sinks, and population regulation. American Naturalist, 132: 652-661.

SAS Institute (1988).- SAS Technical Report: P-179. Additional SAS/STAT procedures. SAS Institute, Inc. Cary, NC 255pp.

SAS Institute (1991).- SAS/STAT Software: the PHREG procedure. Preliminary documentation. SAS Institute, Inc. Cary, NC 59pp.

Schwarz, C.J., Burnham, K.P., and Arnason, A.N. (1988).- Post-release stratification in band-recovery models. Biometrics, 44: 765-785.

Seber, G.A.F. (1982).- The estimation of animal abundance and related parameters. 2nd edition Griffin, London 654 pp.

Van Horne, B. (1983).- Density as a misleading indicator of habitat quality. Journal of Wildlife Management, 47: 893-901.

White, G.C. (1983).- Numerical estimation of survival rates from band-recovery and biotelemetry. Journal of Wildlife Management, 47: 716-728.

Marked Individuals in the
Study of Bird Population
J.-D. Lebreton and P.M. North (eds.)

Survival and Habitat Utilisation: A Case for Compositional Analysis

N.J. Aebischer[1], V. Marcström[2], R.E. Kenward[3] and M. Karlbom[2]

[1]*The Game Conservancy, Fordingbridge, Hampshire, SP6 1EF, United Kingdom,* [2]*Institute of Zoophysiology, Box 560, 751 22 Uppsala 1, Sweden and* [3]*Institute of Terrestrial Ecology, Furzebrook Research Station, Wareham, Dorset, BH20 5AS, United Kingdom.*

Summary. Avian survival rates can vary not only by age and year, but also in relation to individual habitat use. Radiotelemetry makes it possible to determine repeatedly the location of a tagged individual. Its habitat use is then given by the proportion of radiolocations in each habitat type. Radiotelemetry is also a direct method of establishing whether a marked bird is still alive. The importance of different habitat types for survival may be assessed by compositional analysis. This technique recognizes that the sample size is the number of tagged animals, and that radiolocations serve to subsample each individual's habitat use. It allows for the analytical complication that, for a given individual, the proportions describing habitat use sum to one over all habitat types. The principles of compositional analysis are explained and demonstrated by reference to survival and habitat data collected for radiotagged pheasants on the Swedish island of Gotland in 1984 and 1985. There was no effect of age or year on survival, but pheasants that survived made significantly greater use of vegetable crops and less use of pasture than those that died.

1. Introduction

The estimation of survival rates is an important component in understanding the population dynamics of an animal population. Part of the process is untangling how survival rates vary in relation to intrinsic and extrinsic factors. For example, it has long been recognized that avian annual survival rates change with age, an intrinsic factor, and also with year, an extrinsic one. Accordingly, many analytical methods have been developed to take these two factors into account (Seber, 1970, 1971; Brownie *et al.*, 1985; Aebischer, 1988; Pollock *et al.*, 1990).

Another source of variation in survival rates is the habitat occupied by different individuals of the same species. On a very broad scale, Newton (1989) found that territory quality was a major factor associated with sparrowhawk *Accipiter nisus* longevity. Survival of blue tits *Parus caeruleus* was lower in an isolated wood than in one adjacent to human habitations (Dhondt, 1989), and Brittas *et al.* (in press) suggested that differences in the survival of pheasants *Phasianus colchicus* at two sites was linked to habitat differences.

Modern radiotracking techniques (e.g. Kenward, 1987; White and Garrott, 1990) make it possible to measure an individual's habitat use much more precisely than in the past. They provide a means of accurately and repeatedly locating the marked animal. The distribution of radiolocations across a range of narrowly defined habitat types determines habitat use at the level of each individual. Radiotagging also provides a means of determining remotely whether an animal is still alive; if it dies, the body can be found and, often, the cause of death can be established (Kenward, this volume). This paper addresses the question of how to relate an individual's habitat use to its survival based on radiotracking data.

2. Analytical pitfalls

Radiotracking data are in the form of a number of locations situated in a variety of habitat types, for a certain number of marked individuals. For many of them, the time and cause of death is known. The animals may also be classified according to groupings such as age or sex.

In analysing such data, it is tempting to classify the radiolocations by habitat type, survival of the animals and any other groupings, then use a chi-square test or log-linear modelling to assess differential habitat use (e.g. Byers *et al.*, 1984; Heisey, 1985). However, these procedures are fundamentally flawed because they assume that the radiolocations are independent observations. This is only true if each location comes from a separate animal (and the animals are not aggregated). If not, the assumption is wrong at two levels: (1) locations collected for the same individual are likely to be serially correlated (Swihart and Slade, 1985), (2) variation in individual behaviour, caused by each animal's idiosyncratic preferences and past history, means that locations cannot be pooled across individuals. Increasing the sampling time between locations may avoid (1), but (2) is intractable. Furthermore, the greater the number of locations, the more the determination of an animal's habitat use is accurate, but the more correlated are the data. Hence avoiding (1) is counterproductive.

The problem is one of sampling and subsampling. The radiolocations from an individual constitute a subsample of that animal's trajectory through time and space (Aebischer *et al.*, in press). Its habitat use is defined by the proportions of the trajectory contained within each habitat type, and is estimated by the distribution of its radiolocations. The habitat types are usually types of surface cover (e.g. pasture, cropland, woodland), but can also be three-dimensional (e.g. understory, shrub layer, canopy). The mean habitat use by the population is that averaged over all individuals in the population, and the amount of variability is assessed from the variation between individuals.

In an analysis of habitat use, the sample size is therefore the number of individuals, and the unit of data is multivariate, comprising as many values (proportions) as there are habitat types.

A special feature of these data is that for each animal, the proportional habitat use sums to one, i.e. the proportion corresponding to one habitat type is not independent of the proportions relating to the others. It can therefore be misleading to analyse the data for each habitat type in isolation, even apart from the difficulty in determining levels of statistical significance due to multiple testing. What is required is first a global test of differential habitat use by groups of individuals, then, if significant, a means of pinpointing where the differences lie. This is achieved by compositional analysis (Aitchison, 1986).

3. Compositional analysis

A full description of the application of compositional analysis to the general analysis of habitat use is given in Aebischer *et al.* (in press), but the principles are summarised below.

Given D habitat types, an individual's proportional habitat use is estimated by $x_1, x_2, \ldots, x_D$ where x_i is the proportion of radiolocations in habitat i. As described above, $x_1+\ldots+x_D = 1$. Aitchison (1986) called such a set of components a composition, and showed that for any component x_j, the logratio transformation $y_i=\ln(x_i/x_j)$ $(i=1, \ldots, D, i\neq j)$ renders the y_i linearly independent.

An overall difference in habitat use between groups of individuals may be tested by multivariate analysis of variance (Chatfield and Collins, 1980; Kendall, 1980) applied to the logratios y_i. This involves fitting by least squares a multivariate model of the form $\mathbf{Y}=\mathbf{A\Theta}+\mathbf{E}$, where $\mathbf{Y}$ is the logratio data matrix (D-1 columns, one row per marked animal), $\mathbf{A}$ is the regressor matrix, $\mathbf{\Theta}$ the parameter matrix and $\mathbf{E}$ the error (or residual) matrix.

The overall test of significance is based on comparing the fit of two models using, for instance, a generalized likelihood ratio statistic Λ. The first model M_1 with m_1 parameters corresponds to the hypothesis to be tested (differential habitat use), the second one M_2 with $m_2<m_1$ parameters corresponds to the null hypothesis (identical habitat use). If $\mathbf{E}_i$ is the residual sum-of-squares and cross-product matrix resulting from fitting model M_i $(i=1, 2)$, then Λ is the ratio of determinants $|\mathbf{E}_1|/|\mathbf{E}_2|$. Assuming multivariate normality of the residuals, $-N\ln\Lambda$, where N is the number of individual animals, follows an approximate $\chi^2(m_1-m_2)$ distribution (Cox and Hinkley, 1974). If the residuals are not multivariate normal, the distribution of Λ may be derived empirically by data randomisation (Edgington, 1980). The equality of covariance matrices is central to both testing procedures; in particular the latter requires the permutability of the distribution over individuals under the null hypothesis.

Aitchison (1986) demonstrated that the value of Λ does not depend on the component x_j chosen as denominator in the logratio transformation, because the procedure is equivalent to centring the $\ln(x_i)$ data matrix by rows. This means that any one of the D components may be used arbitrarily as the denominator (provided that the same one is used for all individuals), without prejudice to the outcome of the statistical test.

If the test reveals significant differences in habitat use with respect to a factor, for instance survival, it is relevant to ask where the differences lie and attempt to order the habitat types according to their survival value.

To locate the differences, it is necessary to compare the mean logratios for animals that survive with the matching logratios for ones that die (Manly, 1985). With D components, there are D-1 logratios. Using component D as denominator in the logratio transformation, let $s_{1,D}, \ldots, s_{D-1,D}$ represent the mean logratios for the survivors, and $m_{1,D}, \ldots, m_{D-1,D}$ those for the animals that suffer mortality. The inequality $s_{i,D}-m_{i,D}>0$ implies that relative to habitat D, animals that survive make greater use of habitat i than those that die. Equivalently, animals that survive make less use of habitat D relative to habitat i than those that die. If $s_{i,D}-m_{i,D}>0$ for all

$i=1, \ldots, D$-1, then animals that survive make less use of habitat D relative to all other habitat types than those that die, i.e. habitat D is the habitat type relatively least used by survivors. Conversely, $s_{i,D}-m_{i,D}<0$ for all i would imply that habitat D was the habitat type relatively most used by survivors. The argument holds whichever component is used as denominator.

As a result, habitat types can be ranked according to their relative use by survivors by calculating the differences $d_{i,j}=s_{i,j}-m_{i,j}$ for all $i,j=1, \ldots, D$, $i \neq j$, where each component j is in turn used as denominator. The $d_{i,j}$ can be regarded as elements of a DxD matrix with a missing diagonal. The rows of the matrix are indexed by the habitat type used as numerator in the logratio, and the columns by the denominator. Because $s_{i,j}=-s_{j,i}$ and $m_{i,j}=-m_{j,i}$, $d_{i,j}=-d_{j,i}$ and the matrix is antisymmetric. It follows from antisymmetry and the independence property of logratios that each element is independent of the others in the same row or same column. The number of positive elements in each row is an integer between 0 and D-1 which ranks the habitats in order of increasing relative use by survivors: 0 is "worst" and D-1 is "best".

Each element in the ranking matrix is a mean difference for which a standard error can also be calculated. The ratio mean/standard error gives a statistic that can be used to test whether apparent differences are significant, thereby pinpointing where the differences in relative habitat use occur. Its significance levels can be determined by reference to a t distribution if the logratios are normally distributed, or by randomisation if not. Although this implies $D(D$-1) tests, they are not independent (only those within a same row or same column are). If Λ has already indicated that habitat use differs overall between animals that survive and those that die, we recommend applying a standard level of significance by analogy with the protected least-significant-difference method (Carmer and Swanson, 1973; Snedecor and Cochran, 1980).

An alternative method of ordering the habitat types according to survival value is based on the duality between multivariate analysis and discriminant analysis (e.g. Chatfield and Collins, 1980). Let $\mathbf{E}_1$ be the residual logratio matrix corresponding to the hypothesis of differential habitat use by birds that survived and those that died, and $\mathbf{E}_2$ the one calculated assuming no differential use. The solution $\mathbf{a}$ to the matrix equation $\mathbf{E}_1(\mathbf{E}_2)^{-1}\mathbf{a}=\lambda\mathbf{a}$ is an eigenvector whose coefficients $a_1, \ldots, a_{D-1}$ represent the relative survival value of the D-1 logratios; geometrically, they are proportional to the orthogonal projection of the logratios onto the axis linking the centroids of the two groups of birds, i.e. $a_i=k(s_{i,D}-m_{i,D})$ for $i=1, \ldots, D$-1. To obtain coefficients which allow the habitat types to be ordered, $\mathbf{a}$ can be converted to centred logratios $c_1, \ldots, c_D$ by the transformation $c_D=-(\Sigma a_i)/D$, $c_i=a_i+c_D$ ($i=1, \ldots, D$-1). The c_i ($i=1, \ldots, D$) represent scores that order the habitat types on a continuous axis; negative scores denote habitat types whose use is negatively linked to survival and vice versa.

The above data manipulations, model fitting and hypothesis testing based on multivariate normality are performed by most large statistical packages for microcomputers, for instance SYSTAT 5.0 (Wilkinson, 1990). The tests presented here that are based on data permutation were programmed in GENSTAT 5 (Genstat 5 Committee, 1987).

4. Example

The methodology is illustrated using a subset of data from a pheasant study on the Swedish island of Gotland. During the pre-breeding period (February-April) in 1984 and 1985, each of 78 birds was located three times a day during a 10-day tracking session. There were at most two sessions per bird over the three months. The main cause of death was predation by raptors (especially goshawks *Accipiter gentilis*): 19 pheasants were killed by raptors less than one month after the last tracking session, whereas 59 survived beyond the end of May. Our question was whether raptor predation was related to the pheasants' habitat use. For each bird, habitat use was determined by the distribution of its radiolocations across 4 habitat types, using RANGES IV (Kenward, 1990). The habitat types were (1) pasture, (2) woodland, (3) vegetable crops (including potatoes, legumes and soft fruit), and (4) the remainder. As the sample of pheasants not only covered two years, but also involved members of either sex belonging to one of two age classes (first-year or older), there were three additional factors which needed to be taken into account.

The first step for compositional analysis was to transform the four-part composition describing each pheasant's habitat use to three logratios; the fourth habitat type was used as the denominator in the transformation. Some pheasants did not make use of pasture, and others were not found in vegetable crops (only three were in neither). These zero proportions were replaced by 0.001 to enable logarithms to be taken; in biological terms it was equivalent to admitting that use was possible, but that 30 radiolocations were too few to detect it.

The next step was to establish whether habitat use changed in relation to any of the four factors, year, age, sex or death. Model fitting proceeded in two stages. The full model M_1 initially included all four factors and their interaction terms, and was compared to a model M_2 without interaction terms (null hypothesis: no interactions between factors). Model M_2 was then compared to four restricted models from which each factor in turn was excluded (null hypothesis: the excluded factor had no effect). As visual inspection suggested that at least one of the logratios was not normally distributed, the level of significance of Λ was determined by reference to 999 randomisations of the data within the appropriate model structure. There was no evidence of interaction between factors ($\Lambda=0.712$, $P=0.316$), and the tests of differential habitat use with respect to each factor gave $\Lambda=0.980$ ($P=0.710$) for year, $\Lambda=0.961$ ($P=0.454$) for age, $\Lambda=0.961$ ($P=0.474$) for sex and $\Lambda=0.810$ ($P=0.005$) for death. It was clear that habitat use differed with respect to only one factor, namely whether or not the pheasant was killed by a raptor. Ignoring all factors except death and testing for differential habitat use yielded $\Lambda=0.819$ ($P=0.002$).

Given that surviving birds used the habitat differently from those that died, the next step was to find out where exactly the difference lay by constructing a ranking matrix based on mean pairwise differences between logratios. The matrix of mean logratio differences over their standard errors is given in Table 1, as is a simplified matrix where the differences have been highlighted by replacing positive values by "+", negative ones by "-" and statistically significant ones by "+++" and "---" respectively. Relative to all other habitat types, pheasants that survived made

Table 1. Pinpointing differences in relative habitat use, and ranking habitat types according to their survival value to pheasants. (a) matrix of standardised logratio differences (mean logratio differences over the standard error of the difference), and (b) simplified matrix where each value in (a) is replaced by its sign and tripled if significant at $P<0.05$.

(a) Matrix of standardised logratio differences

Numerator	Denominator				Rank
	Pasture	Woodland	Vegetables	Other	
Pasture		-2.41	-3.29	-2.03	0
Woodland	2.41		-2.18	0.31	2
Vegetables	3.29	2.18		3.65	3
Other	2.03	-0.31	-3.65		1

(b) Simplified matrix

Numerator	Denominator				Rank
	Pasture	Woodland	Vegetables	Other	
Pasture		---	---	---	0
Woodland	+++		---	+	2
Vegetables	+++	+++		+++	3
Other	+++	-	---		1

significantly greater use of vegetable crops and significantly less use of pasture. The habitat ranking defined by the number of positive values in each row was Pasture <<< Other < Woodland <<< Vegetables (where <<< denotes a significant difference between two adjacent ranks).

The scoring method of ordering habitat types in terms of survival value gave, based on the logratios, the coefficients -1.501, 1.706 and 0.146. Transformed to centred logratios, the scores for Pasture, Vegetables, Woodland and Other were -1.588, 1.618, 0.058 and -0.088 respectively. The corresponding ranking was Pasture < Other < Woodland < Vegetables, as found by the matrix method.

5. Assumptions

There are four main assumptions underlying the approach outlined above.

5.1. The estimate of habitat use obtained from the distribution of radiolocations is representative of true habitat use

This is essentially a sampling problem. The regime used to determine the position of a radiotagged individual must ensure that all parts of the animal's trajectory have an equal chance of being sampled. This requires either random sampling, or else sampling at regularly spaced intervals; in the latter case, the interval must be such that it does not sample repeatedly the same point of any activity cycle that an animal may exhibit (such as a 24-hour one). It assumes also that wearing the transmitter does not alter the habitat use of the marked animal, and that neither does having its position determined (although in our example, any bias should have been the same for all categories).

5.2. Each individual provides an independent measure of habitat use

Multivariate analysis of variance requires independence among the (multivariate) data items. Lack of independence could arise through monitoring animals that move around together, in a flock or herd. In this case, the sampling unit is the group of animals, not the individuals. It could also arise through monitoring the same individual over several years, and counting each year as a separate source of data. This is acceptable only if it can be shown that within-individual variation is the same as between-individual variation; in the majority of cases, this is unlikely to be the case.

5.3. The logratios must follow a multivariate normal distribution if standard tests of significance are used

As mentioned already, if this condition does not hold, statistical tests may still be performed and evaluated by means of data randomisation. This is the procedure adopted here, even though it was found that the levels of significance obtained assuming multivariate normality were remarkably similar to those obtained through randomisation, and would have resulted in exactly the same conclusions. It must be noted, though, that the assumption of a constant underlying covariance matrix for all individuals remains.

5.4. Each individual makes use of all habitat types

Ideally, none of the proportions describing any individual's habitat use should be zero, so that a situation of dividing by zero or taking a logarithm of zero never arises when applying the logratio transformation. In practice, as seen in the example, the situation does arise, especially when the number of radiolocations is small and the number of habitat types is large.

One rather drastic way of coping with the problem is to eliminate all individuals that do not use all habitat types. Another method is to group little-used habitat types with others, so reducing the number of habitat types. Often, judicious grouping results in relatively little loss of information for a relatively large (but not necessarily total) loss of null proportions. The approach adopted here is to view any remaining zero proportions as a failure of the sampling regime to detect that a habitat type is used when the use is very small, and to replace the zero with a small number. The replacement value must be smaller than the smallest non-null proportion recorded so as to represent effectively a use too small to be detected.

In the case of the Gotland pheasants, the smallest recorded proportion was 0.033, so zeros were replaced by 0.001, a value an order of magnitude smaller. The effect on the conclusions of choosing 0.001 as the replacement value was examined by repeating the analysis with replacement values of 0.01, 0.0001 and 0.00001 (Table 2). In all cases, the test for differential habitat use by pheasants which survived and those that died was significant at $P<0.01$. Again in all cases, the ranking procedure ranked vegetable crops as highest and pasture as lowest in survival value. The only change was in the placing of woodland versus other; the difference was cosmetic as their ranks never differed significantly.

Table 2. Analysis of the Gotland pheasant data when a habitat proportion of 0 (habitat type not used) is replaced by 0.01, 0.001, 0.0001 and 0.00001. (a) Overall test of differential habitat use by birds that survived and those that died; (b) habitat rankings (<<< indicates a significant difference in rank at P<0.05).

(a) Overall test of differential habitat use

Replacement value	Test for differences between pheasants that died and those that survived	
0.01	$\Lambda=0.839$	$P=0.006$
0.001	$\Lambda=0.819$	$P=0.002$
0.0001	$\Lambda=0.814$	$P=0.001$
0.00001	$\Lambda=0.812$	$P=0.001$

(b) Habitat ranking

Replacement value	Habitat ranking in terms of survival value (in order of increasing rank)						
0.01	Pasture	<<<	Woodland	<	Other	<<<	Vegetables
0.001	Pasture	<<<	Other	<	Woodland	<<<	Vegetables
0.0001	Pasture	<	Other	<	Woodland	<<<	Vegetables
0.00001	Pasture	<	Other	<	Woodland	<<<	Vegetables

There remains the possibility that a bird did not use a certain habitat type because that habitat type was, for some reason, not available to it. Reference to a map of the study area shows if this is the case. Within the framework of compositional analysis, the technique to apply in such a situation is considerably more complicated than those described above, and is explained in Aebischer *et al.* (in press).

Another analytical technique that is applicable to compositional data containing zero proportions is correspondence analysis (Benzécri, 1973; Hill, 1973). This accounts for the unit-sum constraint by reciprocal averaging of rows and columns (bi-centring). The logarithmic transformation of compositional analysis is replaced by a linear transformation, which accepts null proportions, but which may suffer from curvature problems (Hill and Gauch, 1980). Correspondence analysis can be extended to MANOVA-type models, testing being carried out by permutation (Ter Braak, 1987; Lebreton *et al.*, 1991).

6. Discussion

There are several theories of avian population dynamics in which an interaction of survival with habitat use plays an important role. For instance, Watson and Jenkins (1968) postulated that territoriality forces many red grouse into suboptimal habitat where they die. Brown (1969) proposed the 'buffer zone' theory of population regulation, whereby optimal habitat is occupied first, then suboptimal habitat, then any remaining individuals become floaters. Although Brown considered optimality in relation to breeding success, it is equally applicable in terms of the likelihood of survival. Part of understanding population processes is therefore knowing if and how survival varies with habitat use. From the practical point of view, such knowledge may have useful implications for species conservation or management (Bernstein *et al.*, 1991).

The case for compositional analysis when relating survival to habitat use rests on the following statistical considerations.

(1) The populations being sampled are made up of the individuals that survived and those that died. The sample size is the number of individuals in each category. The variable to be measured is habitat use, which is multivariate if there are more than one habitat types.

(2) An individual's habitat use is given by the proportion of time it spends in each habitat type, or, equivalently, the proportion of its trajectory through time and space contained in each habitat type. In the absence of continuous monitoring, these proportions may be estimated by the distribution of radiolocations of tagged individuals. The radiolocations, which constitute within-individual subsampling, are not suitable for testing habitat differences at the population level.

(3) The proportions describing habitat use are not independent, because they sum to one for each individual. It is therefore incorrect and misleading to analyse them separately.

Compositional analysis, by transforming the sets of non-independent proportions to sets of unconstrained logratios, resolves (3). By applying multivariate analysis of variance to test for differences in habitat use between animals that survived and those that died, and also to take into account any other potentially influential factors such as age or sex, it correctly addresses the structure of the data and of the population.

There are several implications for the design of radiotracking experiments (see also Kenward, in press). In particular, the sample size, i.e. the number of marked animals, must be adequate to ensure that mean habitat use is determined with sufficient accuracy. We recommend tagging at least 30 individuals if approximately two-thirds are expected to survive the study period, and more if the proportion surviving is likely to be higher. The detection of differences in habitat use in relation to survival also depends on the amount of individual variation in habitat use. The more the individuals in the same category differ in their habitat use, the greater the sample size necessary to reduce the error on the estimate of mean habitat use. There is obviously a trade-off between tagging more animals and collecting more radiolocations per animal. In our experience, it is more important to overcome individual variation than to increase the accuracy with which an individual's habitat use is determined, so that 40 animals with 30 locations each are more useful than 20 animals with 60 locations each.

In most instances, background knowledge of the biology of the animal will be useful in interpreting the results of the analysis. For example, the analysis assumes that an animal's habitat use is unconstrained by the presence of other animals and that all habitat types are equally available to it. If the animals are territorial, or if weaker individuals are pushed into suboptimal habitat by the stronger ones, then differences in habitat use between survivors and non-survivors may reflect differences in individual fitness. In either case, some habitat types may genuinely not be available to some individuals and constitute true missing values. This situation may be handled by taking the availability of habitat types into account by multivariate analysis of covariance, or by considering zero proportions as missing values (see Aebischer *et al.*, in press, where the problem of defining availability is also discussed). Alternatively, the use of correspondence analysis with respect to instrumental variables (Lebreton *et al.*, 1991) may be applicable.

In the example of the Gotland pheasants, 78 birds had been tagged over two years. There was no evidence of a year effect (nor of a sex or age effect) on habitat use during the prebreeding period, so that the data could be pooled to look at the relationship between habitat use and survival, based on 19 birds that were killed by raptors and 59 that survived. Compositional analysis revealed significant differences in habitat use between the two categories of pheasant. In order of survival value, the 'best' habitat type was vegetable crops, and the 'worst' was pasture. Breeding male pheasants establish territories along woodland edges or bordering areas of cover (Lachlan and Bray, 1976). The males guard the females when they are feeding on adjacent open ground at dawn and dusk (Hill and Robertson, 1988). Territory density is higher along edges of arable or vegetable crops than next to pasture (Woodburn and Robertson, 1990). It is possible that vegetable crops provide a higher density of quality feed than pasture. It seems that pheasants on pasture spend more time feeding than those on arable or vegetable crops (Robertson, pers. comm.), so may be exposed to raptor predation for a longer period each day. Certainly, a more detailed study of the birds' activity budgets is required; this too is a suitable candidate for compositional analysis!

Acknowledgements. Our thanks go to P. A. Robertson for useful discussions and helpful criticisms of the manuscript, to J.-D. Lebreton for drawing attention to the scoring method of ordering habitat types and the connection with correspondence analysis, and to an anonymous referee. The pheasant work from which the example was taken was funded by the Swedish Sportsmen's Association and the Natural Environment Research Council.

References

Aebischer, N.J. (1988).- Estimating time-specific survival rates when the reporting rate is constant. Acta Ornithologica, 23: 35-38.

Aebischer, N.J., Robertson, P.A. and Kenward, R.E. (in press).- Compositional analysis of habitat utilization from animal radiotelemetry data. Ecology.

Aitchison, J. (1986).- The Statistical Analysis of Compositional Data. Chapman and Hall, London, UK.

Benzécri, J.-P. (1973).- L'Analyse des Données: II, L'Analyse des Correspondances. Dunod, Paris, France.

Bernstein, C., Krebs, J.R. and Kacelnik, A. (1991).- Distribution of birds amongst habitats: theory and relevance to conservation. *In* Bird Population Studies: Relevance to Conservation and Management, C.M. Perrins, J.-D. Lebreton and G.J.M. Hirons (eds), pp. 317-345. Oxford University Press, Oxford, UK.

Brittas, R., Marcström, V., Kenward, R.E. and Karlbom, M. (in press).- Survival and breeding success of reared and wild ring-necked pheasants. Journal of Wildlife Management.

Brown, J.L. (1969).-. Territorial behavior and population regulation in birds: a review and re-evaluation. Wilson Bulletin, 81: 293-329.

Brownie, C., Anderson, D.R., Burnham, K.P. and Robson, D.S. (1985).- Statistical Inference from Band Recovery Data - A Handbook. 2nd edition. USDI Fish and Wildlife Service, Research Publication No. 156, Washington, USA.

Byers, C.R., Steinhorst, R.K. and Krausman, P.R. (1984).- Clarification of a technique for analysis of utilization-availability data. Journal of Wildlife Management, 48: 1050-1053.

Carmer, S.G. and Swanson, M.R. (1973).- An evaluation of ten pairwise multiple comparison procedures by Monte Carlo methods. Journal of the American Statistical Association, 68: 66-74.

Cox, D.R. and Hinkley, D.V. (1974).- Theoretical Statistics. Chapman and Hall, London, UK.

Chatfield, C. and Collins, A.J. (1980).- Introduction to Multivariate Analysis. Chapman and Hall, London, UK.

Dhondt, A.A. (1989).- Blue tit. *In* Lifetime Reproduction in Birds, I. Newton (ed.), pp. 15-33. Academic Press, London, UK.

Edgington, E.S. (1980).- Randomization Tests. Dekker, New York, USA.

Genstat 5 Committee (1987).- Genstat 5 Reference Manual. Clarendon Press, Oxford, UK.

Heisey, D.M. (1985).- Analyzing selection experiments with log-linear models. Ecology, 66: 1744-1748.

Hill, D.A. and Robertson, P.A. (1988).- The Pheasant: Ecology, Management and Conservation. Blackwell, London, UK.

Hill, M.O. (1973).- Reciprocal averaging: an eigenvector method of ordination. Journal of Ecology, 61: 237-249.

Hill, M.O. and Gauch, H.G. (1980).- Detrended correspondence analysis, an improved ordination technique. Vegetatio, 42: 47-58.

Kendall, M. (1980).- Multivariate Analysis. 2nd edition. Griffin, London, UK.

Kenward, R.E. (1987).- Wildlife Radio Tagging: Equipment, Field Techniques and Data Analysis. Academic Press, London, UK.

Kenward, R.E. (1990).-. RANGES IV. Software for Analysing Animal Location Data. Institute of Terrestrial Ecology, Natural Environment Research Council, Cambridge, UK.

Kenward, R.E. (in press).- Quantity versus quality: programming for robust analyses of data from radio-tracking. *In* Wildlife Telemetry: Remote Monitoring and Tracking of Animals, I.G. Priede and S.M. Swift (eds). Ellis Horwood, Chichester, UK.

Lachlan, C. and Bray, R.P. (1976).- Habitat selection by cock pheasants in spring. Journal of Applied Ecology, 13: 691-704.

Lebreton, J.-D., Sabatier, R., Banco, G. and Bacou, A.M. (1991).- Principal component and correspondence analyses with respect to instrumental variables: an overview of their role in studies of structure - activity and species - environment relationships. *In* Applied Multivariate Analysis in SAR and Environmental Studies, J. Devillers and W. Karcher (eds), pp. 85-114. ECSC, EEC, EAEC, Brussels, Belgium.

Manly, B.F.J. (1985).- The Statistics of Natural Selection on Animal Populations. Chapman and Hall, London, UK.

Newton, I. (1989).- Sparrowhawk. *In* Lifetime Reproduction in Birds, I. Newton (ed.), pp. 279-296. Academic Press, London, UK.

Pollock, K.H., Nichols, J.D., Brownie, C. and Hines, J.E. (1990).- Statistical inference for capture-recapture experiments. Wildlife Monographs, 107: 1-97.

Seber, G.A.F. (1970).- Estimating time-specific survival and reporting rates for adult birds from band returns. Biometrika, 57: 313-318.

Seber, G.A.F. (1971).- Estimating age-specific survival rates from bird-band returns when the reporting rate is constant. Biometrika, 58: 491-497.

Snedecor, G.W. and Cochran, W.G. (1980).- Statistical Methods. 7th edition. Iowa State University Press, Ames, Iowa, USA.

Swihart, R.K. and Slade, N.A. (1985).- Testing for independence of observations in animal movements. Ecology, 66: 1176-1184.

Ter Braak, C.J.F. (1987).- The analysis of vegetation-environment relationships by canonical correspondence analysis. Vegetatio, 69: 69-77.

Watson, A. and Jenkins, D. (1968).-. Experiments on population control by territorial behaviour in Red Grouse. Journal of Animal Ecology, 37: 595-614.

White, G.C. and Garrott, R.A. (1990).- Analysis of Wildlife Radiotracking Data. Academic Press, London, UK.

Wilkinson, L. (1990).- SYSTAT: The System for Statistics. Systat Inc., Evanston, Illinois, USA.

Woodburn, M.I.A. and Robertson, P.A. (1990).- Woodland management for pheasants: economics and conservation effects. *In* The Future of Wild Galliformes in the Netherlands, J.T. Lumeij and Y.R. Hoogeveen (eds), pp. 185-197. Gegevens Koninklijke Bibliotheek, Den Haag, Netherlands.

Reviews

Marked Individuals in the
Study of Bird Population
J.-D. Lebreton and P. M. North (eds.)

An Overview of Software for Terrestrial Vertebrate Population Dynamics

Jean-Dominique Lebreton, Anne-Marie Reboulet and Gérard Banco

Centre d'Ecologie Fonctionnelle et Evolutive, C.N.R.S., BP 5051, 34 033 Montpellier Cedex 1 France

Summary. This paper attempts to review the software available for vertebrate population dynamics, with an emphasis on methods for estimating and comparing vital rates. Its ultimate goals are to propose reasonable procedures for population biologists, and to define foreseeable lines for software developers. The following points are discussed:

- some approaches to data management;
- specialized software for recoveries (ESTIMATE, BROWNIE, MULT);
- specialized software for recaptures when survival estimation is of concern, with or without estimation of population size (RELEASE, JOLLY, JOLLYAGE, SURGE, POPAN, RECAPCO, SURPH); we present briefly a menu-based software package named CR, which bridges RELEASE, JOLLY, JOLLYAGE and SURGE.
- the particular case of capture rates equal to one, which covers radio-tracking data (EGRET);
- specialized software for recaptures with no mortality when population size estimation is of concern (closed population models) (CAPTURE);
- the use for various kinds of models of general purpose statistical software (GLIM, SAS) and of a major research and development program (SURVIV);
- software for predictive models of population dynamics (ELEFAN, RAMAS, and some limited availability programs).

1. Introduction

It is now hardly conceivable for a biometrician to develop any new quantitative approach without providing adequate software to implement its use. In the study of bird, and more generally, of vertebrate populations, it is thus not surprising that the recent explosion of methods (Nichols, 1992) was accompanied by an explosion in software development.

Our purpose is to review the software available for vertebrate population dynamics, with an emphasis on methods for estimating and comparing vital rates (see also the tables given by North, 1990, and Lebreton et al., 1992). We will attempt to use a common notation for the models, and to provide some basic comparisons between programs. The ultimate goals are to help population biologists to use reasonable procedures, sometimes using several programs in combination, and to attempt to define foreseeable lines for software developers. We summarise in Appendix 1 a menu-based software package named CR, which bridges various existing programs for capture-recapture data. The two main traps into which we might well fall are to be apologetic for our own production, and to be rapidly outdated in our recommendations and anticipations. We will certainly not manage to avoid these traps totally.

Table 1. An outline of software for Vertebrate population dynamics

NAME	KEY REFERENCE [1]	TOPICS; features	TYPE[2]
B-RING	Coker, Du Feu (1990)	DATA MANAGEMENT	SP
MULT	Conroy et al. (1988)	RECOVERIES; many models	SP
ESTIMATE BROWNIE	Brownie et al.(1985)	RECOVERIES; most mod.also in MULT	SP
EGRET	S.E.R.C.(non dated)	SURVIVAL MODELS; Radio-track.data	SP
GLIM	Cormack (this volume) (1985, 1989)	RECAPTURES; Anal.of residuals	GPSS
SAS	Burnham (1989)	RECAPTURES; nonlinear proced.	GPSS RD
POPAN3	Arnason et al.(1992)	RECAPTURES; data handling, simul.	SP
RECAPCO	Buckland (1980), Buckland Garthwaite (1990)	RECAPTURES; Bootstrap	SP
CR	This paper (App. 1) Cezilly et al.(1992)	RECAPTURES; Menu system for RELEASE, SURGE...	SP
RELEASE	Burnham et al.(1987)	RECAPTURES; GOF tests, simul.	SP
SURGE	Pradel et al.(1990) Pradel, Lebreton(1991)	RECAPTURES; high flexibility	SP
SURPH [3]	Skalski et al. (this volume)	RECAPTURES; Proport.hazards	SP
JOLLY	Pollock et al.(1990)	RECAPTURES; Open pop. models	SP
JOLLYAGE	Pollock et al.(1990)	RECAPTURES; Age-depend.models	SP
CAPTURE 2CAPTURE	Otis et al.(1978) Rexstad,Burnham (1992)	RECAPTURES;closed populat., simul.	SP
SURVIV	White (1983) White, Garrott (1990)	RECAPTURES RECOV; high flexibility	RD
RAMAS	Ferson, Akcakaya(1990)	DYNAMIC MODELS;	SP
ULM	Legendre (1992)	DYNAMIC MODELS; high flexibility	SP
ELEFAN	Gayanilo et al.(1988)	DYNAMIC MODELS;	SP

[1] Contact addresses are given in Appendix 2

[2] GPSS: General Purpose Statistical Software, SP: Specialized program, RD : Research and Development Program

[3] Presently available only on Workstation

There are at least two competing criteria with which to classify existing software:

1°) The type of models or functions covered: data management, survival models based on dead recoveries, survival models based on capture-recapture data (including or not population size estimation), closed population models based on capture-recapture data, predictive population models;

2°) The general orientation: general purpose statistical program (with adaptation to particular models), research and development program, specialized software;

Some other features, less relevant to classification, are important in practice :

1°) The level and type of user-friendliness : command language, menu system, full screen abilities, on-line help, flexibility within the class of models considered, simulation abilities, quality of documentation.

2°) The accessibility : availability of executable code on floppy disks for PC compatibles is now standard, but the development of workstations and of networks will rapidly induce other ways of accessing software. However, on workstations, it is not unusual when implementing a program to have to rewrite some interface code.

Table 1 gives a list of the programs considered. The selection considers primarily programs widely used and available for PCs, but is certainly incomplete. The power of PCs is quite suitable for most calculations, and the spread of the Windows environment will lift the 640 Kb limitation.

Using a somewhat arbitrary plan, we will discuss:

- some approaches to data management (§ 2);
- specialized software for survival estimation from recoveries (ESTIMATE, BROWNIE, MULT) (§ 3);
- specialized software for survival estimation from recapture data, with or without estimation of population size (POPAN, SURGE, RELEASE, JOLLY, JOLLYAGE, CR, RECAPCO, SURPH) (§ 4);
- software for the particular case of capture rates equal to one, which covers radio-tracking data (EGRET) (§ 5);
- specialized software for population size estimation from recaptures when there is no mortality (closed population models) (CAPTURE) (§ 6);
- general purpose statistical software (GPSS) (GLIM, SAS) and a major research and development (RD) program (SURVIV) (§ 7);
- Software for predictive models of population dynamics (ELEFAN, RAMAS, and some limited availability programs) (§ 8).

2. Capture-Recapture Data Management

The basic data, for recoveries as well as for recaptures, are individual histories, i.e. a statistical array with individuals as rows (statistical units) and occasions of capture as columns. Permanent categorical variables are usually available on the same individuals (Table 2A).

Table 2. Recapture history data:

A) 5 individual recapture histories over 7 capture occasions; 1 symbolizes capture, 0 non-capture. The accompanying variable is for sex (1=males, 2=females).

B) the same data when each individual is characterized by its weight (in three classes: 1, 2 or 3) measured at each capture. 0 symbolizes non-capture.

A

individual	recapture hist.	sex
1	1 1 1 1 0 0 0	1
2	0 1 1 0 0 0 0	2
3	1 1 0 0 0 0 0	1
4	0 0 0 1 0 0 1	2
5	1 1 1 1 1 1 0	1

B

individual	recapture hist.	sex
1	2 2 3 2 0 0 0	1
2	0 3 2 0 0 0 0	2
3	1 1 0 0 0 0 0	1
4	0 0 0 2 0 0 1	2
5	1 2 1 1 1 1 0	1

Individual variables measured at the time of each capture make the data structure more complex (Table 2B). Data concerning recapture over several sites (Hestbeck et al., 1991) are of this latter kind, site number being a particular categorical variable measured only when there is capture. Further qualitative attributes, such as mark number, are usually handled in parallel with these numerical data for better security in data retrieval.

We strongly advocate the use of standard Data Base Management Systems (DBMS) to handle such data (see also Stafford et al., 1986). The basic information from field records is usually a recapture event, i.e. the combination of a date, a site and a mark number. Building a file of such events is the most logical way to handle data. The conversion to recapture histories requires first a selection (e.g., spring resightings only) and a grouping of events concerning the same mark number, e.g. by indexing by it. The transcription into a rectangular array may then require an interface program. For instance, the DBase III+ programming language is adequate for such operations if the number of different individuals is limited (a few thousands). Use of a relational DBMS on workstation or mainframe may be recommended when there are over 10 000 individuals. At the borderline, the speed is considerably improved by translating DBase III+ programs into the C language, using an automatic translator (Pradel, pers.comm.) or by using one of the compilable DBMS languages compatible with Dbase III+.

Interface programs for such data conversions are usually considered as specific to a particular application, and are not distributed. As a consequence, there is certainly a lot of duplication in technical work, which could be limited by the development of adequate recapture and recovery data handling tools. We view B-RING, a data handling program for bird ringers developed by the British Trust for Ornithology, as an attempt in this direction. POPAN3 also has good data handling possibilities. CR offers various interface programs (Appendix 1).

3. Specialized Software for Analysis of Recoveries

A flexible set of models for estimating survival from recoveries of dead individuals is presented by Brownie et al. (1985). The corresponding programs BROWNIE and ESTIMATE have been the reference software for survival models based on recoveries.

More recently, MULT (Conroy et al., 1989) provided a more comprehensive set of models, including most of those considered in BROWNIE and ESTIMATE. Data entry consists of recovery tables; there is a comprehensive user's guide; the dialogue is based on a classical user-friendly menu (no full screen ability).

Table 3. An example of treatment with MULT: Blue Jay *Cyanocitta cyaneus*, region 6 (Henny, 1972, p.60). Under ESTIMATE, model M1 considers time dependent recovery and survival rates, model M2 considers time dependent recovery and constant survival rate. These models could thus be denoted as (F_t, S_t) and (F_t, S). Model M1 is clearly preferable to M2.

```
FINAL PARAMETER VALUES FOR MODEL ESTIMATE  M1
   param.     Estimate     S.error    CONFIDENCE INTERVAL
   MEAN F     0.0108       0.0008     0.0093 -      0.0123
-------------------------------------------------------------
   S ( 1)     0.5714       0.1439     0.2895   -    0.8534
   S ( 2)     0.5283       0.1230     0.2872   -    0.7695
   S ( 3)     0.2945       0.0664     0.1644   -    0.4245
   S ( 4)     1.2687       0.2952     0.6900   -    1.8473
   S ( 5)     0.9742       0.2470     0.4901   -    1.4583
   S ( 6)     0.6025       0.1715     0.2665   -    0.9386
   S ( 7)     0.5470       0.1685     0.2167   -    0.8774
   S ( 8)     0.3944       0.1220     0.1552   -    0.6336
   S ( 9)     0.9229       0.3011     0.3326   -    1.5131
   S (10)     0.7238       0.3582     0.0216   -    1.4259
-------------------------------------------------------------
   MEAN S     0.6828       0.0460     0.5926   -    0.7729
-------------------------------------------------------------
MODEL M2
CONSTRAINTS:S(I)=S - CONSTANT SURVIVAL RATE
ESTIMATES CONVERGED (DISTANCE=0.1097E-11 ON ITER. 4)
FINAL PARAMETER VALUES FOR MODEL ESTIMATE  #M2
   param.     Estimate     S.error    CONFIDENCE INTERVAL
   MEAN F       0.0099      0.0006      0.0087   - 0.0111
     S          0.6389      0.0261      0.5878   - 0.6900

GOODNESS OF FIT AND LIKELIHOOD STATISTICS:
-------------------------------------------------------------
      NUMBER OF   CHI    DEGREES OF              LOG
MODEL PARAMETERS SQUARE   FREEDOM     PROB    LIKELIHOOD
-------------------------------------------------------------
M 1        21    19.9610     23        0.6443  -2099.9338
M 2        12    60.3217     31        0.0012  -2115.2769
=============================================================
COMPARISON OF MODELS:
  M O D E L          M O D E L        CHI-SQ    DF    PROB
ESTIMATE# M 2 VS ESTIMATE# M 1      30.6860   9    0.0003
```

Models in MULT are classified in 7 sets: ESTIMATE, COVAR, INTERVAL, BROWNIE, TWO-TIME, REW-IND, REW-DIR, with respectively 3, 2, 3, 3, 7, 13 and 16 models. The 3 models in ESTIMATE consider time dependent or constant survival and recovery rates for adults. BROWNIE extends the same 3 models to handle simultaneously animals marked as young. COVAR considers two models in which time dependence is reduced to dependence on an external covariate. INTERVAL extends the time dependent models to sampling intervals of unequal length. TWO-TIME considers 7 models in which the basic interval (say, a year) is split into two "seasons", with various hypotheses on time-dependence. REW-IND and REW-DIR consider respectively 13 and 16 models for experiments with reward bands, also under various hypotheses for time-dependence. An example of treatment with MULT is given in Table 3.

Presently, the main shortcoming for the analysis of recoveries is that the set of models encompassed is finite. More complex models can be handled with SURVIV (see 7; below) which may be difficult to use for the intermittent user. A menu-driven program encompassing a flexible set of models, notably including models which require iterative fitting, is still needed for the analysis of recoveries.

4. Capture-Recapture Data : Specialized Software for Survival Estimation and Open-Population Models

The framework is the Cormack-Jolly-Seber (CJS) model (Cormack, 1964; Jolly, 1965; Seber, 1965) and its recent generalizations (Lebreton et al., 1992). The programs considered here are POPAN3, RECAPCO, JOLLY, JOLLYAGE, RELEASE, SURPH and SURGE. The first four consider open population models, i.e. they include population size estimation, while the last three are strictly oriented towards survival estimation. The survival part of open population models can nevertheless be used even when population size estimation is not feasible or is not of concern (Brownie, 1987, Clobert and Lebreton, 1987).

The use of SAS (PROC NLIN), GLIM, and SURVIV is discussed later (§ 7).

All survival models in the CJS framework can be described by a common notation (Lebreton et al., 1992) summarizing the linear predictor for the apparent survival probabilities Φ and that for the recapture probabilities p as in generalized linear models. In this notation the CJS model is denoted as (Φ_t, p_t). The model in which, for instance, survival is constrained to be constant over time is denoted as (Φ, p_t). Most such models can be fitted using SURGE, in particular models using covariates (Clobert et al., 1987) and those concerning several groups of individuals (Pradel et al., 1990. Lebreton et al., 1992). SURGE has no full screen ability, and fitting complex models (e.g. those with additive effects) requires some training. Up to now, identifiability problems are not automatically detected. Moreover, the deviance given by SURGE is defined up to an additive constant and cannot be used for GOF tests.

Appendix 1 and Table 4 summarize the treatment of data concerning the Dipper *Cinclus cinclus* using models derived from model (Φ_t, p_t) with CR, RELEASE, JOLLY, and SURGE. Further details and interpretations on this example are given by Lebreton et al. (1992). JOLLY gives bias-corrected estimates (optional in RELEASE), rarely needed in practice, for model (Φ_t, p_t), and population size estimation -if needed and feasible - under model (Φ_t, p_t), (Φ, p_t), (Φ_t, p), and (Φ, p), i.e. under various constraints of constancy on parameters. For model (Φ_t, p_t), the RELEASE GOF test is more complete than the JOLLY GOF test. SURGE makes it possible to fit intermediate models, e.g. with covariates. Multiple groups are handled by RELEASE and SURGE (Lebreton et al., 1992) which are thus recommended for this class of models, as long as only survival is of concern.

The method used in RECAPCO to provide estimates of survival in the range 0-1 is outdated by the logit transformation implemented in SURGE. However, RECAPCO is the only program that currently provides bootstrapped confidence intervals for some basic models. SURPH, and, more recently, a working version of SURGE (Viallefont, 1992), allow the use of a complementary log-log transformation. This transformation is particularly relevant for survival, since additive group effects correspond then to proportional hazards across groups. The latest versions of RELEASE provide a goodness-of-fit test for a model with immediate trap-dependence, denoted as (Φ_{g*t}, p_{g*m*t}) (Pradel, this volume). Such models can be fitted with SURGE (Pradel, this volume).

Table 5 summarizes the treatment of data concerning the Black-headed Gull using age-dependent models derived from model (Φ_{a2*t}, p_t) by JOLLYAGE and SURGE. The notation Φ_{a2*t} means that survival varies over two-age groups and time independently, while the notation p_t means that the capture probability varies over time. Convergence problems often prevents fitting the model (Φ_{a2}, p_t) using JOLLYAGE. Features unique to JOLLYAGE are the explicit estimates used for model (Φ_{a2*t}, p_t), an incomplete goodness-of-fit test of this model, and the estimation of population size.

POPAN3 and RELEASE have good simulation abilities. This is a key feature for the design of complex studies. Simulation should be used prior to initiation of field work to predict the precision of estimates and the power of tests. Power can also be calculated from expected values obtained from RELEASE (PROC EXPECT) (Lebreton et al., 1992, p.82).

Using JOLLYAGE and RELEASE for initial goodness-of-fit tests and simulation, and SURGE for stepdown selection of reduced models, seems to us to be the most recommendable strategy for survival estimation from capture-recapture data. These programs together with interface programs have been made accessible from the menu system CR (Appendix 1).

Table 4. Treatment of the Dipper *Cinclus cinclus* data in Lebreton et al. (1992) by SURGE, RELEASE, and JOLLY.

model (Φ_t, p_t), SURGE RESULTS: DEVIANCE = 656.950

survival probabilities ESTIMATE	S.e.	recapture probabilities ESTIMATE	S.e.
.7181	.1592	.6962	.1697
.4346	.0688	.9130	.0587
.4781	.0599	.9007	.0522
.6261	.0597	.9324	.0459
.5985	.0559	.5306	.0507

Same point estimates from RELEASE. SURGE estimates of survival and confid. intervals can be forced in range

Bias corrected estimates from JOLLY or RELEASE :

survival probabilities ESTIMATE	S.e.	recapture probabilities ESTIMATE	S.e.
.7133	.1547	.7010	.1643
.4342	.0687	.9241	.0720
.4778	.0597	.9137	.0578
.6256	.0592	.9015	.0535
.5982	.0560	.9330	.0455

Although the overall GOF test (RELEASE) seems to reject the model (X^2_{12}= 43.7947, P=.0000), this is not the case when only components with enough data are considered (X^2_5=1.772, P=.880). Only the latter test is given by JOLLY

model (Φ, p_t), SURGE results: DEVIANCE = 664.4802

survival probabilities ESTIMATE	S.e.	recapture probabilities ESTIMATE	S.e.
.5531	.0275	.7851	.1241
		.8905	.8918
		.8750	.7095
		.9083	.4751
		.9406	.3929
		.9639	.1025

Same point estimates can be obtained from JOLLY (mod.B) with upper limit of confidence intervals out-of-range

model (Φ_t, p), SURGE results: DEVIANCE = 659.730

Model (Φ, p) (D) : SURGE DEVIANCE = 666.838

	survival probabilities ESTIMATE	S.e.	recapture probabilities ESTIMATE	S.e.
SURGE	.5602	.0252	.9026	.0286
JOLLY	.5602	.0248	.9026	.0281

Results of tests differ slightly, because SURGE uses G^2 tests without pooling (Likelihood Ratio Tests). JOLLY uses quadratic X^2 with pooling.

TEST	SURGE		JOLLY	
D vs A	X^2_9 = 9.838	P =.3688	X^2_8= 8.294	P =.4053
B vs A	X^2_4 = 7.530	P= .1104	X^2_2= 7.104	P =.0287
D vs B	X^2_5 = 2.358	P =.7977	X^2_3= 1.535	P =.6743

Table 5. Treatment of the Black-headed Gull *Larus ridibundus* data in Clobert et al.(1987) by JOLLYAGE and SURGE.

Model (Φ_{a2*t}, p_t) = A2
FIRST YEAR SURVIVAL (note unidentifiable parameters)

i	PHI'	SE(PHI')	SE'(PHI')	95% confidence interval
1	0.5000	0.3307	0.3062	-0.1482 - 1.1482
		...		
5	Div/0.	Div/0.	Div/0.	Div/0. - Div/0.
6	Div/0.	Div/0.	Div/0.	Div/0. - Div/0.
7	0.6617	0.2209	0.2071	0.2287 - 1.0946
		...		

ADULT SURVIVAL (abridged results : mean only)

	0.8960	0.0597	0.0731	0.7789 - 1.0131

CAPTURE RATES

2	0.2500	0.2096		-0.1609 - 0.6609
3	0.2000	0.1033		-0.0024 - 0.4024
		...		
10	0.3537	0.0990		0.1597 - 0.5477

Test of the hypothesis that adult and young survival and capture prob. are the same (from Pollock 1981; Biometrics 37:521-529): X^2_2 = 2.16, P = 0.3401
Contingency table GOF test (Pollock et al.1985)
X^2_3 = 2.94, P = 0.4012

Model (Φ_{a2}, p) = D2

Parameter	Estimate	s.e.	95% confidence interval
JOLLYAGE			
PHI	0.8857	0.0232	0.8403 - 0.9312
PHI'	0.5586	0.0568	0.4473 - 0.6700
p	0.1399	0.0150	0.1105 - 0.1693
SURGE			
PHI	0.8857	0.0227	0.8331 - 0.9233
PHI'	0.5586	0.0564	0.4470 - 0.6646
p	0.1399	0.0148	0.1133 - 0.1714

Model (Φ_{a2}, p_t) = B2
JOLLYAGE: Convergence failure after 3 iterations
SURGE: DEVIANCE = 1872.773

PHI	0.8464	0.03653	0.7606 - 0.9053
PHI'	0.5986	0.07246	0.4523 - 0.7292

OVERALL RESULTS
JOLLYAGE: GOODNESS OF FIT TESTS

A2	X^2_3 = 2.94	P= 0.4012
D2	X^2_{20}=32.67	P= 0.0366

There is thus some evidence of time dependence in capture probability. SURGE makes it possible to select model (Φ_{a2}, p_t) and then to model capture probability as a function of effort (Clobert et al., 1987). JOLLYAGE provides population size estimates, although this is irrelevant to these data based on resightings.

5. Capture-Recapture Data: Specialized Software for Survival Estimation When Probability of Capture is 1

Although it is frequently assumed that the capture probability equals 1, this can be safely done almost only with radio-tracking data.

The programs mentioned above can in theory handle such data: estimates of capture probabilities will be equal to 1. However, this kind of model has a long tradition in human medical research, and most developments come from this field. GLIM and SAS (see § 7) can be used to fit various parametric and semi-parametric models, such as the COX proportional hazards model. In particular they can be used to fit the discrete time COX model (Aitkin et al. 1989, p.312 ff).

EGRET (S.E.R.C.) seems to be an interesting user-friendly package for classical epidemiological models (with a name which is attractive to bird population biologists !) although we did not test it. Further programs for radio-tracking data can be found in Kenward (1987) and White and Garott (1990).

6. Capture-Recapture Data: Specialized Software for Closed Population Models

CAPTURE, following the monograph by Otis et al. (1978; see also White et al., 1982), has no competitor. Because of the low power of some tests between models, it seems recommendable to run all models and then to have a close look at the results. A user-friendly front-end, named 2CAPTURE, makes the use of CAPTURE easy (Rexstad and Burnham, 1992). The latest version includes profile likelihood confidence intervals and the estimators developed by Chao (1989).

7. GPS Software and RD Software

As mentioned above, GLIM is a natural tool for models with capture probabilities equal to 1. Moreover, the analogy of capture recapture data with multidimensional contingency tables has been extended by Cormack (1981, 1989, this volume), to handle open population models of the CJS type. The model-oriented language in GLIM is fairly user-friendly, at least for the dedicated user. Variation in capture probabilities (over time, across groups...) can be addressed as interaction terms in the model. Variation in survival appears in compound effects which are less straightforward. Although PC implementations of GLIM impose severe limitations on the number of occasions, GLIM provides in a simple way a lot of information on the quality of fit (residuals), and on parameter identifiability. Survival models for recoveries can also be handled by GLIM (Lebreton, unpublished). SAS (PROC CATMOD) provides similar tools.

Moreover, SAS (PROC NLIN) can be used to handle nearly any model in a research and development context. Burnham (1989) provides guidelines to run CJS type models with SAS.

SURVIV (White, 1983) is a most respected research and development program which provides a framework within which many models can be fitted and simulated by specifying the probabilities of cells of multinomial distributions. It requires an adequate knowledge of model structure, and needs to be recompiled for each use. Front-ends to SURVIV, building the entry file to SURVIV, can be used to explore new classes of models. This has been the case for the multisite capture-recapture models considered by Hestbeck et al. (1991) (Hines, pers. comm.). and for some recent recovery models (Morgan and Freeman, 1989).

Some features of POPAN3 are useful for research and development, e.g. to build specific contingency tables from the data for running specific tests.

8. Predictive Models of Population Dynamics

Because they are frequently needed for demographic analysis after parameters have been estimated, we mention here briefly some of the available software for predictive population models (see Table 1).

RAMAS/age offers a user friendly framework for LESLIE type models. The array of models encompassed is fairly limited, particularly with respect to density-dependence and stochasticity. One may also cite in this context ELEFAN, which considers methods largely specific to fish stock modelling. At first sight it seems somewhat heterogeneous, with non-robust methods included.

LESLIE (Danchin et al., in prep.) is a user-friendly program for Leslie matrix models, with comprehensive output. ULM is a user-friendly program for discrete time population models with good graphic features. It covers a large array of discrete time growth models.

Many programs for analytical calculations can be used to run or analyze discrete time population models in a research and development context. One may mention at least DERIVE (explicit mathematical calculations) and CALMAT (Matrix calculations).

9. Discussion

After this broad overview we will discuss in turn the user's problems and the developer's problems.

The user's problems concern first data handling: we strongly recommend the use of a computer for any kind of data conversion ! It is also recommended to build a toolbox from existing tools in a coherent way and not to reprogram classical data handling tools. Choosing, obtaining, and implementing one or several programs is the next step. The implementation phase is always a source of surprises, because each program assumes some specific environment. For implementation on PCs for example, we recommend that help is sought from somebody knowledgeable of DOS.

The next step - running the program and interpreting the output - is by far the most difficult. The documentation is never as good as expected, and the concepts often difficult. Population biologists should not underestimate the time needed for this phase.

The developer's problems are somewhat symmetrical. At least three recommendations seem important to us:

- Tutorials, serving as intermediate steps between user's guides and articles, are badly needed;
- Programs should be interconnected as much as possible, to make the user more comfortable, and to limit duplication of technical work;
- Research and development should take place in a standard computer environment. The PC-AT computers and more recent PC-compatible computers induced a revolution for the kind of computing needed in animal population dynamics. Even in a PC environment, particular attention should be paid to the portability of programs. For instance most PCs under European norms, e.g. for keyboards, will have less memory available than under American ones and the implementation of large programs is likely to be troublesome.

Some lines of development are so clear that they need no recommendation. The user-friendliness will increase with the availability of full screen environments such as Windows. Pieces of software, with relevant help files, will be made available at a particular computer network address. Such a use of networks is already classical for users of workstations (Internet Ftp facility), not so much for users of PCs. Contrary to the usual one-time distribution, this type of distribution makes the maintenance (upgrading, correcting bugs, adding new options) of software easy.

Among many lines of development, it seems to us that developing flexibility for recoveries, for closed population models, and for multisite recapture models is an exciting prospect. One may consider in particular additive models of the kind developed for recapture models (Lebreton et al., 1992). Many improvements in robustness will come from simulation, resampling methods, graphical tools, model selection tools... There is still much to do to be able to run models of individual histories allowing estimation of any kind of demographic rate.

Acknowledgments. We thank P.M. North, W. Peach and G.C. White for very useful comments on an earlier version of this paper.

References

Aitkin, M., Anderson, D., Francis, B. and Hinde, J.(1989).- Statistical modelling in GLIM. Clarendon Press, Oxford.

Arnason, A. N. et al. (1992).- POPAN-PC User's Manual. Charles Babbage Research Centre, St-Norbert, Manitoba, Canada.

Brownie, C. (1987).- Recent models for mark-recapture and mark-resighting data. Biometrics, 43: 1017-1019.

Brownie, C., Anderson, D.R., Burnham, K.P. and Robson, D.S. (1985).- Statistical inference from band recovery data - a handbook. U.S. Fish and Wildlife Service, Resource publication n°156, 2nd edition, 256 p.

Buckland, S.T. (1980).- A modified analysis of the Jolly-Seber capture-recapture model. Biometrics, 36: 419-435.

Buckland, S.T. and Garthwaite, P.H. (1990).- Quantifying precision of mark-recapture estimates using the bootstrap and related methods. Biometrics, 47: 255-268.

Burnham, K. P. (1989).- Numerical Survival rate estimation for Capture-Recapture Models using SAS PROC NLIN. pp 414-435 in *Estimation and Analysis of Insect Populations*. McDonald, L., Manly, B.F.J., Lockwood, J. and Logan, J. (Eds) Lecture Notes in Statistics. Springer-Verlag, Berlin.

Burnham, K.P., Anderson, D.R., White, G.C., Brownie, C. and Pollock, K.H.(1987).- Design and analysis methods for fish survival experiments based on release-recapture. American Fisheries Society Monograph n° 5, 437 p.

Cezilly, F., Pradel, R. and Lebreton J.D. (1992).- Working with CR version 1.0. Mimeographed document, C.E.F.E., C.N.R.S., Montpellier, France.

Chao, A. (1989).- Estimating population size for sparse data in capture-recapture experiments. Biometrics, 45: 427-438.

Clobert, J. and Lebreton, J.D. (1987).- Recent models for mark-recapture and mark resighting data. A response to C.Brownie. Biometrics, 43: 1019-1022.

Clobert, J., Lebreton, J.D. and Allainé, D.(1987).- A general approach to survival rate estimation by recaptures or resightings of marked birds. Ardea, 75: 133-142.

Coker, D., Du Feu, C. et al.(1990).- B-RING User guide. The British Trust for Ornithology, Thetford, U.K.

Conroy, M.J., Hines, J.E. and Williams, B.K.(1988).- Procedures for the analysis of band recovery data and user instructions for program MULT. U.S. Fish and Wildlife Service, Resource publication n°175, 58 p.

Cormack, R.M.(1985).- Examples of the use of GLIM to analyse capture-recapture studies. pp. 243-273 in North, P.M and Morgan, B.J.T. (Eds) Statistics in Ornithology. Springer Verlag, Berlin.

Cormack, R.M.(1989).- Loglinear models for capture-recapture. Biometrics, 45 : 395-413.

Ferson, S. and Akcakaya, H.R. (1990).- RAMAS/age user manual. Exeter Software, Setauket, New-York, U.S.A.

Gayanilo, F.C.Jr, Soriano, M. and Pauly, D. (1988).- A draft guide to the compleat ELEFAN. International Center for Living Aquatic Resources Management, Manila, Philippines.

Henny, C.J. (1972).- An analysis of the population dynamics of selected avian species with special reference to changes during the modern pesticide era. U.S. Fish and Wildlife Service Wildlife Research Report 1.

Hestbeck, J.B., Nichols, J.D. and Malecki, R. A.(1991).- Estimates of movement and site fidelity using mark-resight data of wintering Canada geese. Ecology, 72: 523-533.

Kenward, R. (1987).- Wildlife radio-tagging; equipment, field techniques and data analysis. Academic press, London.

Lebreton, J.D., Burnham, K.P., Clobert, J. and Anderson, D.R.(1992).- Modelling survival and testing biological hypotheses using marked animals : a unified approach with case studies. Ecological Monographs, 62: 67-118.

Legendre, S. (1992) Unified Life Models. Software for Population Dynamics. Mimeographed doc., Ecole Normale Supérieure, Laboratoire d'Ecologie, Paris, France.

Morgan, B.J.T. and Freeman, S.N. (1989).- A model with first-year variation for ring-recovery data. Biometrics, 45: 1087-1101.

North, P.M. (1990).- Analysis of avian ring recovery and live recapture data: where have we come from Wageningen to Sempach. The Ring, 13: 11-22.

Nichols, J.D.(1992).- Capture-Recapture Models; using marked animals to study population dynamics BioScience, 42: 94-102.

Otis, D.L., Burnham, K.P., White, G.C. and Anderson, D.R.(1978).- Statistical inference from capture data on closed animal populations. Wildlife Monographs, 62 : 1-135.

Pollock, K.H., Hines, J.E. and Nichols, J.D. (1985).- Goodness-of-fit tests for open capture-recapture models. Biometrics, 41 : 399-410.

Pollock, K.H., Nichols, J.D., Brownie, C. and Hines, J.E. (1990).- Statistical inference for capture-recapture experiments. Wildlife Monographs, 107 : 1-97.

Pradel, R., Clobert, J. and Lebreton, J.D.(1990).- Recent developments for the analysis of capture-recapture multiple data sets. An example concerning two Blue Tit populations. The Ring, 13, 193-204.

Pradel, R. and Lebreton, J.D.(1991).- User's Manual for Program SURGE version 4.1. Mimeographed document, C.E.P.E./C.N.R.S., Montpellier, France.

Rexstad, E. and Burnham, K.P. (1991).- User's Guide for Interactive Program CAPTURE. Department of Biology and Wildlife, University of Alaska, Fairbanks, U.S.A..

S.E.R.C. (non dated).- EGRET, state-of-the-art epidemiological computing. Statistics and Epidemiology Research Corporation, Seattle, Washington, U.S.A..

Stafford, S.G., Alaback, P.B., Waddell, K.L. and Slagle, R.L. (1986).- Data management procedures in ecological research. pp.93-113 in Research data management in the ecological sciences, Michener, W.K. (ed). Belle W. Baruch library in Marine sciences nr 16. University of South Carolina Press.

Viallefont, A. (1992).- Robustesse et flexibilité des modélisations en capture-recapture; application aux Oies des neiges (*Anser caerulescens caerulescens*). Unpublished D.E.A. report, University of Montpellier II, Montpellier, France.

White, G.C.(1983).- Numerical estimation of survival rates from band-recovery and biotelemetry data. J.Wildl.Manag., 47 : 716-728.

White, G.C. and Garrott, R.A.(1990).- Analysis of wildlife radio-tracking data. Academic press, San Diego.

White, G.C., Anderson, D.R., Burnham, K.P. and Otis, D.L. (1982).- Capture-recapture and removal methods for sampling closed populations. Los Alamos National Lab., report LA-8787-NERP, 235 p.

APPENDIX 1: the CR Menu System (an outline of the Dipper data analysis in table 4)

Menu

DATA ANALYSIS	Files	System	End

Data Preparation starts from file ED.RH (Biomeco Format)

ED.RH

```
294 7
$
$
1 1 1 1 0 0 0
1 1 0 0 0 0 0
1 1 1 1 1 1 0
......
0 0 0 0 0 0 1
```

menu option:

Files

```
BIOMECO format to RELEASE format
BIOMECO format to SURGE   format
BIOMECO format to JOLLY   format
.....
```

ED.REL

```
PROC TITLE EUROPEAN DIPPER MALES/FEMALES;
PROC CHMATRIX OCCASIONS= 7  GROUPS= 1 ;
1111000  1 ;
1100000  1 ;
1111110  1 ;
.......
0000001  1 ;
GLABEL( 1)=GROUP1;
PROC STOP ;
```

ED.JLY

```
TITLE=EUR/DIPPER M/F
PERIODS=7
TYPE=CAP/REL
FORMAT=( 7(1X,2I1),1X,I4)
11 11 11 11 00 00 00  1
11 11 00 00 00 00 00  1
11 11 11 11 11 11 00  1
11 11 00 00 00 00 00  1
....                  .
00 00 00 00 00 00 11  1
```

ED.SUR

```
 6
22 49 52 45 41 46
11  6  5  3  2  0
20  8  7  4  2
22 11  5  2
26 16 11
25 16
23
 9  6  2  2  1  2
29 11  2  3  2
29 12  6  3
16 11  7
16  9
23
```

Data Treatment:

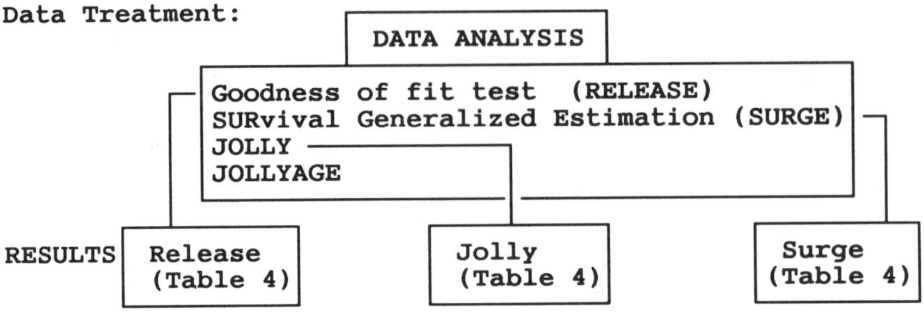

APPENDIX 2: Contact address for the software cited (in alphabetical order)

B-RING W. Peach, British Trust for Ornithology, The Nunnery, Nunnery place, Thetford IP24 2PU, U.K.

BROWNIE Office of Bird Migr. Manag., Patuxent Wildlife Research Centre, Laurel, MD 20708, U.S.A.

CALMAT Exeter Software, 100 North Country road, Bldg. B., Setauket, N.Y. 11733, U.S.A.

CAPTURE, 2CAPTURE Colorado Cooperative Fish and Wildlife Research Unit, 201 Wagar Building, Colorado State Univ., Fort-Collins, Co 80523,U.S.A.

CR J.D. Lebreton, C.E.F.E. C.N.R.S., BP 5051, 34033 Montpellier cedex, France

DERIVE Soft Warehouse, 3615 Harding Avenue, Suite 505, Honolulu, Hawaii 96816-3735, U.S.A.

EGRET Statistics and Epidemiology Research Corporation (S.E.R.C.), 909 Northeast 43rd street, Seattle, Washington 98105, U.S.A.

ELEFAN International Center for Living Aquatic Resources Management, Manila, Philippines.

ESTIMATE Office of Bird Migr.Manag., Patuxent Wildlife Research Centre, Laurel, Md 20708, U.S.A.

GLIM Numerical Algorithm Group, Wilkinson House, Jordan Hill road, Oxford OX2 8DR, U.K.

JOLLY J.E. Hines, Patuxent Wildlife Research Centre, Laurel, Md 20708, U.S.A.

JOLLYAGE J.E. Hines, Patuxent Wildlife Research Centre, Laurel, Md 20708, U.S.A.

LESLIE J.D. Lebreton, C.E.F.E. C.N.R.S., BP 5051, 34033 Montpellier cedex, France

MULT J.E. Hines, Patuxent Wildlife Research Centre, Laurel, Md 20708, U.S.A.

POPAN3 Charles Babbage Research Center, PO Box 272, St-Norbert Postal station, Winnipeg, Manitoba R3V IL6, Canada.

RAMAS Exeter Software, 100 North Country road, Bldg. B., Setauket, N.Y. 11733, U.S.A.

RECAPCO S.T. Buckland, Scottish Agricultural Statistics Service, Aberdeen Unit, Macaulay Land Use Research Institute, Craigiebuckler, Aberdeen AB9 2QJ, Scotland.

RELEASE G.C. White, Dpt of Fishery and Wildlife Biology, Colorado State University, Fort-Collins, Co 80523, U.S.A.

SAS SAS institute, Box 8000, Cary, NC 27511, U.S.A.

SURGE R. Pradel, J.D. Lebreton, C.E.F.E. C.N.R.S., BP 5051, 34033 Montpellier cedex, France

SURPH J.R. Skalski, Centre for Quantitative Science, University of Washington, HR-30, Seattle Wa 98195, U.S.A.

SURVIV G.C. White, Dpt of Fishery and Wildlife Biology, Colorado State University, Fort-Collins, C0 80523, U.S.A.

ULM S. Legendre, Laboratoire D'Ecologie, Ecole Normale Supérieure, 46 rue d'Ulm, 75230 Paris cedex 05, France

Marked Individuals in the
Study of Bird Population
J.-D. Lebreton and P. M. North (eds.)

Trends in the Analysis of Recovery and Recapture Data

David R. Anderson[1], Mark A. Wotawa[2], and Eric A. Rexstad[3]

[1]*Colorado Cooperative Fish & Wildlife Research Unit, U. S. Fish & Wildlife Service, Fort Collins, CO 80523 USA*

[2]*Colorado Cooperative Fish & Wildlife Research Unit, Colorado State University Fort Collins, CO 80523 USA*

[3]*Institute of Arctic Biology and Department of Biology & Wildlife, University of Alaska-Fairbanks, Fairbanks, AK 99775 USA*

Summary. This paper reviews trends in the analysis of ring recovery and capture-recapture data and makes several predictions about trends into the remainder of the decade. The number and size of data sets have increased very substantially and now many taxonomic groups are represented by marked subpopulations. Interest is shifting from parameter estimation to hypothesis testing and structural models.

There has been an explosion of new statistical theory. The key issue in the comprehensive analysis of recovery or recapture data is proper model selection. The Principle of Parsimony is the conceptual guide to model selection and Akaike's Information Criterion (AIC) provides a simple and effective method for practical use. Given a model, a large amount of theory exists for making optimal estimates and tests. Several methods offer improved means to estimate precision of estimates (e.g., empirical variance inflation using quasi-likelihood, and improved confidence intervals using logit-based or profile likelihood intervals). The small sample properties of AIC and other model selection criteria are topics which need additional research.

Goodness-of-fit tests are reviewed and their interpretations given. Small expectations associated with large data sets remain problematic. Heterogeneity in survival often cause little bias in ϕ, but can result in poor model fit. Recent efforts to estimate the magnitude of heterogeneity are reviewed. Additional research on heterogeneity is recommended.

An increasing number of research papers are showing that radio telemetry equipment has had impacts on avian species. These effects frequently impact both survival and reproduction. We suggest caution in attempting to make valid estimates of survival from radio-marked birds.

Relatively few biologists conduct treatment-control survival experiments using samples of marked individuals. A general theory for design and analysis of such experiments exists, but few experiments are being conducted. We strongly recommend that more experimentation be done. Increasing interest in multistrata models, movement parameters, state transition models, and post-stratification has taken place since the 1989 EURING meetings. These results are reviewed briefly and more research results are recommended.

Three brief examples are given (northern spotted owls ringed throughout their range, mallards ringed during the winter in eastern Colorado, and gadwalls ringed in North Park, Colorado). These data sets are used to illustrate a variety of subtle points and forecast future analysis trends. Finally, trends in computer hardware and software are given with a discussion of future requirements and present limitations.

1. Introduction

Rigorous statistical theory began in the mid-1960s for open population capture-recapture surveys (e.g., Cormack, 1964, Jolly, 1965, Seber, 1965, but also see Darroch, 1959 and Leslie et al., 1953) and in the early 1970's for ring recovery surveys (Seber, 1970 and Robson and Youngs, 1971, but also see Seber, 1962). During these times, a typical data set extended for perhaps 4-8 years; even by the mid-1970's it was fairly unusual to have more than 15 years of sample data on marked populations. Until the mid-1970's data were simple, covariates were infrequently measured or modeled, and few treatment-control experiments were conducted using sampling of marked populations (but see Bellrose, 1959). Seber's book summarizing then current methods was first released nearly 20 years ago (1973).

Currently there are perhaps hundreds of high quality data sets representing a wide variety of taxonomic groups. It is not uncommon to see data sets covering more than 25 years. Further, the numbers released each year are often large or the capture probability (p) is relatively large (i.e., >0.4). Surveys over the past 2-4 decades often provide "multiple data sets" for analysis (e.g., data for 2-3 age classes, for both sexes, for several areas or colonies over a k year period). As foreseen by Jolly (1965:229-230) the analysis of such data sets allows some parameters to be in common across data sets, while other parameters are unique to a class. Continuous or discrete external covariates are often available (e.g., North and Morgan, 1979). The combination of advanced statistical methods, sophisticated software and large and complex data sets allows biologists to address increasingly interesting questions from data on marked populations. Statistical inference from multiple data sets allows the potential to identify increased structural complexity and is a topic of current research. The theory for the analysis of multiple data sets falls under the term "meta-analysis" (see Hedges and Olkin, 1985, but Burnham et al., 1987 and Lebreton et al., 1992 treat this subject in detail).

As statistical theory has become more sophisticated, data sets have become more complex, and biological questions have become more penetrating, a synthesis was required before more progress was likely. This synthesis was provided by the realization that the key analysis and inference issue was one of proper model selection (Lebreton et al., 1992 and references therein). An exciting and practical paradigm casts the model selection problem into a simple optimization framework (see Burnham and Anderson, 1992). A contemporary analysis strategy is based on building several models (perhaps >80 for large data sets) hypothesized to represent the data. Knowledge and hypotheses about the biology of the species and the sampling should drive the model building. Akaike's Information Criterion (AIC) is computed for each model, and the model with the minimum AIC is selected as a basis for inference (Akaike, 1973). Often the data do not support selection of one model that is clearly the "best" but a small subset of good models can be identified. This procedure is based on the Principle of Parsimony and is easy to compute and interpret (see e.g., Pradel et al., 1990 and Szymczak and Rexstad, 1991).

During the past few years there has been renewed interest in the estimation of survival probabilities (ϕ) (e.g., Clobert et al., 1985) and waning interest in the estimation of population size (N). In addition, more work has been directed toward estimating structural relationships and testing their significance than only point estimates of parameters. AIC, likelihood ratio tests and goodness-of-fit tests are critical tools in the analysis of complex data sets (e.g., Lebreton et al., 1992).

Once a model has been chosen as the basis for inference, then many analytic procedures are available; most exist in a likelihood framework (see Burnham et al., 1987, North, 1990, Pollock et al., 1990, and Seber, 1992). Point estimation using maximum likelihood (ML) (or some equivalent method such as iteratively reweighted least squares, Burnham, 1989) is well known, given a particular model. Seber (1986 and 1992) provide recent reviews of the literature while Nichols (1992), Pollock et al. (1990) and Lebreton et al. (1990) provide conceptual overviews of the field. North (1990) and Perrins et al. (1991) contain new results; both theory and application. Lebreton et al. (1990) extend estimation to allow for age-specific breeding probabilities. In general, the state of the art in the analysis of avian data is quite good (but see Dobson, 1990).

2. Estimation of Precision Given a Model

Modern methods for the analysis of data from marked animal populations involve sophisticated models containing both structural and variance components. Aggressive model building, starting with a global model, and proper model selection should produce a model that incorporates the significant structural features of the data; otherwise, bias in the estimators is likely. Alternatively, if the model has too much structure or too many parameters, then precision is unnecessarily lost and "effects" may be inferred, but are not justified by the data. AIC is a method for balancing these concepts in arriving at a "proper" model. However, the binomial sampling variance assumed by most models currently in use may often be inadequate as a model for the variation in the empirical data. Thus, uncertainty in the estimates may be underestimated by the model-based estimators of the sampling variances and covariances. The structure of the model might be adequate, but the binomial sampling variance is inadequate. This has led to the use of quasi-likelihood methods, bootstrapping and improved confidence interval procedures.

One application of quasi-likelihood allows a variance inflation factor c to be estimated as

$$\hat{c} = \chi^2/df$$

where χ^2 is the goodness-of-fit test statistic for the global model and df is the associated degrees of freedom of this test. Then, theoretical, model-based sampling variances and covariances are "inflated" as $\hat{c} \cdot \widehat{var}(\hat{\phi})$ and $\hat{c} \cdot \widehat{cov}(\hat{\phi}_i, \hat{\phi}_j)$, respectively (see Williams, 1982 and Burnham et al., 1987). Often $\hat{c}$ ranges from a little over 1 to perhaps 3. Likelihood ratio tests, AIC and profile likelihood intervals should be adjusted for this extra-binomial variation if we judge $\hat{c}>1$ (see Lebreton et al., 1992:84-85).

The concepts of bootstrapping (Efron, 1979) are well known, but have not yet seen wide use in the analysis of capture-recapture and ring recovery data (but see Buckland, 1984 and Buckland and Garthwaite, 1991). For the large data sets now available, the computational burden associated with bootstrapping is a limitation. Still, bootstrapping will see further use in the near future as a method to obtain improved estimates of precision.

ML estimators are asymptotically normally distributed, however with sample sizes typically encountered, the sampling distribution often has a long right tail and therefore achieved confidence interval coverage is frequently below the nominal level. Burnham et al. (1987:214) provides a simple way to set improved intervals based on the logistic transformation, $logit(\hat{\phi})=log_e(\hat{\phi}/(1-\hat{\phi}))$, provided that $0<\hat{\phi}<1$. This transformed quantity is more normally distributed than $\hat{\phi}$. Then $\widehat{se}(logit(\hat{\phi}))=\widehat{se}(\hat{\phi})/[\hat{\phi}(1-\hat{\phi})]$ and an approximate $(1-\alpha)$% confidence interval is computed from the lower and upper bounds, $\hat{\phi}_L$ and $\hat{\phi}_U$, as

$$\hat{\phi}_L = \frac{\hat{\phi}}{\hat{\phi}+(1-\hat{\phi})\cdot C} \qquad \text{and} \qquad \hat{\phi}_U = \frac{\hat{\phi}}{\hat{\phi}+(1-\hat{\phi})/C},$$

where

$$C = exp\left(\frac{z_{\alpha/2} \cdot \hat{se}(\hat{\phi})}{\hat{\phi}(1-\hat{\phi})}\right).$$

This method is simple to compute and software should implement this procedure.

Profile likelihood intervals (Barndorff-Nielson, 1986) will see increased use in the future. This procedure is computer intensive for models of high dimension, but represents the best omnibus method for deriving confidence intervals with the nominal achieved coverage. This method is useful for setting confidence intervals even in the case where $\hat{\phi}>1$ or equal to 0. Furthermore, endpoints of profile likelihood intervals always lie within [0,1] for ϕ and $\hat{p}$. Profile likelihood intervals can be adjusted to allow for extra-binomial variation by incorporating an estimate of c. If the degrees of freedom in the goodness-of-fit test are large (*i.e.*, >20-30), then $\hat{c}\cdot\chi^2$ can be used, resulting in a wider confidence interval. If the degrees of freedom are not large, then the F distribution is required and profile likelihood intervals must be based on $\hat{c}\cdot F_{1,\,df,\,1-\alpha}$.

3. Goodness-of-Fit Testing

Proper model selection involves two overlapping components: 1) selecting a parsimonious model that 2) fits the data. AIC has been suggested for the former and below, we describe a useful approach in evaluating goodness-of-fit within a capture-recapture framework.

TESTs 2 and 3 are ideal for testing the fit of capture recapture data and these are easily computed in program RELEASE (Burnham et al., 1987:64-67 and associated references). TESTs 2 and 3 are presented as series of 2x2 contingency tables as a result of pooling across cells of the m_{ij} matrix. Fisher's exact test may be preferred over the usual Pearson χ^2 if cell counts of these tables are small, which in practice is common. Here, one can compute the value of χ^2 that produced the given P value and sum the k χ^2 statistics to obtain an overall test. Alternatively, one can use Fisher's method and reject the null hypothesis if $-2\sum log_e(P_i) \geq C$ where C is obtained from the upper tail of the chi-square distribution with $2k$ degrees of freedom (Hedges and Olkin, 1985).

Examination of the residual patterns of the individual cells of the recapture (or recovery) matrix as described by Freeman and North (1990) and Freeman and Morgan (1990) is also suggested. If a global model is rejected, these observations in combination with TESTs 2 and 3 provide a means of determining whether the lack of fit is due to the "structure" of the model or alternatively if no pattern is recognized, a possible failure of one of the inherent variance assumptions (e.g., parameter heterogeneity or dependence of individual fates).

The former case, an improper structure, suggests that important variables have been excluded, thereby restricting the interpretations of variables which were included; this indicates more model building is necessary to account for additional variability. If additional structural components are not considered *a priori*, the hypothetico-deductive approach of modern science dictates a new experimental design for valid inference.

The alternative explanation for poor fit, failure of the variance assumptions, can be mitigated by adjusting the variances, covariances, AIC, profile likelihood intervals, and likelihood ratio tests by the value of $\hat{c}$. Models are becoming available to assess the failure of assumptions (e.g., Burnham et al., 1987) and are recommended for comprehensive analysis. The current trend of larger data sets with many capture occasions possibly coupled with "marginal" amounts of data, result in many cells with very small expected values. This magnifies the effect of outliers or heterogeneity on lack of fit (*cf* Freeman and North, 1990, Freeman and Morgan, 1990). Because future surveys are likely to be properly designed and incorporate substantial components of variation, but also inevitably introducing heterogeneity, we envision increasing use of this latter approach.

4. Heterogeneity

It has long been recognized that the estimation of survival rates in avian populations is predicated on the convenient but untenable assumption that all members of an age and sex class have the same probability of survival (Brownie et al., 1985:10). A decade ago, Pollock and Raveling (1982) and Nichols et al. (1982) addressed the effect of violating of this assumption upon survival rate estimators. These authors concluded that heterogeneity in survival probabilities did not bias estimators substantially, but did cause negative bias in variance estimates. A similar conclusion was reached in recent paper by Barker (1992) with regard to achieved confidence interval coverage.

Burnham and Rexstad (In press) developed extensions to model M_2 (Brownie et al., 1985) in which the assumption of homogeneity is relaxed and estimation includes a survival rate, recovery rates and a measure of parameter heterogeneity. Heterogeneity was measured as the variance of the distribution of survival probabilities in the ringed cohort at the time of ringing, σ^2. These models are shown to have reasonable properties and possible extensions to models of heterogeneity and time-variations in survival probabilities are discussed by Burnham and Rexstad (In press). Rexstad and Anderson (in prep.) applied the heterogeneity-only models to a number of data sets of mallards (*Anas platyrhynchos*) and found evidence of heterogeneity in mallards ringed as adults. They were unable to detect differences in heterogeneity between sexes or between winter-ringed and fall-ringed birds. Additional work remains to be done in this area, particularly with regard to age-specific models, allowing the incorporation of large numbers of birds ringed as young into the estimation of heterogeneity. It must be noted that heterogeneity is a second-order estimation problem and it is difficult to obtain precise estimates of heterogeneity (σ^2) from ring-recovery data. In addition, the small sample distribution of certain test statistics is a subject in need for more research.

Researchers concerned with the effect of heterogeneity on estimates of treatment effects are encouraged to conduct simulation experiments using the EXPECT and SIMULATE options in program RELEASE (Burnham et al., 1987). Investigations by these authors suggest these estimators are robust with respect to heterogeneity in survival or recapture probabilities. However, the ease with which researchers can conduct simulations tailored to their specific experiments argues strongly for these investigations prior to actual field studies.

5. Problems with Radio Telemetry for the Study of Avian Populations

Radio telemetry has been used to estimate survival in avian populations (see Pollock et al., 1989). Alternatively, Lakani (1990) has suggested use of telemetered birds to independently estimate the survival rate of young, and thus make the remaining parameters in the life table model identifiable (see Anderson et al., 1985).

Many papers have appeared in the past 2-4 years showing evidence that radio transmitters often have detrimental effects on birds. Space limitations prevent a comprehensive review of this literature, but the reader is directed to recent papers by Marks and Marks (1987) and Gessaman and Nagy (1988). There is already a substantial literature on these issues with some papers appearing in the early 1970s. Papers in the past five months include Burger et al. (1991), Hooge (1991), Paton et al. (1991), Sodhi et al. (1991) and Foster et al. (1992). Individual studies have sometimes shown insignificant differences in survival, but the power of these tests is often quite low. Taken in the aggregate, it seems that radio telemetry equipment often impacts both survival and reproduction in birds. We suggest caution in the use of current telemetry equipment and attachment methods in the study of survival of avian species.

Multi-strata and Post-Stratification Models

Studies and questions involving multiple strata are becoming increasingly necessary in the biological sciences. Initial applications focused primarily on estimation of population size, then more recently on estimation of survival rates for individual populations. Currently, increasing interest in movement rates among strata has led to extensions of capture-recapture and recovery methods to estimate movement parameters. Following Brownie et al. (submitted), studies involving multiple strata generally fall within two major categories. The first include those in which strata are assigned at the time of first capture and remain fixed throughout the study (e.g., sex or area) or change strata in a predictable manner (e.g., age), while the second include those in which individuals may change strata through time with some unknown probability. The former approach is addressed extensively by Lebreton et al. (1992) for capture-recapture methods; analogous methods apply for recovery data (see also Brownie et al., 1985:144-152 and 170-182). Schwarz et al. (1988) also considers fixed strata although class assignment is made at the time of a terminal recovery (i.e., post-stratification). These fixed attribute approaches are useful as they account for and reduce heterogeneity of recovery and survival rates, however they do not account for immigration and emigration.

The class of models incorporating movement among strata was first introduced by Chapman and Junge (1956) and Darroch (1961) for a closed population and 2 capture occasions. Open model extensions followed which allowed 3 or more capture occasions and the identification of additional parameters (Iwao, 1963, Arnason, 1972a, 1972b, 1973, Seber, 1982:555). More recently, Schwarz et al. (submitted:appendix B) gave maximum likelihood estimates of Arnason's model, while Brownie et al. (submitted) provided reduced parameter versions (see also Nichols et al., 1992). Schwarz et al. (submitted) also provide analogous concepts under a recovery framework. Additional applications were considered by Hestbeck et al. (1991) and Brownie et al. (submitted); they generalize Arnason's and Schwarz's models by relaxing the Markovian assumption to include "memory" of prior strata.

Some major limitations of models considering movement among strata exist: primarily, confounding of movement parameters with recovery or survival rates and the amount of data necessary to maintain sufficient precision with additional parameters. With additional assumptions or independent estimates of some parameters useful interpretations of movement parameters can be "derived" (see e.g., Hestbeck et al., 1991, Nichols et al., 1992, Brownie et al. submitted, Schwarz, 1988). A unified approach involving both recovery and recapture data suggested by Schwarz (1988) and illustrated by Burnham (1992) can overcome identifiability problems, although these techniques are both complex and data intensive. We feel that increasingly large data sets addressing more complex questions, associated with greater habitat fragmentation and other human disturbances, will generate a great need for creative models incorporating transitions among multiple strata (Gilpin and Hanski, 1991).

6. Design of Experiments

Burnham et al. (1987) provide a general theory for treatment-control experiments using sample data from marked populations. This theory and associated software (RELEASE) has seen some use in fisheries issues and studies involving the assessment of the effect of pesticides. Until recently, few biologists have considered survival experiments in a capture-recapture or ring recovery framework (Anderson et al., 1987).

Statistical theory is available for experiments using either batch or unique marks and sampling terminal recoveries, multiple recaptures, or a combination. Theory is based on a set of nested models for V treatment-control groups, sampled over k occasions. An intensive array of statistical hypothesis tests are provided to assess model assumptions and AIC aids in model selection. Many formulae are available to aid in the experimental design and the importance of replication is stressed (Rexstad et al., 1990). Program RELEASE contains options (SIMULATE and EXPECT) to assist in designing experiments and evaluating properties of specific models. We forecast a substantial growth in the number of survival experiments conducted on marked animal populations during the remainder of the 1990s.

6.1 Example 1: Northern Spotted Owl

Anderson and Burnham (1991) report on a unique effort to examine trends in the demographic parameters of the northern spotted owl, a threatened subspecies under the U. S. Endangered Species Act. Capture-resighting data were available on five large study areas throughout the range of the owl (i.e., NW California, W Oregon, and W Washington). The data involved 2-4 age classes and two sex classes collected for 5-7 years over the five areas. The objective was to estimate age- and time-specific survival for females from the capture history matrix for each area. Age-specific fecundity of females was estimated from accompanying data. These rate estimates were used in a Leslie-Lefkovitch matrix to estimate the finite rate of population change (λ) and its standard error ($se(\hat{\lambda})$) (Anderson et al., 1990).

The analysis involved six biologists working on the owl and six analysts with expertise in capture-recapture analysis, software and statistical theory. This team interacted in an intensive workshop to share information and expertise. During this time, they examined perhaps 80-90 models with the empirical data. In particular, a meta-analysis was conducted on the five data sets for adult females whereby 64 models were built and the log-likelihood ($log_e(\mathcal{L})$), degrees of freedom, and AIC were tabulated.

The AIC selected model was (ϕ_T, P_{g*T}) which represents a significant negative trend (T) in adult female survival (specifically, a linear trend in $logit(\phi)$). The estimated capture probabilities ($\hat{p}$) increased as a linear trend on $logit(p)$ and differed significantly by group (g=area). No significant difference in $\hat{\phi}$ was detected across groups. The decrease in $\hat{\phi}$ for adult females is alarming for a Threatened subspecies. Further, the meta-analysis resulted in $\hat{\lambda}$=0.93 with $\hat{se}(\hat{\lambda})$=0.03, indicating a significant annual decline in this population.

This example depicts several trends in data analysis. First, a group effort comprised of several biologists and analysts might become a more common occurrence. Second, on significant social/biological issues, several data sets might be available for analysis and review. In these cases it seems compelling to consider a comprehensive analysis. Third, in issues likely to be subjected to judicial hearings, it is critical to be exacting in terms of objective model selection, estimation methods used, goodness-of-fit tests, proper estimation of empirical variances, etc. We expect to see more of these issues in the court systems by the end of the 1990s.

6.2 Example 2: Colorado Mallards

This example represents an ongoing analyses of the survival of mallards ringed during winter (post-harvest) in eastern Colorado. The far-sighted nature of the original study design was evident, for relatively equal numbers of mallards were marked with uniquely numbered rings by sex and age (subadults and adults) for 21-24 years in 7 previously identified "management units." A limited sport harvest provided the necessary recovery information for up to 26 years.

The need for parsimonious modeling of such a large and complex set of data is obvious; a global model incorporating all of the above factors would contain over 1200 parameters. Parsimony was achieved using a combination of likelihood ratio tests or their asymptotic equivalents (Brownie et al., 1985) and AIC. Models outside the limitations of programs BROWNIE and ESTIMATE (Brownie et al., 1985), and MULT (Conroy et al., 1989) required the algorithms of program SURVIV (White, 1983) which was run on a SUN/SPARC work-station with the virtual memory of a UNIX operating system. Without *Windows* or OS-2, the 640K memory limitation of DOS-based PCs could not accommodate the large dimensions needed to analyze this set of data. Even if memory was not a limitation, the number of models and their dimensions would impose a major time constraint if run on personal computers.

An interesting problem arose concerning "spatial scale" when testing among the seven areas. A compromise was necessary when pooling areas to maintain sufficient power and "sub-population" integrity. Parameter constraints across areas were based upon *a priori* knowledge of mallard ecology and physiographic distinctions. A series of testing and pooling resulted in a hierarchy of nested spatial scales, each of which could be applied to subsequent tests concerning attributes other than area.

A second consequence of the large dimensions was poor fit of the global model for males in all areas (and all spatial scales). Initial examination of the "standard normal deviates" provided by program BROWNIE failed to indicate any striking pattern, suggesting structural integrity. Additionally, the off diagonal location and positive residual value of cells contributing greatest to the lack of fit might suggest individual heterogeneity exists. The superior fit of models incorporating heterogeneity and relatively small values of $\hat{c}$ (approximately 1.1-1.4) Provide support for this supposition. If the lack of fit is not a result of poor structure, quasi-likelihood methods are justified to adjust appropriate variance estimates and associated tests.

Initial results of the "best" parsimonious models for males suggests a negative time trend in survival rates and time-specific recovery rates. Because of the important implications concerning harvest rates and compensatory mortality, 5 harvest-related covariates in addition to time were incorporated into the model on a logit scale and AIC was used to select a subset of models to evaluate their influence on recovery and survival rates. Recovery rates would be directly influenced by harvest variables because they contain an inherent, non-identifiable harvest parameter. Therefore, in this case, recovery rates were modeled first, and only the best subsets were used in modeling survival rates; this reduces the set of models under consideration to only a fraction of the excessive number of all possible subsets. Weather covariates are also being investigated in a similar manner for their influence on mallard survival. Interpretation of parameter estimates under models including covariates would follow similarly accepted multiple regression practices, including the evaluation of collinearity and outliers.

This study exemplifies several trends we expect to see in the application of capture-recapture and recovery methods: large, complex, multi-strata data sets requiring a large array of models, global models with inadequate fit, and the evaluation of external covariates. Theory and software now exist to design experiments and analyze data which can resolve problems such as spatial scale and test hypotheses concerning important biological processes such as compensatory mortality; we highly recommend an approach involving experimental treatments (see Anderson et al., 1987, Burnham et al., 1987).

6.3 Example 3: Colorado Gadwall

An example we feel is indicative of future analyses is presented by Szymczak and Rexstad (1991) in their estimation of survival rates of gadwalls (*Anas strepera*). This species is lightly exploited, such that ring recoveries from hunter-killed birds are few, leading to imprecise estimates of survival rates. However, an extensive ringing effort for >10 years resulted in a large number of ringed birds recaptured in subsequent ringing periods. A hybrid analysis incorporating both terminal recoveries and live recaptures was employed (also see Burnham, 1992 and this volume).

The model developed attempted to estimate time- and sex-specific survival, recovery, recapture, and fidelity rates (probability of returning to the ringing site, given the individual was alive). The fidelity rates were key to the analysis in that if fidelity rates could be assumed to be unity, then both terminal recoveries and live recaptures could be used to estimate (common) survival rates (Burnham, this volume). Another interesting feature was a joint analysis of the sexes; this was done in an effort to find a parsimonious model in which annual male and female survival rates were assumed to be highly related. If this assumption was tenable, this would lead to a large reduction in the number of parameters to be estimated (see Lebreton et al., 1992).

With the large parameter space under consideration (66 parameters in the most general model), parsimony was highly desirable. The number of candidate models was also large because of the potential for time- and sex-specificity for four sets of parameters. AIC was used to select a model for inference. Likelihood ratio tests were used to explicitly test the assumptions that fidelity rates were unity and that female survival rates could be modeled as functions of male survival rates (i.e., parallel on a logit scale). The minimum-AIC model assumed fidelity equal to unity, recovery and recapture probabilities sex-specific, and survival rates time-specific and linked across sexes. Coefficients of variation for survival rates were on the order of 10%. This example illustrates the use of extensive model-building and model selection as critical components of data analysis.

7. Computer Hardware and Software

The practical application of advanced statistical theory over the past 15 years has been extensive because of advances in computer hardware and the availability (usually without cost to the user) of specialized software. We see this trend continuing and increasingly necessary. Computational requirements will increase sharply in the decade ahead. There will be increasingly large data sets, much more model building (perhaps a 10-50 fold increase), more parameters to estimate, and more comprehensive analyses of multiple data sets. Models allowing individual covariates (Skalski et al., this volume) will require substantial computing demands. In addition, several newer techniques are computer intensive (e.g., profile likelihoods and bootstrapping).

By the time this paper appears, there will be a substantial need for more than the 640K of RAM that is currently supported on IBM PC compatible hardware. While computer speed is also a growing limitation, the more important limiting resource is the amount of RAM available. Operating systems such as OS-2 and *Windows* offer a solution to the present limitation on available RAM. Current software must be recompiled to run under these new operating systems. In particular, the few software packages with advanced, user-friendly interactive capability face significant conversion problems in adapting to a new operating system.

Both application and system software will be significant limitations in the future in comparison to hardware. Within 3–5 years, perhaps the research community will have progressed to SPARC/RISC machines running UNIX, but the relearning and reprogramming required to make this change may be distressing. Still, the rest of this decade will see sharply increasing demands for RAM and computer speed to cope with the sophisticated analytical methods available.

It might be possible to bring much of the *existing* software into one, integrated system through user-friendly main menu. The software supporting this menu would reformat several data types and call various software routines as indicated by the user. This "cobbled-up" system might be imperfect in several ways, but these could be transparent to the user. The user could enter data and select several analysis routines and examine output. The fact that the various routines required differing data formats and some were written in C while others were in FORTRAN or BASIC would be handled by the frontend software through a sophisticated menu. Critical to this approach would be a user friendly menu offering context-sensitive *help*.

Perhaps the input to SURVIV (White, 1983) could be automated via appropriate interface software (such work is independently in progress by C. Brownie and associates and R. Dorazio, pers. comm.). SURVIV has seen intensive use as a research tool, but would see additional use if complex models could be entered more readily. Program SURGE (Lebreton et al., 1992) might see even further development; alternative link functions, more goodness-of-fit testing, and an interactive front end would be all significant additions. An integrated software system must be well documented, allow a great deal of analysis flexibility and must offer the user detailed output. Advanced graphics capabilities may be available in specialized software before the end of the decade. The ability to examine outliers, graph Monte Carlo generated sampling distributions or plot estimated survival functions of external variables might become more widespread if it can be done easily. Certainly a research need is to find meaningful graphic displays that will aid in interpretation of capture-recapture data, including residual variation.

Finally, some thought should be given to the explosion of new theory and new applications appearing continually. Perhaps fewer than 15 people throughout the world have a reasonable understanding of, say, 90% of the statistical methods and software available in the early 1990s on capture-recapture and ring recovery surveys. This fact focuses the need for more advanced workshops and conferences to spread the knowledge base. More synthesis and overview papers must be published. We believe the research community in this field should consider the development of an expert system to aid in the transfer of information and technology to others. Such a system would be dynamic, continually expanding, and more useful with each expanding year.

8. Conclusions

The rest of the 1990s will see a tremendous increase in model building for the analysis of complex data sets. Increased modeling of fundamental biological parameters as functions of external variables will be required. Additional research on model selection, heterogeneity, and the use of individual covariates is needed. Bootstrapping to obtain empirical variance estimates will increase and all these matters will require greatly increased computing power and memory. More team work including statisticians and biologists will be the trend. User-friendly software must continue to be developed and become more widely available.

Acknowledgments - We appreciate the review comments offered by Drs. K. P. Burnham and G. C. White.

References

Akaike, H. (1973).- Information theory and an extension of the maximum likelihood principle. Pages 267-281 *in* B.N. Petran and F. Csáki, eds. International symposium on Information theory. Second edition. Akadémiai Kiadi, Budapest, Hungary.

Anderson, D.R., Burnham, K.P., Nichols, J.D., and Conroy, M.J. (1987).- The need for experiments to understand population dynamics of American black ducks. Wildlife Society Bulletin, 15:282-284.

Anderson D.R., Burnham, K.P., and White, G.C. (1985).- Problems in estimating age–specific survival rates from recovery data of birds ringed as young. Journal of Animal, Ecology, 54:89-98.

Anderson, D.R., Bart, J., Edwards, T.C. Jr., Kepler, C.B., and Meslow, E.C. (1990).- 1990 status review: northern spotted owl (*Strix occidentalis caurina*). U.S. Department of the Interior, Fish and Wildlife Service. Portland, OR., Unpublished report. 95pp.

Anderson, D.R. and Burnham, K.P. (1991).- Demographic analysis of northern spotted owl populations. Appendix C. *in* Recovery plan for the northern spotted owl. U.S. Fish & Wildlife Service 662pp.

Arnason, A.N. (1972a).- Parameter estimates from mark-recapture experiments on populations subject to migration and death. Research Population Ecology, 8:97-113.

Arnason, A.N. (1972b).- Prediction methods and variance estimates for the parameters of the triple-catch--two population model with migration and death. University of Manitoba, Department of Computer Science, Science Report No. 76, 30pp.

Arnason, A.N. (1973).- The estimation of population size, migration rates and survival in a stratified population. Research Population Ecology, 15:1-8.

Barker, R. (1992).- Effect of heterogeneous survival on bird-banding model confidence interval coverage rates. Journal of Wildlife Management, 56:111-116.

Barndorff-Nielson, O.E. (1986).- Inference on the full or partial parameters based on the standardized log likelihood ratio. Biometrika, 73:307-322.

Bellrose, F.C. (1959).- Lead poisoning as a mortality factor in waterfowl populations. Illinois Natural History Survey Bulletin, 27:235-288.

Brownie, C., Anderson, D.R., Burnham, K.P., and Robson, D.S. (1985).- Statistical inference from band recovery data–A handbook. U. S. Fish and Wildlife Service, Resource publication 156. Washington, D.C., USA. 305pp.

Brownie, C., Hines, J.E., Nichols, J.D., Pollock, K.H., and Hestbeck, J.B. Submitted. (1992).- Capture-recapture studies for multiple strata including non-Markovian transition probabilities. Biometrics.

Buckland, S.T. (1984).- Monte Carlo confidence intervals. Biometrics, 40:811-817.

Buckland, S.T., and Garthwaite, P.H. (1991).- Quantifying precision of mark-recapture using the bootstrap and related methods. Biometrics, 47:255-268.

Burger, L.W., Ryan, M.R., Jones, D.P., and Wywialowski, A.P.. (1991).- Radio transmitter bias estimation of movement and survival. Journal of Wildlife Management, 55:693-697.

Burnham, K.P. (1989).- Numerical survival rate estimation for capture-recapture models using SAS PROC NLIN. Pages 416-435 *in* L. McDonald, B. Manly, J. Lockwood, and J. Logan, (eds.) Estimation and Analysis of Insect Populations. Springer-Verlag. New York, USA.

Burnham, K.P. (1991).- On a unified theory for release-resampling of animal populations. Pages 11–35 *in* Proceedings of the 1990 Taipei Symposium in Statistics, M.T. Chao and P.E. Cheng (eds.) Institute of Statistical Science, Academia Sinica. Taipei, Taiwan, R.O.C.

Burnham, K.P., and Anderson, D.R. (1992).- Data-based selection of an appropriate biological model: The key to modern data analysis. Pages 16-30 *in* Wildlife 2001: Populations, McCullough, D., and R. Barrett (eds.). Elsevier Sci. Publ., Ltd., New York.

Burnham, K.P., Anderson, D.R., White, G.C., Brownie, C., and Pollock, K.H. (1987).- Design and analysis methods for fish survival experiments based on release-recapture. American Fisheries Society Monograph 5. Bethesda, Maryland, USA. 437pp.

Burnham, K.P. and Rexstad, E.A. (In press).- Modeling heterogeneity in survival rates of banded waterfowl. Biometrics.

Chapman, D.G. and Junge, C.O., Jr. (1956).- The estimation of the size of a stratified animal population. Annals of Mathematical Statistics, 27:375-389.

Clobert, J., Lebreton, J.D., Clobert-Gillet, M., and Coquillart, H. (1985).- The estimation of survival in bird populations by recaptures or resightings of marked individuals. Pages 197-213 *in* B.J.T.Morgan, and P.M. North, (eds.) Statistics in Ornithology. Springer-Verlag, New York, USA.

Conroy, M.J., Hines, J.E., and Williams, B.K. (1989).- Procedures for the analysis of band-recovery data and user instructions for program MULT. U.S. Fish & Wildlife Service, Resource Publication No. 175. 61pp.

Cormack, R.M. (1964).- Estimates of survival from the sighting of marked animals. Biometrika, 51:429-438

Darroch, J.N. (1959).- The multiple-recapture census: Estimation when there is immigration or death. Biometrika, 46:336-351.

Darroch, J.N. (1961).- The two-sample capture-recapture census when tagging and sampling are stratified. Biometrika, 48:241-260.

Dobson, A. (1990).- Survival rates and their relationship to life-history traits in some common British birds. Pages 115-146 *in* Power, D.M., (ed.) Current ornithology. Vol.7, Plenum Press, N.Y., USA.

Efron, B. (1979).- Bootstrap methods: Another look at the jackknife. Annals of Statistics, 7:1-26.

Freeman, S.N., and Morgan, B.J.T. (1990).- Studies in the analysis of ring-recovery data. Ring, 13:271-288.

Freeman, S.N., and North, P.M.. (1990).- Estimation of survival rates of British, Irish and French grey herons. Ring, 13:139-166.

Foster, C.C., Forsman, E.D., Meslow, E.C., Miller, G.S., Reid, J.A., Wagner, F.F., Carey, A.B., and Lint, J.B. (1992).- Survival and reproduction of radio-marked adult spotted owls. Journal of Wildlife Management, 56:91-95.

Gessaman, J.A., and Nagy, K.A. (1988).- Transmitter loads affect flight speed and metabolism of homing pigeons. Condor, 90:662-668.

Gilpin, M., and Hanski, I. (1991).- Metapopulation dynamics: empirical and theoretical investigations. Academic Press, London. 366pp.

Hedges, L.V., and Olkin, I. (1985).- Statistical methods for meta-analysis. Academic Press. N.Y., USA.

Hestbeck, J.B., Nichols, J.D., and Malecki, R.A. (1991).- Estimates of movement and site fidelity using mark-resight data of wintering Canada geese. Ecology, 2:523-533.

Hooge, P.N. (1991).- The effects of radio weight and harnesses on time budgets and movements of acorn woodpeckers. Journal of Field Ornithology, 62:230-238.

Iwao, S. 1963. On a method for estimating the rate of population interchange between two areas. Research Population Ecology, 5:44-50.

Jolly, G.M. (1965).- Explicit estimates from capture-recapture data with both death and immigration–stochastic model. Biometrika, 52:225-247.

Lakani, K.H. (1990).- Choice of models for estimating age-specific bird survival rates using data from ringing recoveries. Ring, 13:23-36.

Lebreton, J.D., Burnham, K.P., Clobert, J., and Anderson, D.R. (1992).- Modeling survival and testing biological hypotheses using marked animals: A unified approach with case studies. Ecological Monographs, 62:67-118.

Lebreton, J.D., Hemery, G., Clobert, J., and Coquillart, H. (1990).- The estimation of age-specific breeding probabilities from recaptures or resightings in vertebrate populations. I. Transversal models. Biometrics, 46:609-622.

Leslie, P.H., Chitty, D., and Chitty, H. (1953).- The estimation of population parameters from data obtained by means of the capture-recapture method. III. An example of the practical application of the method. Biometrika, 40:137-169.

Marks, J.S., and Marks, V.S. (1987).- Influence of radio collars on survival of sharp-tailed grouse. Journal of Wildlife Management, 51:468-471.

Nichols, J.D. (1992).- Capture-recapture models. BioScience, 42:94-102.

Nichols, J.D., Sauer, J.R., Pollock, K.H., and Hestbeck, J.B. (1992).- Estimating transition probabilities for stage–based population projection matrices using capture-recapture data. Ecology, 73:306-312.

Nichols, J.D., Stokes, L.S., Hines, J.E., and Conroy, M.J. (1982).- Additional comments on the assumption of homogeneous survival rates in modern bird banding estimation models. Journal of Wildlife Management, 46:953-962.

North, P.M. (1987).- Ring recovery analytical methods: a brief review. Pages 7-12 *in* North, P.M., (ed.) Ringing recovery analytical methods. Acta Ornithologica, 23.

North, P.M. (ed.) (1990).- The statistical investigation of avian population dynamics using weather data from ringing recoveries and live recaptures of marked birds. Ring, 13.

North, P.M., and Morgan, B.J.T. (1979).- Modelling heron survival using weather data. Biometrics, 35:667-681.

Paton, P.W.C., Zabel, C.J., Neal, D.L., Steger, G.N., Tilghman, N.G., and Noon, B.R. (1991).- Effects of radio tags on spotted owls. Journal of Wildlife Management, 55:617-622.

Perrins, C.M., Lebreton, J.D., and Hirons, G.J.M. (eds.) (1991).- Bird population studies: Relevance to Conservation and Management. Oxford University Press, Oxford. 683pp.

Pollock, K.H., Nichols, J.D., Brownie, C., and Hines, J.E. (1990).- Statistical inference for capture-recapture experiments. Wildlife Monograph, 107. 97pp.

Pollock, K.H., and Raveling, D.H. (1982).- Assumptions of modern band-recovery models with emphasis on heterogeneous survival rates. Journal of Wildlife Management, 46:88-98.

Pollock, K.H., Winterstein, S.R., and Conroy, M.J. (1989).- Estimation and analysis of survival distributions for radio-tagging animals. Biometrics, 45:99-109.

Pradel, R.J., Clobert, J., and Lebreton, J.D.. (1990).- Recent developments for the analysis of capture-recapture multiple data sets. An example concerning two blue tit populations. Ring, 13:193-204.

Rexstad, E.A., and Anderson, D.R. (In press).- Heterogeneous survival rates of mallards (*Anas platyrhynchos*). Canadian Journal of Zoology, 70:

Rexstad, E., Burnham, K.P., and Anderson, D.R. (1990).- Design of survival experiments with marked animals: a case study. American Fisheries Society. Symposium 7:581-587.

Robson, D.S., and Youngs, W.D. (1971).- Statistical analysis of reported tag-recaptures in the harvest from an exploited population. Biometrics Unit, Cornell Univ., Ithaca, N.Y. BU-369-M.

Schwarz, C.J. (1988).- Post-release stratification and migration models in band-recovery and capture-recapture models. Unpublished PhD dissertation. Univ. of Manitoba. 384pp.

Schwarz, C.J., Burnham, K.P., and Arnason, A.N.. (1988).- Post-release stratification in band-recovery models. Biometrics, 44:765-785.

Schwarz, C.J., Schweigert, J.F., and Arnason, A.N. (Submitted).- Estimating migration rates using tag-recovery data. Biometrics.

Seber, G.A.F. (1962).- The multi-sample single recapture census. Biometrika, 49:330-349.

Seber, G.A.F. (1965).- A note on the multiple recapture census. Biometrika, 52:249-259.

Seber, G.A.F. (1970).- Estimating time-specific survival and reporting rates for adult birds from band returns. Biometrika, 57:313-318.

Seber, G.A.F. (1973).- The estimation of animal abundance and related parameters. First edition. Macmillan, New York, USA. 506pp.

Seber, G.A.F. (1982).- The estimation of animal abundance and related parameters. Second edition. Macmillan, New York, USA. 654pp.

Seber, G.A.F. (1986).- A review of estimating animal abundance. Biometrics, 42:267-292.

Seber, G.A.F. (1992).- A review of estimating animal abundance II. Reviews of the International Statistics Institute.

Sodhi, N.S., Warkentin, I.G., James, P.C., and Oliphant, L.W. (1991).- Effects of radiotagging on breeding merlins. Journal of Wildlife Management, 55:592-600.

Szymczak, M.R., and Rexstad, E.A. (1991).- Harvest distribution and survival of a gadwall population. Journal of Wildlife Management, 55:592-600.

White, G.C. (1983).- Numerical estimation of survival rates from band-recovery and biotelemetry data. Journal of Wildlife Management, 47:716-728.

Williams, D.A. (1982).- Extra-binomial variation in logistic linear models. Applied Statistics, 31:144-148.

Taxonomic index

Subject index

Entries in CAPITAL for Software